ACS SYMPOSIUM SERIES **655**

Enzymes for Pulp and Paper Processing

Thomas W. Jeffries, EDITOR
Forest Service, U.S. Department of Agriculture

Liisa Viikari, EDITOR
VTT Biotechnology and Food Research

Developed from a symposium sponsored
by the Cellulose, Paper, and Textile Division

American Chemical Society, Washington, DC

Library of Congress Cataloging-in-Publication Data

Enzymes for pulp and paper processing / Thomas W. Jeffries, editor, Liisa Viikari, editor.

p. cm.—(ACS symposium series, ISSN 0097–6156; 655)

"Developed from a symposium sponsored by the Cellulose, Paper, and Textile Division at the 211th National Meeting of the American Chemical Society, New Orleans, Louisiana, March 24–28, 1996."

Includes bibliographical references and indexes.

ISBN 0–8412–3478–7

1. Wood-pulp—Biotechnology. 2. Enzymes—Industrial applications. 3. Papermaking—Chemistry.

I. Jeffries, T. W. (Thomas William), 1947– . II. Viikari, Liisa, 1949– . III. Series.

TS1176.6 B56E59 1996
676—dc20

96–45741
CIP

This book is printed on acid-free, recycled paper.

Copyright © 1996

American Chemical Society

All Rights Reserved. The appearance of the code at the bottom of the first page of each chapter in this volume indicates the copyright owner's consent that reprographic copies of the chapter may be made for personal or internal use or for the personal or internal use of specific clients. This consent is given on the condition, however, that the copier pay the stated per-copy fee through the Copyright Clearance Center, Inc., 222 Rosewood Drive, Danvers, MA 01923, for copying beyond that permitted by Sections 107 or 108 of the U.S. Copyright Law. This consent does not extend to copying or transmission by any means—graphic or electronic—for any other purpose, such as for general distribution, for advertising or promotional purposes, for creating a new collective work, for resale, or for information storage and retrieval systems. The copying fee for each chapter is indicated in the code at the bottom of the first page of the chapter.

The citation of trade names and/or names of manufacturers in this publication is not to be construed as an endorsement or as approval by ACS of the commercial products or services referenced herein; nor should the mere reference herein to any drawing, specification, chemical process, or other data be regarded as a license or as a conveyance of any right or permission to the holder, reader, or any other person or corporation, to manufacture, reproduce, use, or sell any patented invention or copyrighted work that may in any way be related thereto. Registered names, trademarks, etc., used in this publication, even without specific indication thereof, are not to be considered unprotected by law.

PRINTED IN THE UNITED STATES OF AMERICA

Advisory Board

ACS Symposium Series

Robert J. Alaimo
Procter & Gamble Pharmaceuticals

Mark Arnold
University of Iowa

David Baker
University of Tennessee

Arindam Bose
Pfizer Central Research

Robert F. Brady, Jr.
Naval Research Laboratory

Mary E. Castellion
ChemEdit Company

Margaret A. Cavanaugh
National Science Foundation

Arthur B. Ellis
University of Wisconsin at Madison

Gunda I. Georg
University of Kansas

Madeleine M. Joullie
University of Pennsylvania

Lawrence P. Klemann
Nabisco Foods Group

Douglas R. Lloyd
The University of Texas at Austin

Cynthia A. Maryanoff
R. W. Johnson Pharmaceutical Research Institute

Roger A. Minear
University of Illinois at Urbana–Champaign

Omkaram Nalamasu
AT&T Bell Laboratories

Vincent Pecoraro
University of Michigan

George W. Roberts
North Carolina State University

John R. Shapley
University of Illinois at Urbana–Champaign

Douglas A. Smith
Concurrent Technologies Corporation

L. Somasundaram
DuPont

Michael D. Taylor
Parke-Davis Pharmaceutical Research

William C. Walker
DuPont

Peter Willett
University of Sheffield (England)

Foreword

THE ACS SYMPOSIUM SERIES was first published in 1974 to provide a mechanism for publishing symposia quickly in book form. The purpose of this series is to publish comprehensive books developed from symposia, which are usually "snapshots in time" of the current research being done on a topic, plus some review material on the topic. For this reason, it is necessary that the papers be published as quickly as possible.

Before a symposium-based book is put under contract, the proposed table of contents is reviewed for appropriateness to the topic and for comprehensiveness of the collection. Some papers are excluded at this point, and others are added to round out the scope of the volume. In addition, a draft of each paper is peer-reviewed prior to final acceptance or rejection. This anonymous review process is supervised by the organizer(s) of the symposium, who become the editor(s) of the book. The authors then revise their papers according to the recommendations of both the reviewers and the editors, prepare camera-ready copy, and submit the final papers to the editors, who check that all necessary revisions have been made.

As a rule, only original research papers and original review papers are included in the volumes. Verbatim reproductions of previously published papers are not accepted.

ACS BOOKS DEPARTMENT

Contents

PULP BLEACHING

OTHER APPLICATIONS

Preface

ENZYMES PERMEATE OUR LIVES in ways we never realize. They are used to prepare our foods and beverages, clean our clothes, and diagnose our illnesses. They are used for chemical syntheses and are incorporated into electromechanical devices. They are the protein machines behind life itself and, through the tools of biotechnology, they are being adapted to virtually all aspects of biomaterials handling.

Enzyme technology is driven by the need for economic, efficient, ecological processing. Cost-effective large-scale applications are made possible by the capacity for producing novel enzymes in large quantity through biotechnology. Manufacture, stabilization, packaging, and distribution are done on a scale that has taken enzymes from the shelf of exotic specialty reagents into the holding tank of bulk commodities. Manufacturing expertise is increasing as new and more efficient hosts for protein expression are developed. The enzyme industry is constantly seeking new markets; its biggest challenges lie in the discovery of new applications.

The pulp and paper industry is emerging as one of the largest markets for enzyme applications in the world. The demand for paper increases globally as standards of living rise and the need for clean, efficient processing is ever greater. Increased pulp yield, improved fiber properties, enhanced paper recycling, and reduced processing and environmental problems are all consequences of enzyme applications in the pulp and paper industry. Acceptance of these technologies is growing as knowledge and industry expertise increase.

We have long known the enzymes that act on paper: cellulases, hemicellulases, ligninases, and lipases. Only recently, however, have we begun to explore the potentials for their application in large scale, such as for bleach enhancement, contaminant removal, pitch removal, fiber modification, and lignin degradation. Enzymes are highly selective in their action. That is their bane and beauty. Conditions must be exact for their use, but the results are precise. Effects can be subtle and subject to interference from the harsh environments often found in industrial settings. For these reasons, some view enzyme applications skeptically and want more evidence of value before moving forward.

The symposium on which this book is based was sponsored by the Cellulose, Paper, and Textile Division at the 211th ACS National Meeting which took place in New Orleans, Louisiana, from March 24–28, 1996. The papers in this volume are exploratory. They probe the bounds of

possibility. They report the discovery of new biocatalysts and novel uses for those we know well. The authors are diverse, coming from the Americas, Europe, Asia, Africa, and Australia. They work in industry, academia, and government, and their expertise ranges from fundamental biochemistry and genetics to large-scale application and marketing. The processes are nascent. They take many different forms, and in few instances have optima been attained.

This volume is not meant to be definitive. In an emerging field as large and rapidly changing as the present one, it can only be a snapshot of progress. With time we will know of its success.

THOMAS W. JEFFRIES
Institute for Microbial and Biochemical Technology
Forest Products Laboratory
Forest Service, U.S. Department of Agriculture
One Gifford Pinchot Drive
Madison, WI 53705

LIISA VIIKARI
VTT Biotechnology and Food Research
P.O. Box 1501
FIN–02044 VTT, Finland

September 3, 1996

Overviews

Chapter 1

Roles for Microbial Enzymes in Pulp and Paper Processing

T. Kent Kirk and Thomas W. Jeffries

Institute for Microbial and Biochemical Technology, Forest Products Laboratory, Forest Service, U.S. Department of Agriculture, One Gifford Pinchot Drive, Madison, WI 53705

Microbial enzymes are enabling new technologies for processing pulps and fibers. Xylanases reduce the amount of chemicals required for bleaching; cellulases smooth fibers, enhance drainage, and promote ink removal; lipases reduce pitch; lignin-degrading enzymes remove lignin from pulps. Several of these processes are commercial, and others are beginning to emerge. In the future, enzyme-based processes could lead to cleaner and more efficient pulp and paper processing. The papers in this book describe fundamental and applied aspects of xylanase, cellulase, lignin-degrading enzymes, and lipase with a view toward the development of novel processes, unusual enzymatic activities and elucidation of underlying mechanisms.

Paper manufacture is one of the largest industries in the United States. In 1995, the US produced more than 82 million metric tons of paper, paperboard and secondary products with a shipped value of $166 billion (*1*) An industry of this magnitude supports a large, rapidly changing technical base, and because it draws heavily upon timber and water resources, it is subject to close scrutiny from environmental interests.

Several decades ago researchers realized that because paper is composed of natural polymers – cellulose, hemicelluloses, and lignin – microbial enzymes and organisms might be useful in its processing. Only in the last decade, however, have microbial enzymes been used commercially in the pulp and paper industry, and microorganisms, though long employed in waste treatment, are only now beginning to be used in other processing steps.

The main reason for the slowness in using enzymes in pulp and paper processing is that the substrates – wood and pulps – are difficult to degrade. Because it is the lignin that is removed from wood in chemical pulping, and from pulps in bleaching, the research focus historically has been on lignin-biodegrading systems. Lignin likely evolved in part as a deterrent to microbial degradation, and it continues to be an impediment to biotechnological processing of wood and pulps. Its degradation by isolated enzymes remains difficult.

The potential for environmentally benign, efficient lignin removal spurred research that led to the discovery of lignin-degrading enzymes in the early 80s, and to extensive

This chapter not subject to U.S. copyright
Published 1996 American Chemical Society

investigations into their activities through the present. Even so, having ligninolytic enzymes in hand, and knowing how they function, has not – until very recently – shown promise for practical applications.

During the past ten years the number of possible applications of enzymes in pulp and paper manufacture has grown steadily, and several have become, or are approaching, commercial use. These include enzymatic bleaching with xylanases, pitch removal with lipases, and freeness enhancement with cellulases and hemicellulases. Others such as contaminant removal and fibrillation of recycled fibers by cellulases could be commercial soon. As we gain more experience with these systems, one of the biggest barriers to their use – a reticence by industry – is diminishing. Enzymes are considered exotic, and paper makers usually cannot use them in the same manner as inorganic chemicals. Enzyme manufacturers often lack sufficient industry contacts to transfer technology to the mills, and traditional suppliers of paper chemicals face loss of some markets with increased use of enzymes. All of these factors weigh against rapid widespread application, but the industry has continued to develop. Increasingly, pulp and paper companies are employing microbiologists and biochemists to guide them in the use of biotechnology.

Our purpose in this chapter is to provide a historical framework for applications and fundamental studies in this field. Several recent in-depth reviews of individual applications (*2-5*) present summaries of the latest developments. Historically, development of enzyme applications in pulp and paper began with studies of the use of cellulases to facilitate fiber beating, and progressed through current efforts to apply lignin-degrading enzymes (Table I).

Table I. Milestones in the development of enzyme technology for pulp and paper processing

Year	Development	Researchers and reference
1959	Pulp fibrillation by cellulases	Bolaski and Gallatin (*6*)
1984	Enzymatic beating with xylanases	Comtat, Mora and Nóe (*7*)
1984	Hemicellulose removal from dissolving pulps by xylanases	Paice and Jurasek (*8*)
1986	Prebleaching with xylanases	Viikari, Ranua, Kantelinen, Sundquist and Linko (*9*)
1988	Enhanced drainage with cellulases	Fuentes and Robert (*10*)
1988	Decreased vessel picking with cellulases	Uchimoto, Endo and Yamagishi (*11*).
1989	Depitching pulp with lipase	Irie, Matsukura and Hata (*12*).
1991	Deinking with cellulases and xylanases	Kim, Ow and Eom (*13*)
1993	Pulp delignification with laccase	Call and Mülke (*14*)
1996	Bleaching with manganese peroxidase	Harazono, Kondo and Sakai (*15*)

Fibrillation and Strength Enhancement

Beating and refining are mechanical processes that enhance fibrillation and inter-fiber bonding. Properly applied, microbial enzymes can enhance or restore fiber strength, reduce beating times, and increase inter-fiber bonding through fibrillation.

Pulp fibrillation by cellulases was recognized as a means to enhance strength properties as early as 1959 (*6*). Cellulases from the fungus *Aspergillus niger* were used to enhance fibrillation, thereby improving the strength of paper by increasing fiber-fiber contact. It was principally applied to cotton linters and other non-wood pulps. A process patented in 1968 used cellulases from the white-rot fungus *Trametes suaveolens* to reduce refining or beating time (*16*). Nomura reported that cellulase plus cellobiase added to pulps facilitated fibrillation without strength loss (*17*) Jokinen et al have also described the use of cellulases to improve fibrillation of pulps (*18*).

The principal challenge in using enzymes to enhance fiber bonding is to increase fibrillation without reducing pulp viscosity. Viscosity decreases when cellulases cleave cellulose chains, lowering the degree of cellulose polymerization (number of glucose residues per chain) and destroying fiber integrity. In one attempt to get around this problem, researchers fiberized pulp using "cellulase-free" xylanase from mutants of *Sporotricum pulverulentum* and *Sporotricum dimorphosphorum*. (*7, 19-21*) Relatively mild xylanase treatments removed less than 2% of the total fiber weight while improving fibrillation and fiber bonding and decreasing beating times. This decreased the drainage rate (increased the Schopper-Riegler (SR) value), and the water retention value. However, at the same time it decreased viscosity, and decreased breaking length drastically. Both of these negative effects are attributable to residual cellulase activity, which these workers had attempted to inhibit. Today, many commercial, cellulase-free xylanase preparations are available through cloning and over expression, so their application in pulp fibrillation might be practical. In fact, some of the side benefits of enzymatic bleaching and deinking are *increased* pulp viscosity and drainage (*22*).

The exact mechanism by which enzymatic pulp fibrillation occurs is still not understood, and this area deserves more basic research.

Drainage

The drainage rate of a pulp – that is, the rate of water loss during paper formation – determines the speed of paper machine operation. Drainage rates of secondary or recycled fibers are particularly important because they tend to be much lower than drainage rates of primary (virgin) fibers, so that using large quantities of secondary (recycled) fiber slows the papermaking process. In the United States, where manufacturers mainly use primary fibers, paper machines tend to be designed for such stocks. Increasing the recycled fiber content requires operating the machines at lower rates.

Fuentes and Robert discovered that cellulases can improve the drainage rates of recycled fibers (*23*). In one study, treating recycled fibers with cellulases and xylanases reduced the SR value (i.e., it increased drainage) by 18 to 20%. Commercial enzymes were used to treat batches of recycled fibers on both laboratory (*24*) and pilot scales (*25*). Pulp strength was reduced, but it could be improved by refining the pulp before enzyme treatment. Starch sizing likewise improved mechanical properties. Cellulases are presently being used commercially to enhance drainage rates of recycled fibers in France, and one might expect to see wider application .

Current research on cellulases is refining the understanding of how they function (e.g., *26, 27*). Considerable progress has been made recently in understanding cellulase structures and functions. Research is showing that cellulases vary in mode of action, pointing to the opportunity to optimize cellulase selection for specific uses. In applied research, we need to better understand the relationships between fibrillation and drainage.

Modification of Other Pulp Properties

Enzymes can improve paper properties in specific ways. For example, some hardwoods contain large vessels that make a rough paper surface. During printing, these vessels lead to "picking", which prevents complete contact of the ink with the paper surface, and an imperfect image. This can be particularly important in image copy, and it is of increasing significance as vessel-rich eucalyptus pulps are used in larger quantities. Cellulase treatment can reduce "picking," (*11*).

Xylanase from the fungus *Schizophylum commune* reduced xylan 22% in a delignified mechanical aspen pulp (*8*). Purified xylanase from the fungus *Trichoderma harzianum* reduced the xylan content of unbleached kraft pulp by 25% (*28*), and a

cloned xylanase from the bacterium *Bacillus subtilis* reduced the xylan content of bleached hardwood kraft pulp 20% (*29*). The use of xylanase, however, has not become commercial, in part because xylan removal is incomplete. More recently obtained xylanases might prove more effective, and this application should be reexamined.

Retting

To date, pulping wood with isolated enzymes has not been accomplished, and is not to be expected, because enzymes cannot penetrate the lignified cell walls. Enzymes can, however, pulp herbaceous fibers. Microbial retting is an ancient process dating to the beginnings of civilization. Traditional retting uses mixed microbial populations–mainly soft-rot bacteria–introduced with crude inocula. Fibers that are retted include flax, jute, and coconut hulls (*30*). In this process microbial pectinases (pectin-depolymerizing enzymes) release cellulosic fibers from fiber bundles.

Contemporary practice uses selected microbial strains or isolated enzymes. The chief disadvantages of the traditional method is the bad odor that develops in the retting tanks, during handling, and in the discharge of the effluent; its uncontrolled nature; and its slowness. Enzymatic retting is faster than traditional retting, readily controlled, and produces fewer odors, but further development is required to make it competitive with the traditional methods. Commercial enzymes such as cellulases, hemicellulases, pectinases and other polysaccharidases have been applied to flax at various levels and compared to traditional retting methods (*31*).

Pectinolytic enzymes secreted by soft-rot bacteria (*32*) also cause maceration of woody bast fibers derived from the phloem of plants. These fibers, used to make cordage, matting and various fabrics, are long, strong, and commonly stiff. Pectinolytic and xylanolytic enzymes can help soften them. Alkaline presoaking enhances enzymatic activity. A combined alkali-enzyme process increased fibrillation, decreased the Canadian standard freeness value, decreased the shives content, and improved sheet formation (*33*). Enzymatic pulps prepared with pectinolytic enzymes produce bulkier paper with higher opacity and better printability than pulps prepared from the same stock solely by an alkaline process. Chemical and enzyme retting have both been carried out on a semi-industrial scale, and the characteristics of fibers produced by these two methods are not significantly different (*34*).

Combinations of cellulases, xylanases and pectinases have been used to soften and smooth the surfaces of jute-cotton blended fabrics (*35*). By obtaining an optimum balance of enzymes, it is possible to lower the dosing rate and improve efficiency.

In recent years, a few fundamental studies have been initiated on the enzymatic retting process. These employ purified enzymes on defined substrates, and characterization of the resulting products. A purified pectinase from an *Aspergillus* released three size classes of polysaccharides from flax. To ensure maximum strength of the thread manufactured from retted flax, only a small fraction of the pectins belonging to the fiber bundles need to be hydrolyzed. Some advantage might be gained, therefore, in using enzyme preparations with better specificity (*36*).

In developing nations, and particularly in countries where forest stands are endangered from over exploitation, better use might be made of herbaceous fibers for paper production. Such feedstocks should be amenable to enzymatic pulping, and the resulting processes should give higher yields with fewer environmental problems. Clearly, however, much more work needs to be done in this area before enzymatic pulping of herbaceous fibers will see wide application.

Pitch Control

Certain types of wood pulps, including sulfite pulps and various mechanical pulps – especially from pines – have high pitch contents. Pitch is a term used collectively for hy-

drophobic components of wood: resin and resin acids, triglycerides, waxes, etc. Pitch causes numerous problems in pulp and paper manufacture, including deposits in and on equipment, adverse effects on water absorption by the pulps, holes and tearing of the paper due to sticky deposits on dryer rolls, discoloration and hydrophobic spots in the paper. (*12*). Current methods for controlling pitch include "seasoning" the wood before pulping (which allows the pitch to deteriorate by the action of indigenous microbes), and adding talc or other chemicals to the pulp to coat the resin and deactivate its surface. Seasoning takes a long time, and often leads to discoloration, and the chemicals cost money. The most troublesome components of resins are triglycerides, fatty acid esters of glycerol. These can be removed by solvent extraction or by strong alkali, but the former process is not practical, and the latter results in yield losses and discoloration. Neither is used.

Hata and coworkers at Jujo Paper Company reported in 1990 that lipases can reduce pitch problems by lowering the triglyceride content of groundwood pulp (*37*). A lipase obtained from *Candida cylindrica*, when added to the groundwood stock chest reduced pitch problems and talc consumption considerably. Mill trials showed that the number of defects in the paper decreased along with the frequency of machine cleaning. The dosage of chemicals was reduced, and the time used for the traditional "seasoning" of the chips (to control pitch) was greatly shortened, resulting in cleaner chips and lowered bleach consumption. Perhaps most importantly, lipase improved pulp properties. Use of lipases went on-line in Jujo mills shortly after 1990, and reportedly is now being used by other companies in Japan. Other work showed that incubating CTMP in the presence of lipase speeds water absorption while it increases the strength and the specific volume of the resulting paper (*38*)

In Austria, Messner and coworkers (*39*) showed that lipases are also effective in reducing pitch problems associated with sulfite pulps. Pitch in sulfite pulps can also present problems during paper manufacture. Deposits on exposed parts of the paper machine such as air foils or machine wire can degrade the product and impair production. Chlorinated pitch triglycerides formed during bleaching are particularly troublesome. Fischer and Messner found that treating unbleached sulfite pulps with lipase followed by alkaline extraction removed most of the triglycerides (*44*). Because the enzyme absorbs rapidly to pulp fibers, it is not possible to recycle it for reuse. On the other hand, its absorptive properties enable application at low pulp consistencies, and the enzymes remain active and attached during various treatments and washing stages (*40*). The lipase process has been scaled-up in a 12-ton per day pulp trial, and has been shown to remove 90% of the triglycerides in three h with stirring at 37°C (*41*).

Bleaching Chemical Pulps

The kraft process accounts for 85% of the total pulp production in the United States, and is the single major process world-wide. Bleached kraft pulp is a major, relatively high-value component of the total production of kraft paper. Kraft pulping removes most of the lignin, and dissolves and degrades hemicelluloses without severely damaging cellulose. The kraft process results in excellent pulp from a wide variety of wood species. Unfortunately, kraft pulping also generates large quantities of chromophores. Chromophores are composed of residual lignin and carbohydrate degradation products. They are hard to extract because they are physically entrapped in and covalently bound to the carbohydrate moieties in the pulp matrix (*42-44*). Manufacturers use elemental chlorine (Cl_2) and chlorine dioxide (ClO_2) to bleach the chromophores, and then they extract them, along with residual lignin, to make white (“bright”) pulp. Because of consumer resistance and environmental regulation of chlorine in bleaching, pulp makers are turning to oxygen, ozone and peroxide bleaching, even though they may be more expensive and less effective than Cl_2.

The biggest success story in the use of enzymes in the pulp and paper industry is hemicellulases (mainly xylanases) as aids in pulp bleaching ("enzymatic pre-bleaching"). Viikari et al. (*9*) discovered that treating kraft pulps with fungal xylanases decreases the amount of bleach chemical required to attain a given brightness. Mill trials followed, confirming the laboratory results, and since the early 1990's xylanases have been used commercially in Scandinavia and Canada – and more recently in Chile. Following the initial discovery of the effectiveness of xylanase, reports confirmed that xylanases can reduce chemical demand in bleaching (*45, 46*). At least 15 patents or patent disclosures dealing with enzymatic treatments to enhance bleaching of kraft pulps were submitted between 1988 and 1993. In addition, there have been numerous research publications and presentations at conferences. The reason for this high interest lies in the economic importance of kraft pulping and the regulatory pressure against chlorine. Another motive for pulp makers to use xylanase for bleaching is to save chemical costs. Hemicellulase treatments are effective on both hardwoods and softwoods, but they affect hardwood kraft pulps more (*47*). Although promising results were obtained initially (*48, 49*), mannanases are not as effective as xylanases, even with softwood pulps, apparently because of limited accessibility of the substrate (*49*). Mannanase treatments have very little effect on handsheet properties. The mannanase from *Trichoderma reesei* has, however, been shown to act synergistically with xylanases to enhance pulp bleaching (*50*). Most proceedings' reports and many of the patent disclosures have been reviewed in earlier publications (*3, 51-54*). We concentrate here on recent findings on mechanisms of efficacy and enzyme characteristics.

Mechanism. The mechanism of hemicellulase prebleaching is not completely understood. One hypothesis suggests that precipitated xylan blocks or occludes extraction and that xylanase increases accessibility (*50, 55, 56,*). This model is based on reports that xylan reprecipitates on the fiber surfaces (*57-59*). More recently, Suurnäkki et al. (*60*) found that no extensive relocation of xylan to the outer surface occurs during pulping, so the occlusion model might not be a sound premise. The second model suggests that lignin or chromophores generated during the kraft cook react with carbohydrate moieties (*43*). Hemicellulases liberate residual lignin by releasing xylan-chromophore fragments, thereby increasing their extractability.

Skjold-Jørgensen et al. (*61*) found that xylanase treatment decreased the demand for active chlorine (aCl_2) for a batch kraft pulp by 15%, but decreased aCl_2 of pulp from a continuous process by only 6 to 7%. They also showed that DMSO extraction of residual xylan does not lead to an increase in bleachability, but that xylanase treatment does. This indicates that it is the DMSO-insoluble xylan fraction that is chemically bound to the chromophores. Paice et al. (*62*) have shown that the prebleaching effect on black spruce pulp is associated with a drop in the degree of polymerization, even though the xylan content decreases only slightly. Prebleaching thus appears to be associated with xylan depolymerization, even though not necessarily with solubilization of the xylan-derived hemicellulose components. Senior and Hamilton (22) have shown that xylanase treatment and extraction change the reactivity of the pulp by enabling a higher chlorine dioxide substitution to achieve a target brightness and that they raise the brightness ceiling of fully bleached pulps.

Enzyme Characteristics. Effective xylanases should have several properties. First, they should be stable on kraft pulps. Some xylanase preparations non-specifically absorb to pulp fibers and are inactivated by degradation products from kraft pulping (*63*). Second, they should have a neutral to alkaline pH optimum. Residual alkali leaks out of the pulp during enzyme treatment, and the pH of even well washed pulp stocks can shift upwards dramatically. Third, they should have good thermal stability (*64*). The pulp is hot (75°C) when it first comes out of the stock washers, and heat-tolerant enzymes generally have higher turnover numbers. Fourth, factors affecting the

interaction of the enzymes with the pulps are important. These include the effective molecular weight, net ionic properties, and specific action pattern (*65*). Finally, they should not be contaminated with cellulases. If cellulases are present, pulp viscosities decrease. Without cellulase, xylanase treatment increases viscosity, because some lower molecular weight xylans are removed (*66*). Even so, excess xylan removal can reduce burst strength and long span tensile strength by reducing inter-fiber bonding even though it does not weaken the fibers themselves (*67*).

Differences in kinetic properties, substrate specificities, and effects on pulp bleachability have been observed with various pure xylanase isozymes. Certain xylanases release chromophores more than others when used at the same activity levels. Four xylanases from *Streptomyces roseiscleroticus* released chromophores and reduced the kappa number of hardwood and softwood kraft pulps. Some resulted in greater kappa reduction (a measure of residual lignin) and others released more chromophores. Characterization of the chromophoric materials by reverse-phase HPLC indicated that compounds absorbing strongly in the visible region were relatively hydrophobic (*68*). More recently, the release of chromophoric groups has been reported with other *Streptomyces* xylanases (*69*), and the release of chromophores has been shown to correlate linearly with increased brightness (*70*). With improved knowledge of substrate specificity and interaction, it may be possible to identify enzymes that have more specific effects in releasing chromophores without substantially releasing xylan.

Xylanases can be classified structurally into two major groups: Family F or 10, and Family G or 11 (*71*). Family F xylanases are relatively high molecular weight and Family G are relatively low molecular weight. The Family G enzymes can be further divided into those with high and low isoelectric points (pI). Liberation of reducing sugars from purified xylan by the pI 5.5 xylanase from *T. reesei* correlates well with its bleaching effect on fibers, but the behavior of the pI 9.0 xylanase is more complex. It appears to be affected not only by its catalytic activity, but also by electrostatic interactions with its substrate (*72*). The effect of the pI 9.0 xylanase on bleachability of pulp increased more at pH 7.0 than would be expected from its activity on purified xylan.

Binding of *Trichoderma* xylanases to polysaccharides is affected by the pH and the ionic strength (*73*). Fibers carry a net negative charge at neutral pH due to the presence of carboxylic acid groups, so interaction of the enzyme with the fiber is affected by charge on the protein. Enzymes are totally bound to fibers when the pH is below their pI, but are mainly unbound at pH values above the pI. A protein with a pI of 9.0 would absorb to the fibers at pH 7, but a protein with a pI of 5.5 might not absorb at all. This effect of electrostatic interaction also seems to be modulated by the effects of counter ions (*74*). A more fundamental explanation of these differences is found in the structure of the *Trichoderma* xylanases. The pI 5.5 xylanase possesses a smaller, tighter substrate pocket and a lower pH optimum than the pI 9.0 xylanase (*75*). It also exhibits a fifteen-fold higher turnover number (*76*), and a three-fold lower K_m. The pI 9.0 xylanase has a more open structure, and a wider pH range, and tends to produce larger oligosaccharides. The difference in pI is attributable to the presence of more lysine and arginine residues in the pI 9.0 xylanase. These are found in a particular region of the enzyme of the enzyme, and they possibly interact with the glucuronic acid side chains of xylan (*77*). Thus it appears that the disproportionately greater ability of the pI 9.0 enzyme to enhance bleaching could be due to its specific interaction with charged groups. Current research is focused on obtaining improved xylanases that exhibit unusual thermal stability, (*78-80*) alkaline activity, (*81*) or high specificity for chromophore release (*68, 82*).

Enzymatic Enhancement of Contaminant Removal

Printing and writing grade papers used in offices are among the most valuable papers manufactured and sold. They amount to about 25% (20 million tons per year) of U.S.

paper production, and recycling is a rapidly growing segment of the paper industry. Less than 10% of office waste papers is recycled back into printing and writing grades, but this could increase several fold if technical problems can be resolved (*83*. About 88% of un-segregated office waste paper (OWP) is composed of chemical fibers (*84*).

Mixed wastepapers present technical and economic challenges to the paper recycler, and of the wide variety of fibers and contaminants present in the paper stock, toners and other non-contact polymeric inks from laser-printing processes are the most difficult to deal with (*85*). Toners and laser printing inks are synthetic polymers with embedded carbon black; they do not disperse readily during conventional repulping processes. Moreover, they are not readily removed during flotation or washing. Because of these problems, recycled papers contaminated with toners have a relatively low value. Conventional deinking uses surfactants to float toners away from fibers, high temperatures to make toner surfaces form aggregates, and vigorous, high intensity dispersion for size reduction. Most of the deinking chemicals and high-energy dispersion steps are expensive. The high-energy dispersion step is both capital- and energy-intensive and can also reduce fiber length.

Cellulases were reported in 1991 to be effective in removing conventional inks from newsprint, in research done in Korea (*13*); a Japanese patent was filed slightly earlier (*86*) Prasad and coworkers also reported enzymatic deinking of newsprint in 1992 (*87*). Microbial enzymes have also been shown to enhance the release of toners from office waste. When cellulases and xylanases were applied to xerographic-printed papers in a medium consistency mixer, they released toner particles and facilitated subsequent flotation and washing steps. In comparison to the control treatment (water only), enzymes released 95% more residual toner particles from recycled fibers. The amounts of enzymes required were highly cost effective with conventional deinking chemicals (*88, 89*). This approach employs relatively low doses of neutral or alkaline-active cellulases along with surfactant and mechanical action at high consistency. The toners are released from the surfaces of the fibers and removed by subsequent flotation. This process has been scaled up in pilot plant trials and has proved to be effective (*90*). Obviously, use of cellulases with pulps must be done carefully to avoid excessive depolymerization of the cellulose. Woodward et al. reported in 1994 that cellulases can be used to separate un-inked from inked fibers (*91*). In this approach, relatively low consistency pulp suspensions are used, and ink particles are separated during recirculation.

Alkaline lipases will facilitate the removal of soy lipid-based offset printing inks (*92, 93*). At present, soy ink-printed materials comprise only a small fraction of the total recycled paper, but this application may find increased value as soy-based inks become more widely used.

Applications of Other Polysaccharidases

Starch-hydrolyzing enzymes such as amylases have potential where starch needs to be removed. Thus, amylases were shown 12 years ago to be useful in freeing pulp fibers from the microcapsules in pressure-sensitive carbonless copying paper waste (*94*). Any process step in which starch hydrolysis and removal is an objective might be facilitated with amylases. Pectinases have been shown to effectively eliminate the cationic demand caused by pectin in peroxide bleach waters (*95*). These enzymes also show potential for reducing energy demand in debarking spruce logs before chipping (*96*).

Lignin-Oxidizing Enzymes

In 1974 and 1975 we showed at the Forest Products Laboratory that synthetic ^{14}C-labeled lignins are mineralized (oxidized to CO_2) by lignin-degrading (white-rot) fungi (*97*). In 1976 we subjected some of the radioactive lignins to kraft cooking and sulfite cooking, and further subjected the resulting kraft lignins and lignin sulfonates to

chlorine bleaching. Interestingly, the white-rot fungi could still mineralize these lignins (*98*). This told us first, that the lignin-degrading machinery of the fungi is clearly quite non-specific, because the pulping and bleaching processes severely modify the already heterogeneous lignin polymer. Second, the results told us that the chemical pulping and bleaching procedures do not produce modified lignins that are inherently environmentally recalcitrant. Third, the results suggested that we might use white-rot fungi to clean up chlorine bleach plant effluents. This turned out to be true, and we worked on that application for several years (*99*). Finally, the results indicated that if we could ever isolate the enzymes responsible for fungal lignin oxidation, we might be able to use them to bleach chemical pulps. We reasoned that the fungi themselves should be able to delignify chemical pulps, and we showed this to be the case in 1978 (*100*). Canadian workers have pursued this line of work, i.e., bleaching with living fungus cells (*101*). In any event, for nearly 20 years scientists have been aware of the potential of lignin-degrading enzymes in bleaching chemical pulps.

During those 20 years, lignin-oxidizing enzymes were discovered and characterized extensively (see *102* for a review). As understood today, there are three components of the lignin-oxidizing enzyme system: manganese peroxidase (MnP), lignin peroxidase (LiP), and laccases. In the presence of H_2O_2, MnP oxidizes Mn^{2+} to Mn^{3+}, which can oxidize phenolic units in lignin. A system composed of MnP, Mn^{2+}, H_2O_2 and unsaturated lipids, however, also oxidizes non-phenolic units (*103*) and depolymerizes lignin. The physiological significance of this system is not yet known. Lignin peroxidase, in the presence of H_2O_2, oxidizes both phenolic and non-phenolic units in lignin, and has also been shown to depolymerize the polymer (*104*). Laccase oxidizes phenolic units in lignin in the presence of O_2. In the presence of certain substrate "mediators," laccase also can oxidize non-phenolic units (*105*).

Soon after the discovery of lignin-degrading enzyme systems, various laboratories attempted to bleach kraft pulps with them; the first report was by Linko and coworkers, and was presented as part of the same paper that included the first xylanase results, at the First International Conference on Biotechnology in the Pulp and Paper Industry, in Stockholm, in 1986 (*50*). Until quite recently, the lignin-oxidizing enzymes have not shown much promise in bleaching trials. Now two enzymes have given encouraging results: laccase and manganese peroxidase (*14, 106*). To date, the third lignin-oxidizing enzyme, lignin peroxidase, has shown little prospect. However, the amount of research that has been conducted on the practical use of these lignin-degrading enzymes has been minimal, indicating that more effort is justified.

Future Directions

In order for enzymes to have significant impacts on pulp and paper processes, they will need to be effective in consistent manners under various operating conditions. Pulp substrates are more recalcitrant and variable than the grain starch or soluble pectin substrates found in food processing applications, and the environmental factors can be more severe. Cellulases, hemicellulases and lignin-degrading enzymes generally function best under slightly acidic or neutral pH, but the most common pulping reactions and recycled fiber processes are alkaline. In addition, the pulp slurries are usually hot. It seems likely, therefore, that future research will be directed toward the discovery or engineering of enzymes that are more robust with respect to pH and temperature. Much progress has already been made in understanding the basic mechanisms of xylanases and cellulases and in engineering their properties. Lignin-degrading enzymes and the mechanisms they employ should receive increased research attention, with emphasis on how they might point to new applications in pulping and bleaching. The new experimental process employing polyoxometalates to bleach pulp oxidatively (*107*), for example, is conceptually based on the role that transition metals play in lignin-degrading enzymes (I. Weinstock and R. Atalla, personal communication).

Literature Cited

1 Anon. 1996. *Statistics of Paper, Paperboard and Wood Pulp*. American Forest and Paper Association, 1111 19th St. Suite 800, Washington, DC. 20036.
2 Kuwahara, M.; Shimada, M., Eds. *Biotechnology in Pulp and Paper Industry*. **1992**. Uni Publ., Ltd., Tokyo, 545 p.
3 Daneault, C.; Leduc, C.; Valade, J.L. *Tappi J.* **1994**, *77*, 125.
4 Paice, M. G.; Bourbonnais, R.; Reid, I. D.; Archibald; Jurasek, L. *J. Pulp Paper Sci.* **1995**, *21*, J280.
5 Welt, T.; Dinus, R. J. *Prog. Paper Recycl.* **1995**, *Feb.*, 36.
6 Bolaski, W.; Gallatin, A.; Gallatin, J.C. United States pat. 3,041,246, 1959.
7 Comtat, J.; Mora, F.; Noé, P. French pat. 2,557,894, 1984.
8 Paice, M.G.; Jurasek, L. *J. Wood Chem. Technol.* **1984,** *4*, 187.
9 Viikari, L.; Ranua, M.; Kantelinen, A.; Sundquist, J.; Linko, M. *Botechnology in the Pulp and Paper Industry,* Stockholm, Sweden, 1986, pp. 67-69.
10 Fuentes, J.L.; Robert, M. European pat. 262040, 1988.
11 Uchimoto, I.; Endo, K.; Yamagishi, Y. Japanese pat.. 135,597/88, 1988.
12 Irie, Y.; Matsukura, T.; Hata, K. European pat. app. EP 374700. 1989.
13 Kim, T.-J.; Ow, S. S.-K.; Eom, T.-J. *Proc. Tappi Pulping Conf.* Tappi Press: Atlanta, **1991**, 1023.
14 Call, H.P. United States pat. 5,203,964, 1993.
15 Harazono, R.; Kondo, R.; Sakai, K. *Appl. Environ. Microbiol.*, **1996**, *62,* 913.
16 Yerkes, W.D. United States pat. 3,406,089, 1968.
17 Nomura, Y. Japanese pat. 126,395/85, 1985.
18 Jokinen, O.; Kettunen, J.; Lepo, J.; Niemi, T.; Laine, J.E. United States pat. 5,068,009, 1991.
19 Mora, F.; Comtat, J.; Barnoud, F.; Pla, F.; Noé, P. *J. Wood Chem. Technol.* **1986**, *6*, 147.
20 Noé, P.; Chevalier, J.; Mora, F.; Comtat, J. *J. Wood Chem. Technol.* **1986**, *6*, 167.
21 Barnoud, F.; Comtat, J.; Joseleau, J.P.; Mora, F.; Ruel, K. *Third International Conference on Biotechnology in the Pulp and Paper Industry*; Stockholm, Sweden 1986. pp. 70-72.
22 Senior, D.J.; Hamilton, J. *Tappi J.* **1993**, *76*, 200.
23 Fuentes, J.L.; Robert, M. French pat. 2604198, 1986.
24 Pommier, J.C.; Fuentes, J.L.; Goma, G. **1989**, *Tappi J. 72* (6), 187.
25 Pommier, J.C.; Goma, G.; Fuentes, J.L.; Rousset, C. *Tappi J.* **1990**, *73* (12), 197.
26 Gilkes, N.R.; Jervis, E.; Henrissat, B.; Tekant, B.; Miller, R.C., Jr.; Warren, R.A. J.; Kilburn, D.G.. *J. Biol. Chem.* **1992**, *267*, 6743.
27 Kleman-Leyer, K.M.; Gilkes, N.R.; Miller, R.C., Jr.; Kirk, T.K. *Biochem. J.* **1994**, *302*, 463.
28 Senior, D.J.; Mayers, P.R.; Miller, D.; Sutcliffe, R.; Tan, L.; Saddler, J.N. *Biotechnol. Lett.* **1988**, *10*, 907.
29 Jurasek, L.; Paice, M.G. *Biomass* **1988**, *15*, 103.
30 Nazareth, S.; Mavinkurve, S.; *Int. Biodeter.* **1987**, *23*, 343.
31 Sharma, H.S.S. *Int. Biodeter.* **1987**, *23*, 181.
32 Kobayashi, Y.; Komae, K.; Tanabe, H.; Matsuo, R. *Biotech. Adv.* **1988**, *6*, 29.
33 Kobayashi, Y.; Matsuo, R.; Hamada, Y.; Ohkawa, A.; Ebuchi, E. *Mokuzai Gakkaishi* **1984**, *30*, 848.
34 Sharma, H.S.S. *Int. Biodeter.* **1987**, *23*, 329.
35 Sreenath, H.K.; Shaw, A.B.; Yang, V.Y.; Gharia, M.M.; Jeffries, T.W. *J. Ferm. Bioengineer.* **1995**, *81*, 18.

36 Morvan, C.; Jauneau, A.; Flaman, A.; Millet, J.; Demarty, M. *Carbohyd. Polym.* **1990**, *13*, 149.
37 Irie, Y.; Matsukura, M.; Usui, M.; Hata, K. Proc. *Tappi Papermakers Conf.* **1990**, Tappi Press: Atlanta, 1.
38 Malgrem, K.; Pedersen, L.S.; Skjold-Jøergensen, S. International pat. app. WO 91/07542, 1991.
39 Fischer, K.; Messner, K. *Tappi J.* **1992**, *75* (2), 130.
40 Fisher, K.; Messner, K. *Enzyme Microb. Technol.* **1992**, *14*, 470.
41 Fischer, K.; Puchinger, L.; Schloffer, K.; Kreiner, W.; Messner, K. *J. Biotechnol.* **1993**, *27*, 341.
42 Yamasaki, T.; Hosoya, S.; Chen, C.-L.; Gratzl, J.S.; Chang, H.-M. *The Ekman Days Int. Symp. Wood Pulping Chem.* Stockholm, Sweden, 1981, Vol. 2, pp. 34-42.
43 Gierer, J.; Wännström, S. *Holzforschung* **1984**, *38*, 181.
44 Jiang, J.-e.; Chang, H.-m. *J. Wood Chem. Technol.* **1987**, *7*, 81.
45 Tolan, J.S.; Canovas, R.V. *Pulp Paper Can.* **1992**, *93*, 39.
46 Scott, B.P.; Young, F.; Paice, M.G. *Pulp Paper Can.* **1993**, *94*, 57.
47 Viikari, L. Kantelinen, A. Poutanen, and Ranua, M. In: Kirk, T.K.; Chang, H.-m. Eds. *Biotechnology in Pulp and Paper Manufacture* Butterworth-Heinemann, Boston. **1990**. pp. 145-151
48 Clark, T.A.; McDonald, A.G.; Senior, D.J.; Mayers, P.R. In Kirk, T.K; Chang, H.-m., Eds. *Biotechnology in Pulp and Paper Manufacture*; Butterworth-Heinemann: Boston, Mass., **1990**, 153-167.
49 Buchert, J.; Kantelinen, A.; Rättö, M.; Siika-aho, M.; Ranua, M.; Viikari, L. In Kuwahara, M.; Shimada, M., Eds. *Biotechology in Pulp and Paper Industry.* Uni Publishers: Tokyo, **1992**, pp. 139-144.
50 Buchert, J.; Kantelinen, A.; Rätto, M.; Siika-aho, M.; Ranua, M.; Viikari, L. In Kuwahara, M., and Shimada, M. Eds. *Biotechnology in Pulp and Paper Industry.* Uni Publishers Co. Ltd., Tokyo, Japan. 1992. pp. 139-144.
51 Trotter, P.C. *Tappi J.* **1990**, *73*, 201.
52 Jeffries, T.W. *ACS Symp Ser.* **1992**, *476*, 313.
53 Jeffries, T.W.; Patel, R.N.; Sykes, M.S.; Klungness, J. *Mat. Res. Soc. Proc.* **1992**, *266*, 277.
54 Shimoto, H.; Sharyo, M.; Kiriyama, T.; Sakaguchi, H. Kami pa Gikyoshi **1991**, *45*, 1316.
55 Kantelinen, A.; Sundquist, J.; Linko, M.; Viikari, L. 1991. In *Proc 6th Int. Symp. Wood Paper Chem.,* Vol. 1, pp. 493-500.
56 Kantelinen, A.; Hortling, B.; Sundquist, J.; Linko, M.; Viikari, L. *Holzforschung,* **1993**, *47,* 318.
57 Yllner, S.; Östberg, K.; Stockman, L. *Svensk Papp.* **1957**, *60*, 795.
58 Axelsson, S.; Croon, I.; Enström, B. *Svensk Papperstidn.* **1962**, *65*, 693.
59 Meller, A. *Holzforschung* **1965**, *19*, 118.
60 Suurnäkki, A.; Heijnesson, A.;, Buchert, J.; Viikari, L.; Westermark, U. *J. Pulp Paper Sci.* **1996**, *22*, J43.
61 Skjold-Jørgensen, S.; Munk, N.; Pedersen, L.S. In Kuwahara, M.; Shimada, M. Eds. *Biotechnology in Pulp and Paper Industry.* Uni Publishers Co. Ltd., Tokyo, Japan. **1992**. pp. 93-99.
62 Paice, M.G.; Gurnagul, N.; Page, D.H.; Jurasek, L. **1992**, *Enzyme Microb. Technol. 14*, 272.
63 Senior, D.J.; Mayers, P.R.; Breuil, C.; Saddler, J.N. In Kirk, T.K. and Chang, H.-m. Eds. *Biotechnology in Pulp and Paper Manufacture* Butterworth-Heinemann, Boston. 1990. pp. 169-182.

64 Viikari, L.; Tenkanen, M.; Rättö, M.; Buchert, J.; Kantelinen, A.; Bailey, M.; Sundquist, J.; Linko, M. 1992. In Kuwahara, M.; Shimada, M. Eds. *Biotechnology in Pulp and Paper Industry.* Uni Publishers Co. Ltd., Tokyo, Japan. **1992**. pp. 101-113.
65 Högman, S.; Jöves, H.; Rosenberg, E.; Shohem, Y. In Kuwahara, M.; Shimada, M. Eds. *Biotechnology in Pulp and Paper Industry.* Uni Publishers Co. Ltd., Tokyo, Japan. **1992**. pp. 107-115.
66 Paice, M.G.; Bernier, R.; Jurasek, L. *Biotechnol. Bioengineer.* **1988**, *32*, 235.
67 Roberts, J.C.; McCarthy, A.J.; Flynn, N.J.; Broda, P. *Enzyme Microb. Technol.* **1990**, *12*, 210-213.
68 Patel, R.N.; Grabski, A.C.; Jeffries, T.W. *Appl. Microbiol. Biotechnol.* **1993**, *39,* 405.
69 Garg, A.P.; McCarthy, A.J.; Roberts, J.C. *Enzyme Microb. Technol.,* **1996**, *18,* 261.
70 Elegir, G.; Sykes, M.; Jeffries, T.W. **1995**, In *Proceedings of the 6th Int. Conf. Biotechnol. Pulp Paper Industry* Vienna, Austria, 23.
71 Davies G.; Henrissat B. *Structure* **1995**, *3*, 853.
72 Buchert, J.; Ranua, M.; Kantelinen, A.; Viikari, L. **1992**, *Appl. Microbiol. Biotechnol.* *37*, 825.
73 Tenkanen, M.; Buchert, J.; Viikari, L. *Enzyme Microb. Technol.* **1995**, *17*, 499.
74 Buchert, J.; Tenkanen, M.; Pitkänen, M.; and Viikari, L. *Tappi J.* **1993**, *76*, 131.
75 Törrönen, R.L.; Messner, M.R.; Gonzalez, R.; Kalkkinen, N.; Harkki, A.; Kubicek, C.P.; *Biotechnology* **1992**, *10*, 1461.
76 Tenkanen, M.; Puls, J.; Poutanen, K.; *Enzyme Microb. Technol.* **1992**, *14*, 566.
77 Törrönen, A.; Rouvinen, J; *Biochemistry* **1995**, *34*, 847.
78 Bernier, R. L.; Kluepfel, D.; Morosoli, R.; Shareck, F. United States pat. 5,116,746 1992.
79 Bergquist, P. L.; Gaibbs, M. D.; Saul, D. J.; Te'o, V. S. J.; Dwivedi, P. P.; Morris, D.; Donald, A.; Kessler, A. **1993**, In *Proceedings, Seventh Int. Symp. Wood Pulping Chem.* Beijing, P. R. China., 596.
80 Nath, D.; Rao, M. *Biotech. Lett.* **1995**, *17*, 557.
81 Yang, V.W.; Zhuang, Z.; Elegir, G.; Jeffries, T.W. *J. Ind. Microbiol.*, **1995**,*15*, 434.
82 Elegir, G.; Sykes, M.; Jeffries, T.W. *Enzyme Microb. Technol.* **1995**. *17,* 954.
83 Brookshaw, C.B. Ed. *Pulp and Paper North American Fact Book.* **1994**, Miller Freeman, San Francisco, 458 p.
84 Iannazzi, F.D.; Strauss, R. In *Proceedings of Waste Paper IV Conference*, Chicago, April 15, 1993.
85 Jaakko Poyry Consulting Inc. and Franklin Associates Ltd., *Progr. Paper Recycl.,* August, 1992.
86 Oji Paper. Japanese pat. JP 2080683, 1990.
87 Prasad, D.Y.; Heitmann, J.A.; Joyce, T.W. *Prog. Paper Recycl.* **1992**, *1*, 21.
88 Jeffries, T.W.; Klungness, J.H.; Sykes, M.S.; Rutledge-Cropsey, K. 1993. *Proc. 1993 TAPPI Recycling Symp.*, pp. 183-188.
89 Jeffries, T.W.; Klungness, J.H.; Sykes, M.S.; Rutledge-Cropsey, K. *Tappi J.* **1994**, *77*, 173.
90 Heise, O.U.; Unwin, J.P.; Klungness, J.H., Fineran, W.G., Jr., Sykes, M.; Abubakr, S. *Tappi J.*, **1996,** *79,* 207.
91 Woodward, J.; Stephan, L.M.; Koran, Jr., L. J.; Wong, K.K.Y.; Saddler, J.N. *Bio/Technology* **1994**, *12*, 905.
92 Lion. Japanese pat. JP 2080684, 1990.
93 Sharyo, M.; Sakaguchi, H.; Kokai T.K. Japanese pat. JP 02160984, 1990.
94 Schulz, W.J. United States pat. 4,450,043, 1984.

95 Thornton, J.W. *Tappi J.* **1994**, *77*, 161.
96 Rättö, M.; Kantelinen, A.; Bailey, M.; Viikari, L. *Tappi J.* **1993**, *76*,125.
97 Kirk, T.K.; Connors, W.J.; Bleam, R.D.; Hackett, W.F.; Zeikus, J.G. *Proc. Nat. Acad. Sci.* **1975**, *72*, 2515.
98 Lundquist, K.; Kirk, T.K.; Connors, W.J. *Arch. Microbiol.* **1977**, *112*, 291.
99 Joyce, T.W.; Chang, H.-m.; Campbell, A.G.; Gerrard, E.D.; Kirk, T.K. *Biotech. Adv.* **1984**, *2*, 301.
100 Kirk, T.K.; Yang, H.H. *Biotechnol. Lett.* **1979**, *1*, 347.
101 Reid, I. D.; Paice, M.G. *FEMS Microbiol. Rev.* **1994**, *13*, 369.
102 Cullen, D.; Kersten, P.J. In Brambl, R.; Marzluf, G.A. Eds. *The Mycota, Vol III*, Springer-Verlag, Berlin, **1996**, P. 295-314.
103 Bao, W.; Fukushima, Y.; Jensen, K.A.; Moen, M.A.; Hammel, K. *FEBS Lett.* **1994**, *354*, 297.
104 Hammel, K.E.; Jensen, K.A.; Mozuch, M.D.; Landucci, L.L.; Tien, M.; Pease, E.A. *J. Biol. Chem.* **1993**, *268*, 12274.
105 Paice, M.G.; Bourbonnais, R.; Reid, I.D.; Archibald; Jurasek, L. *J. Pulp Paper Sci.* **1995**, *21*, J280.
106 Paice, M.G.; Bourbonnais, R.; Reid, I.D. *Tappi J.* **1995**, *78*, 161.
107 Weinstock, I. A., R. H. Atalla, R. S. Reiner, M. A. Moen, K. E. Hammel, C. J. Houtman, C. L. Hill. *New J. Chem.* **1996**, *20*, 269.

Chapter 2

Enzyme-Aided Bleaching of Kraft Pulps: Fundamental Mechanisms and Practical Applications

Liisa Viikari, A. Suurnäkki, and J. Buchert

VTT Biotechnology and Food Research, P.O. Box 1501, FIN–02044 VTT, Finland

Hemicellulase-aided bleaching is one of the few well established, ecomically feasible biotechnical applications in the pulp and paper industry. The enzymatic treatments, based on xylanases and mannanases, introduce modifications in the carbohydrate structures, leading to enhanced delignification. The mechanism is based on the partial depolymerization of hemicelluloses, which impede the chemical removal of residual lignin from pulp fibres. The enzymatic method can be combined to various types of kraft pulping processes and bleaching sequences. The benefits obtained by the enzymatic treatment depend on the type of raw material, pulping process and bleaching sequence. The enzymatic step leads to reductions in chemical consumption and costs and maintains product quality. The method is applied in mill scale in several countries. This review describes the present knowledge about the mechanisms of the method as well as the practical results obtained.

The pulp bleaching technologies entered a new era in late 80s due to growing concern about the formation and release of chlorinated compounds in the recipient. In this context, the hemicellulase-aided bleaching, first introduced in 1985, offered a new type of environmentally safe approach (*1*). The hemicellulase treatment is an indirect bleaching method, rendering the fibres more accessible to bleaching chemicals and leading to more efficient delignification. After the launching of commercial hemicellulases to the markets, the method was soon adopted at many mills. Since then, however, other bleaching technologies have also been developed. These include various chlorine-free, oxygen based bleaching sequences using chemicals, such as peroxide, oxygen and ozone.

The location and structure of hemicelluloses in the fibres affect the delignification of pulps as well as the technical properties of the fibre products. The removal of residual lignin seems to be both physically and chemically restricted by

0097–6156/96/0655–0015$15.00/0
© 1996 American Chemical Society

hemicelluloses in the fibre matrix. Lignin-carbohydrate linkages have been proposed to restrict the chemical removal of residual lignin from pulp fibres (*1*). Due to solubilization and relocation during pulping process, hemicelluloses and lignin may form physically or chemically interlinked matrices in the fibres. Different hypotheses for the mechanism have been presented. It has been suggested that xylanases attack xylan chemically bound to lignin (*1-3*) or degrade reprecipitated xylan, deposited on fibre surfaces during the cook (*4*). The type of xylan attacked has also been suggested to contain chromophores (*5*). Hitherto, all these assumptions have gained some support. However, it is obvious that both the type of raw material (hardwood or softwood) and the pulping conditions profoundly affect the amount and location of hemicelluloses and lignin in the fibres. Recent results in pulp chemistry have also illuminated the fundamentals of hemicellulase-aided bleaching. In addition to xylanases, mannanases capable of attacking pulp glucomannans have been shown to offer great potential in improving the bleachability of new types of low kappa number pulps.

The scientific interest in this method is reflected in the number of papers published during recent years describing numerous xylanases from new sources, as well as bleaching results obtained using various hemicellulases, pulps and bleaching sequences. Several reviews have been published (*6-9*). In this paper, current knowledge on the proposed mechanisms and practical results are reviewed.

Structure of Hemicelluloses in Kraft Pulps

The native hemicellulose structure is heavily modified during pulping processes. In the beginning of conventional sulphate i.e. kraft cooking, xylan in wood is partly solubilized in the alkaline cooking liquid and many of the side groups and acetic acid residues are cleaved off (*10, 11*). It has recently been observed that the majority of the 4-O-methylglucuronic acid side groups in xylan are converted to hexenuronic acid already in the early phases of the kraft cook (*12, 13*).

As the alkali concentration decreases towards the end of the kraft cook, dissolved xylan tends to readsorb on the surface of cellulose microfibrils (*14, 15*). It has recently been suggested that pine kraft xylan reprecipitates evenly to all accessible surfaces to the fibre wall (*16*). In addition to xylan chains, dissolved lignin and covalently bound lignin and xylan have been suggested to reprecipitate to fibre surfaces during cooking (*17, 18*), resulting in relatively high amounts of lignin on the fibre surfaces (*19-21*). The amount of xylan readsorbed during cooking depends on the wood species used in pulping. High amounts of xylan have been found to locate on the surface of birch kraft fibres, probably partly due to readsorption, whereas in pine kraft fibres the concentration of xylan on the fibre surfaces has not been observed to be higher than in the whole fibres (*22*). A large part of wood glucomannan is also dissolved in the beginning of the kraft cook, but due to their instability in alkali, the solubilized polymers are completely degraded in the pulping liquor (*15, 23, 24*).

As a result of the solubilization of hemicelluloses during cooking the distribution and content of xylan and glucomannan in kraft pulp fibres differ from that in the native wood fibres (*25*). In softwood kraft fibres the xylan concentration is generally higher in outer layers, and glucomannan is more concentrated in the middle layers of the fibre. However, due to different analysis methods variations in the

distribution of polysaccharides in softwood kraft fibres have been reported (*26-28*), although there is general agreement that the outer surface layer of hardwood kraft fibres is rich in xylan.

Recently, several modified kraft pulping methods as well as totally new sulphate pulping methods have been introduced (*29*). In the pulps produced by these methods no or little reprecipitation of xylan and lignin is expected to occur due to the relatively constant alkali concentration throughout the cooking process. Consequently, the composition of outer surfaces of pulp fibres is probably different from that of the conventional kraft pulp fibres. In sulphite cooking, hemicellulose is extensively solubilized to mono- and oligomeric compounds and no reprecipitation occurs (*30*). Thus, the distribution of hemicellulose is relatively constant across the pulp fibres.

Hemicellulases

Several species of fungi and bacteria are known to produce the whole spectrum of hemicellulose-degrading enzymes; xylanases and mannanases (*31-36*). Most of the xylanases characterized are able to hydrolyze xylans from various origins, showing differences only in the spectrum of end products. The main products formed from the hydrolysis of xylans are xylobiose, xylotriose and substituted oligomers of two to five xylosyl residues. The chain length and the structure of the substituted products depend on the mode of action of the individual xylanase. Some xylanases, however, show rather strict substrate specificity. The three dimensional structures of several low molecular mass xylanases have recently been determined (*37-40*). The structure of the *Trichoderma reesei* pI 9 xylanase is ellipsoidal, having dimensions of about 30 to 40 Å (*39*). Unlike most cellulases, it does not contain any separate substrate binding domain. Some bacterial xylanases, however, have been found to contain either a cellulose binding domain (*41-43*) or a xylan binding domain (*44, 45*).

Compared with xylanases, mannanases are a more heterogenous group of enzymes. The main hydrolysis products from galactomannans and glucomannans are mannobiose, mannotriose and various mixed oligosaccharides. The hydrolysis yield is dependent on the degree of substitution as well as on the distribution of the substituents (*46*). The hydrolysis of glucomannans is also affected by the glucose/mannose ratio. Recently, the mannanase of *T. reesei* was found to have a multidomain structure similar to that of several cellulolytic enzymes (*47*). The protein contains a catalytic core domain which is connected by a linker to a cellulose binding domain (*48*). Hitherto, no three dimensional structures of mannanases have been published.

Most xylanases studied are active in slightly acidic conditions between pH 4 and 6 and at temperatures below 70°C. More thermophilic and alkalophilic xylanases are of great importance due to the prevailing conditions in pulp processing. Xylanases which are stable and function efficiently at high temperatures are produced by several thermophilic bacteria (*36*). The most thermophilic xylanases hitherto described are produced by the extremely thermophilic bacterium *Thermotoga* sp. (*43, 44*). Several xylanase genes encoding proteins active at temperatures from 75°C up to 95°C (pH 6-8) have been isolated. Thermophilic mannanases have been purified e.g. from *C. saccarolyticus* and *Thermotoga neapolitana*. Xylanases and mannanases with alkali pH optimae have been detected in an alkalophilic *Bacillus* sp. (*36*).

The main enzymes needed to enhance the delignification of both hardwood and softwood kraft pulp have been shown to be endo-β-xylanases (*49, 50*). A positive effect has been achieved with most xylanases studied, independently of the origin of the enzyme. Both fungal and bacterial xylanases have been shown to increase the bleachability (*51*) and several commercial xylanases are available, varying with respect to their pH and temperature optimae.

Mannanases, on the other hand, appear to be more specific with respect to their substrate, and only few mannanases have been shown to hydrolyze glucomannans in softwood pulps. Recently, the effects of purified or partially purified endo-acting β-mannanases from *Caldocellulosiruptor saccharolyticum, Aspergillus niger* and *Trichoderma reesei* on pulp delignification were compared in bleaching (Suurnäkki, A. *et al. Tappi J.*, in press). Of these enzymes, the *T. reesei* mannanase was shown to be most efficient. The mannanase of *B. subtilis* has been shown to be able to solubilize wood mannan but was totally unable to solubilize mannan which was bound to kraft pulp (*52*). The first bleach-boosting mannanase product, produced by *T. reesei*, emerged on the market in 1995.

As compared with xylanases and mannanases, the side-group cleaving enzymes alone or in combination with endoenzymes have had only minor effects on pulp bleachability (*6*). Other purified enzymes which have been studied for improving the bleachability of pulps include individual cellulolytic enzymes (*53*). Only the unspecific endoglucanase I from *Trichoderma reesei,* also exhibiting xylanase activity, was shown to increase the bleachability.

Mechanisms of Hemicellulase-Aided Bleaching

The effect of hemicellulases in bleaching is based on the modification of pulp hemicelluloses, enhancing the removal of lignin in chemical bleaching. It has been proposed that the action of xylanases is due to the partial hydrolysis of reprecipitated xylan (*4*) or to removal of xylan from the lignin-carbohydrate (LC) complexes (*3*). However, these hypotheses are not mutually exclusive, i.e. relocated xylans may contain LC complexes and both mechanisms would allow the enhanced diffusion of entrapped lignin from the fibre wall. Limited removal of pulp xylan is known to increase the leachability of residual lignin from kraft pulps (*54*) and thus also to increase the pulp bleachability during subsequent bleaching stages. In addition, it has been suggested that the hemicellulase treatment removes chromophoric groups from the pulp (*5, 51*). The suggested mechanisms as well as their consequences are presented in Figure 1. The methods used for mechanistic studies have included modified pulping methods, production of model pulps, analysis of degradation products of enzymatic treatments, chemical extractions of lignin and xylans, mechanical peeling, surface analysis by ESCA and different delignification tests.

Relocated Xylan. Both xylan and lignin are dissolved and partially readsorbed on the fibres during pulping. A rather high content of lignin has been observed both in the primary fines and in the surface material of pine kraft fibres (*19, 20*). Xylanases combined with ESCA have been used to determine the xylan content on the outer surfaces of cellulose fibres. In softwood kraft fibres, removal of xylan by xylanases was found to uncover lignin (*21*). The relatively low amount of xylan, observed on the

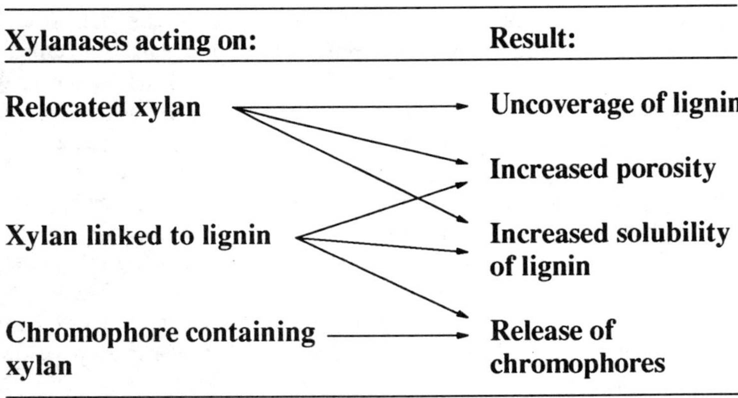

Figure 1. Suggested mechanisms of xylanase-aided bleaching.

outer surfaces of fibres, can obviously be easily removed by xylanases (*22*). Thus, it can be expected that removal of xylan improves the extractability of lignin by exposing lignin surfaces. As dissolved xylan molecules can penetrate most of the pores in cellulose fibres, it can be postulated that xylanases, acting on all available surfaces, enhance the removal of lignin in the whole fibre. In birch kraft pulp, the primary fines and the fibre surface material were found to be considerably richer in both xylan and lignin than the whole pulp (*22*). Therefore, relatively more xylan should be enzymatically removed from the outer surfaces and expose the lignin. The role of reprecipitated xylan in the xylanase-aided bleaching of birch kraft pulps has been confirmed by comparing the effect of xylanase treatment on bleachability of kraft pulps cooked by a batch method and of pulps produced in a flow-through digester and therefore containing only traces of reprecipitated xylan (*4*).

On the other hand, xylanase pretreatment has later also been reported to enhance the bleachability of softwood pulps produced by novel cooking methods with more stable alkali profiles (*55, 56*) and presumably thus containing less reprecipitated xylan than conventional softwood kraft pulps. In addition, the role of reprecipitated xylan in xylanase-aided bleaching has been studied by extraction with dimethyl sulfoxide (DMSO), a chemical which has been claimed to remove reprecipitated xylan selectively from pulps (*57*). This treatment did not improve the bleachability of kraft pulp, whereas xylanase treatment did. Recently, however, Allison *et al.* (*58*) reported that the removal of pulp xylan by DMSO is dependent on the degree of polymerization (DP) of xylan. As the DP of reprecipitated xylan is not known, the role of reprecipitated xylan in xylanase-aided bleaching cannot be conclusively determined by DMSO extractions. Despite the obvious analytical difficulties in determining relocated xylans, it can be expected that the mechanism of xylanase-aided bleaching is not based on hydrolysis of relocated xylan alone.

Lignin Carbohydrate Complexes. Both softwood and hardwood kraft pulps have been reported to contain LC-complexes in which carbohydrates and lignin may be connected to each other by ether or glycosidic linkages (*2, 59*). However, no direct evidence for the type of linkage(s) existing between carbohydrates and lignin has yet

been presented. Increased solubilization of xylan-lignin complexes both from model pulps (*60*) and from kraft pulps (*3, 61*) has been observed by xylanase treatment.

According to GPC analyses, part of the lignin released during the enzymatic treatment appears to be covalently bound to xylan, whereas most of the lignin may be physically interlinked with xylan in the fibre matrix. The action of xylanases on both reprecipitated and LC-xylan in enhancing bleachability suggests that it is probably not only the type but also the location of the xylan that is important in the mechanism of xylanase-aided bleaching. The xylanase of *T. reesei* has been observed to act rather uniformly in all accessible surfaces of kraft pulps (*62, 63*), indicating that the effect of xylanase on bleachability is not only an outer surface phenomenon. The composition of xylan solubilized in limited or extensive treatments has not revealed essential differences, indicating that the structure of xylan is rather similar in all parts of fibres, or that the enzymes are specific to a certain type of xylan (*61*).

Chromophores. It has frequently been observed that xylanase treatment has a slight decreasing effect on the kappa number. This has been explained to be due to removal of lignin fragments or chromophoric structures from pulp (*61*). However, the reduction in the kappa number as measured by permanganate oxidation may be partially due to an artefact. The recently discovered hexenuronic acid (*12*), containing a double bond, may give rise to the consumption of permanganate, increasing the apparent kappa number. Thus, enzymatic removal of xylan containing hexenuronic acid groups can lead to a lower kappa number.

Mode of Action of Mannanases. Compared with xylanase-aided bleaching, the mechanism of mannanase-aided bleaching has attracted only minor interest, probably due to its rather limited effect in most pulp types. However, the mechanism of mannanase-aided bleaching has been assumed to differ from that of xylanase-aided bleaching, due to the different distribution of glucomannan and xylan in pulp fibres (*63, 64*). Unlike in the case of xylanases, no correlation between the amount and composition of enzymatically solubilized glucomannan and the effect on bleachability has been observed. However, the role of the composition and configuration of the outer surfaces of pulp fibres seems to be important in mannanase-aided bleaching. Mannanase treatment was found to enhance the bleachability of pulps produced by modified or continuous pulping methods (*56*, Suurnäkki, A. *et al.*, Tappi J., in press), which are generally considered to contain less reprecipitated xylan and lignin on the fibre surfaces. Furthermore, mannanase treatment was effective in enhancing the bleachability of conventional pine kraft fibres only when the outer surface material of fibres was mechanically removed prior to treatment (*62*). It is possible that underneath the outermost surface layer of kraft fibres and on the surface of modified pulps the glucomannan is more closely located to lignin and that the enzymatic removal of glucomannan therefore increases the leachability of lignin.

Practical Results

The main goal in the enzymatic bleaching of kraft pulps has been to reduce the consumption of chlorine chemicals in the traditional and ECF bleaching processes. However, enzymes can also be used successfully for increasing the brightness of pulp,

which is of key importance in the development of totally chlorine-free (TCF) bleaching sequences. Addition of an enzymatic step to any conventional chemical bleaching sequence results in a higher final brightness value of the pulp. During the past five years, the method has been continuously used in industrial scale together with low-chlorine or totally chlorine-free bleaching methods. The reasons for using enzymes vary depending on the conditions at mill sites. Today xylanases are used both in ECF and TCF sequences. In ECF sequences the enzymatic step is often adopted due to the limiting chlorine dioxide production capacity. The use of enzymes allows bleaching to higher brightness values when chlorine gas is not used. In TCF sequences, the advantage of the enzymatic step is due to improved brightness, maintenance of fibre strength and savings in bleaching costs.

The enzymatic treatment is generally carried out with brown-stock or oxygen delignified pulp prior to the actual bleaching stage. As the first generation enzymes used in pulp treatments have generally been active in the pH range of 5-8 and at moderate temperatures of 40-60°C, adjustment of pH and temperature prior to the treatment has been necessary. Xylanase treatment has been studied to enhance pulp bleachability in traditional and various ECF and TCF bleaching sequences. The benefits attained by enzymatic pretreatment obviously vary between the different bleaching sequences (Suurnäkki, A. *et al.*, Tappi J., in press, *65, 66, 67*). In conventional kraft softwood and hardwood pulps, xylanases have been reported to result in an increase of about 2-6 ISO units in brightness values after modern bleaching sequences and in a decrease in chlorine chemical consumption of about 10-20% in conventional bleaching sequences. In softwood pulps produced by modified pulping procedures, ie. by EMCC and MCC, the effect of xylanase treatment on bleachability has been reported to be less pronounced than in conventional kraft pulps after ECF bleaching (*65, 66*). Oxygen delignification carried out prior to the actual bleaching sequence has been found to enhance the susceptibility of pulp hemicellulose to enzymatic solubilization (*4, 65, 68*). However, when ECF bleaching sequences are used, oxygen delignification has not been found to increase the effect of xylanase treatment on the bleachability of softwood kraft pulps (*69, 70*).

In 1992 more than ten mills worldwide were reported to use xylanases continuously for improved bleaching of kraft pulps. Almost one hundred mill trials have been carried out, about half of them in Europe. Most of the kraft pulp in Europe is produced in Scandinavia, where most of the mill trials have also been performed. Different paper products, including magazine papers (SC, LWC) and tissue papers, manufactured from enzymatically treated pulps have been successfully introduced to the markets.

Conclusions

Hemicellulases were the first group of specific enzymes used in large scale in the pulp and paper industry. The method is also an example of sustainable technology in the traditional chemical industry. The method has clear environmental benefits and is economically attractive. The hemicellulase treatment, together with a chemical extraction, leads to a significant reduction in the residual lignin content of the pulps. The partial hydrolysis of xylan facilitates the extraction of lignin from pulp in higher amounts and with higher molecular weights. However, due to the indirect mode of

action, the effect of hemicellulase-aided bleaching is limited. The improved bleachability is mainly based on the action of endo-β-xylanases, a group of enzymes which can be efficiently produced in industrial scale. In addition to lignin modifying enzymes, new commercial hemicellulases with higher pH and temperature optima should improve the applicability of enzymes.

The two main mechanisms proposed, *i.e.* the action of enzyme in reprecipitated xylan or in LCC-xylan in enhancing the leachability of lignin, are not mutually exclusive and seem both to be valid in the xylanase-aided bleaching of kraft pulps. The relative importance of the two types of xylans in the mechanism seems to depend on the type of pulp. The bleachability of pine kraft pulp has been found to be most affected by the action of xylanase on xylan, physically or chemically linked to lignin, in all accessible pulp surfaces. In birch kraft pulp, the xylanases appear to act most efficiently on the outer surfaces of fibres, possibly mostly on the reprecipitated xylan. In the case of mannanases, the structure of the outer surface of fibres seems to be most important in determining the efficiency of enzyme-aided bleaching. In addition to improvement of the enzymatic application, research on the action of enzymes in cellulose fibres has resulted in new analytical methods and improved knowledge of wood chemistry.

Literature Cited

1. Viikari, L.; Ranua, M.; Kantelinen, A.; Sundquist, J.; Linko, M. *Proc. 3rd Int. Conf. Biotechnology in the Pulp and Paper Industry*; STFI: Stockholm, Sweden, 1986; p 67.
2. Yamasaki, T.; Hosoya, S.; Chen, C.-L.; Gratzl, J. S.; Chang, H.-M. *Proc. Int. Symp. Wood Pulp. Chem.*, Stockholm, 1981; p 2:34.
3. Yang, J. L.; Eriksson, K.-E. L. *Holzforschung* **1992,** *46*, p 481.
4. Kantelinen, A.; Hortling, B.; Sundqvist, J.; Linko, M.; Viikari, M. *Holzforschung* **1993**, *47*, p 318.
5. Wong, K. K. Y.; Clarke, P.; Nelson, S. L. *ACS Symp. Ser.*; 618; 1996; p 352.
6. Viikari, L.; Kantelinen, A.; Sundqvist, J.; Linko, M. *FEMS Microb. Rev.* **1994**, *13*, p 335.
7. Bajpai, P.; Bajpai, P. K. *Process Biochem.* **1992**, *27*, p 319.
8. Onysko, K. A. *Biotech. Adv.* **1993**, *11*, p 179.
9. Denault, C.; Leduc, C.; Valade, J. L. *Tappi J.* **1994**, *77*, p 125.
10. Hamilton, J. K.; Partlow, E. V.; Thompson, N. S. *Tappi* **1958**, *41*, p 803.
11. Croon, I.; Enström, B. F. *Tappi* **1961**, *44*, p 870.
12. Teleman, A.; Harjunpää, V.; Tenkanen, M.; Buchert, J.; Hausalo, T.; Drakenberg, T.; Vuorinen, T. *Carbohydr. Res.* **1995**, *272*, p 55.
13. Buchert, J.; Teleman, A.; Harjunpää, V.; Tenkanen, M.; Viikari, L.; Vuorinen, T. *Tappi J.* **1995**, *78*, p 125.
14. Yllner, S.; Enström, B. *Svensk Papperstidn.* **1956**, *59*, p 229.
15. Yllner, S.; Östberg, K.; Stockman, L. *Svensk Papperstidn.* **1957**, *60*, p 795.
16. Mitikka, M.; Teeäär, R.; Tenkanen, M.; Laine, J.; Vuorinen, T. *Proc. 8th Int. Symp. Wood and Pulping Chem.*, Helsinki, Finland, 1995; Vol. 3, p 231.
17. Simonsson, R. *Svensk Papperstidn.* **1971**, *74*, p 691.

18. Jansson, J.; Palenius, I.; Stenlund, B.; Sågfors, P.-E. *Paperi ja Puu* **1975**, *57*, p 387.
19. Laine, J.; Stenius, P.; Carlsson, G.; Ström, G. *Cellulose* **1994**, *1*, p 145.
20. Heijnesson, A.; Simonson, R.; Westermark, U. *Holzforschung* **1995**, *49*, p 313.
21. Buchert, J.; Carlsson, G.; Viikari, L.; Ström, G. *Holzforschung* **1996**, *50*, p 69.
22. Suurnäkki, A.; Heijnesson, A.; Buchert, J.; Viikari, L.; Westermark, U. *J. Pulp Paper Sci.* **1996**, *22*, p J43.
23. Aurell, R.; Hartler, N. *Svensk Papperstidn.* **1965**, *68*, p 59.
24. Casebier, R. L.; Hamilton, J. L. *Tappi* **1965**, *48*, p 664.
25. Sjöström, E. *Wood Chemistry, Fundamentals and Application,* 2 edn., Academic Press Inc.: San Diego, CA, 1993.
26. Luce, J. E. *Pulp Paper Mag. Can.* **1964**, *65*, p T419.
27. Scott, R. W. *J. Wood Chem. Technol.* **1984**, *4*, p 199.
28. Bachner, K.; Fischer, K.; Bäucker, E. *Das Papier* **1993**, *10A*, p V30.
29. McDonough, T. J. *Tappi J.* **1995**, *78*, p 55.
30. Annegren, G. E.; Rydholm, S. A. *Svensk Papperstidn.* **1959**, *62*, p 737.
31. Biely, P. *Trends Biotechnol.* **1985**, *3*, p 286.
32. Poutanen, K. *Characterization of xylanolytic enzymes for potential applications.* Ph.D. Thesis, Technical Research Centre of Finland, Publications 47, Espoo, Finland, 1988.
33. Eriksson, K.-E.; Blanchette, R. A.; Ander, P. *Microbial and Enzymatic Degradation of Wood and Wood Components.* Springer-Verlag: Berlin, Germany, 1990.
34. Wong, K. K. Y.; Saddler, J. N. *Crit. Rev. Biotechnol.* **1992**, *12*, p 413.
35. Coughlan, M. P.; Hazlewood, G. P. *Biotechnol. Appl. Biochem.* **1993**, *17*, p 259.
36. Viikari, L.; Tenkanen, M.; Buchert, J.; Rättö, M.; Bailey, M.; Siika-aho, M.; Linko, M. In *Bioconversion of Forest and Agricultural Plant Residues*; Saddler, J.N., Ed.; CAB International: Wallingford, UK, 1993; p 131.
37. Okada, H. *Adv. Prot. Des.* **1989,** *12*, p 81.
38. Campbell, R. L.; Rose, D. R.; Wakarchuk, W. W.; To, R.; Sung, W.; Yaguchi, M. *Proceedings of the Second TRICEL Symposium on Trichoderma Cellulases and Other Hydrolases*, Foundation for Biotechnical and Industrial Fermentation Research: Helsinki, 1993; p 63.
39. Törrönen, A.; Harkki, A.; Rouvinen, J. *EMBO J.* **1994**, *13*, p 2493.
40. Törrönen, A.; Rouvinen, J. *Biochemistry* **1995**, *34*, p 847.
41. Hazlewood, G. P.; Gilbert, H. J. In *Xylan and Xylanases;* Visser, J. Beldman, G., Kusters-van Someren, M. A., Voragen, A. G. J., Eds.; Elsevier Science Publishers: Amsterdam, Netherlands, 1992; p 259.
42. Sakka, K.; Kojima, Y.; Kondo, T.; Karita, S.; Ohmiya, K.; Shimada, K. *Biosci. Biotech. Biochem.* **1993**, *57*, p 273.
43. Winterhalter, C.; Liebl, W. *Appl. Environ. Microbiol.* **1995**, *61*, p 1810.
44. Shareck, F.; Roy, C.; Yaguchi, M.; Morosoli, R.; Kluepfel, D.; *Gene* **1991**, *107*, p 75.
45. Irving, D.; Jung, E. D.; Wilson, D. B. *Appl. Environ. Microbiol.* **1994**, *60*, p 763.
46. McCleary, B. V. In *Enzymes in Biomass Conversion*; Leatham, G. F., Himmel, M. E. Eds.; ACS Symp. Ser. 460; American Chemical Society: Washington, 1991; p 437.

47. Tenkanen, M.; Buchert, J.; Viikari, L. *Enzyme Microb. Technol.* **1995**, *17*, p 499.
48. Stålbrand, H.; Saloheimo, A.; Vehmaanperä, J.; Henrissat, B.; Penttilä, M. *Appl. Environ. Microbiol.* **1995**, *61*, p 1090.
49. Paice, M.; Bernier, M.; Jurasek, L. *Biotechnol. Bioeng.* **1988**, *32*, p 235.
50. Tenkanen, M.; Buchert, J.; Puls, J.; Poutanen, K.; Viikari, L. In *Biotechnology in Pulp and Paper Processing;* Kuwahara, M., Shimada, M. Ed.; Elsevier Science Publishers: Amsterdam, Netherlands, 1992; p 547.
51. Patel, R. N.; Grabski, A. C; Jeffries, T. W. *Appl. Microbiol. Biotechnol.* **1993**, *39*, p 405-412.
52. Rättö, M.; Siika-aho, M.; Buchert, J.; Valkeajärvi, A.; Viikari, L. *Appl. Microbiol. Biotechnol.* **1993,** *40*, p 449
53. Buchert, J.; Ranua, M.; Siika-aho, M.; Pere, J.; Viikari, L. *Appl. Microbiol. Biotechnol.* **1994**, *40*, p 941.
54. Hortling, B.; Korhonen, M.; Buchert, J.; Sundqvist, J.; Viikari, L. *Holzforschung* **1994**, *48*, p 441.
55. Allison, R.; Clark, T.; Suurnäkki, A. *Proc. 49th Appita Ann. General Conf.*, Hobart Tasmania, 1995; p 157.
56. Suurnäkki, A.; Kantelinen, A.; Buchert, J.; Viikari, L. *Tappi J.* **1994**, *77*, p 211.
57. Munk, N.; Nissen, A. M.; Vollmond, T.; Lund, H. *Proc. 47th Appita Ann. Gen. Conf.*, Rotorua New Zealand, 1992; Vol. 1, p 257.
58. Allison, R. W.; Clark, T. A.; Ellis, M. J. *Appita* **1995**, *48*, p 201.
59. Gellerstedt, G.; Lindfors, E.-L. *Proc. Int. Pulp Bleaching Conf.*, SPCI: Stockholm, Sweden, 1991; Vol. 1, p 73.
60. de Jong, E.; Wong, K. K. Y.; Windsor, L. R.; Saddler, J. N. In *Biotechnology in Pulp and Paper Industry - Advances in Applied and Fundamental Research;* Srebotnik, E.; Messner, K. Eds.; WUA Universitätsverlag: Vienna, Austria, 1996; p 127.
61. Suurnäkki, A. *Hemicellulases in the bleaching and characterization of kraft pulps.* Ph.D. Thesis, Technical Research Centre of Finland, Publications 267, Espoo, Finland, 1966.
62. Suurnäkki, A.; Heijnesson, A.; Buchert, J.; Westermark, U.; Viikari, L. *J. Pulp Paper Sci.* **1996**, *22*, p J91.
63. Suurnäkki, A.; Heijnesson, A.; Buchert, J.; Tenkanen, M.; Viikari, L.; Westermark, U. *J. Pulp Paper Sci.* **1996**, *22*, p J78.
64. Buchert, J.; Salminen, J.; Siika-aho, M.; Viikari, L. *Holzforschung* **1993**, *47*, 473.
65. Allison, R. W.; Clark, T. A.; Suurnäkki, A. *Proc. 49th Appita Ann. General Conf.*, Hobart, Tasmania, 1995, p 157.
66. Tolan, J. S. *Proc. TAPPI Pulping Conf.* Book 1, Boston, MA, 1992; p 13-17.
67. Pedersen, L. S.; Kihlgren, P.; Nielsen, A. M.; Munk, N.; Holm, H. C.; Choma, P. P. *Proc. Tappi 1992 Pulping Conf.*, Book 1, Tappi Press: Atlanta, USA, 1992; p 31.
68. Saake, B.; Clark, T.; Puls, J. *Holzforschung* **1995**, *49*, p 60.
69. Allison, R. W.; Clark, T. A.; Wrathall, S. H. *Appita* **1993**, *46*, p 349.
70. Allison, R. W.; Clark, T. A.; Ellis, M.J. *Appita* **1995**, *48*, p 201.

Chapter 3

Survey of Mill Usage of Xylanase

J. S. Tolan[1], D. Olson[2], and R. E. Dines[3]

[1]Iogen Corporation, 400 Hunt Club Road, Ottawa, Ontario K1G 3N3, Canada
[2]Fletcher Challenge Canada, Mackenzie, British Columbia, Canada
[3]R. E. Dines Associates, Vankleek Hall, Ontario, Canada

Xylanase enzymes have been used by mills since 1991 to enhance the bleaching of kraft pulp. Canadian mills treated 750,000 tonnes of pulp with xylanase in 1994, representing 8% of Canada's bleached kraft pulp production. This paper describes the results of a survey of the Canadian bleached kraft mills' experiences with xylanase enzymes.

Pulp Production and Bleaching. There are two broad categories of paper produced in roughly equal quantities around the world: high quality paper, which includes paper for writing, books, wrapping, packaging, and other specialized uses, and lower quality paper such as newsprint and that used for telephone directories. The high quality paper is produced from pulp which is primarily made by chemical processes while the low quality paper is primarily based on mechanical action. A general review of pulp and paper manufacturing has been published (*1*). This chapter is concerned with processing to make high quality paper.

Most of the pulp used for high-quality paper is manufactured by the kraft process and is known as kraft pulp. In the kraft process, wood chips are cooked in an alkali cooking liquor to break up the fiber integrity, solubilize the lignin around the fibers, and produce a brown pulp, known as brownstock. The brown colour is from the lignin remaining in the pulp that cannot be removed in the cooking process without destruction of the pulp fibers. Some brownstock is sold commercially without further chemical processing, such as for brown paper bags used in grocery stores. However, the majority of kraft pulp is bleached to varying degrees to improve its whiteness, strength, and other desired properties that depend on the finished product requirements. For example, writing paper is very white, and the bleaching is carried out to achieve the desired whiteness. Wrapping papers must be strong; in this case, the bleaching is carried out to the point where the strength of the paper is increased sufficiently. Photocopy paper is bleached to the point where it efficiently absorbs ink; this corresponds to a high degree of whiteness. A good review of the manufacture of kraft pulp has been published (*2*).

0097–6156/96/0655–0025$15.00/0
© 1996 American Chemical Society

The bleaching of the pulp is carried out in three to six consecutive chemical treatments, called stages. The layout and conditions of the stages differs between mills, but through the early-1980's there was a typical pattern of bleaching stages (*3*). In almost all mills, the first bleaching stage was carried out with chlorine (abbreviated as "C"). Chlorine gas was dissolved in water and then contacted with the pulp as the pulp flowed through a sealed pipe and up a tower, known as the chlorination tower. A typical usage of chlorine was 5% on weight of pulp, and chlorination was carried out a pH of about 2. The stage was designed such that all of the chlorine reacted by the time the pulp reached the top of the tower. The pulp was then washed, and the net result of the chlorination stage was the removal of about 75% of the lignin in the brownstock.

Recently, mills have used chlorine dioxide ("D") or a mixture of chlorine and chlorine dioxide ("$C_{50}D_{50}$") in the chlorination stage. In some mills, oxygen ("O") is used before the chlorination stage.

The second stage was carried out by adding sodium hydroxide to the chlorinated pulp to adjust its pH to about 10 and heating the pulp to about 70°C for 1 hour, then washing. This is the extraction stage ("E") and it results in the removal of lignin to the point where about 10% of the brownstock lignin remains. This is still enough lignin to maintain a brown colour in the pulp.

Recently, mills have included oxygen ("EO") and hydrogen peroxide ("EOP") in the first extraction stage.

The third stage is another acidic oxidation stage analogous to the chlorination stage, but this is carried out with chlorine dioxide and known as the chlorine dioxide stage ("D"). Chlorine dioxide is a gas that is unstable and generated on-site at a kraft mill. It was discovered in 1946 to be more effective as a whitening chemical than chlorine. Chlorine dioxide is used at a level of about 1% on pulp, at pH 4 and 80°C. The stage lasts three hours and is carried out in a tower, from which the pulp exits and is washed with water. At the end of this stage, the pulp is off-white and used for some applications, in which it is referred to as semi-bleached pulp.

The fourth and fifth stages are extraction (E) and chlorine dioxide (D) stages, as described above. After the fifth stage, the pulp is highly white and sold as market pulp.

The interest by the public in environmental implications of processes and the detection of chlorinated compounds in food packaging in the mid-1980's prompted pressure on the industry to decrease or eliminate the use of chlorine in bleaching. Some of the technologies that have been implemented to accomplish this have been the use of modified cooking processes to remove more of the lignin, the use of an oxygen reaction stage before the chlorination stage, the substitution of chlorine dioxide for chlorine in the chlorination stage, and the addition of hydrogen peroxide in the extraction stages. All of these technologies have been implemented by some mills in the past decade.

Xylanase Enzymes for Bleaching. One new technology has evolved to decrease the use of chlorine for bleaching, and that is treatment of the pulp with xylanase enzymes.

Xylanase is used in a mill by adding it to washed brownstock and allowing the enzyme to act on the pulp in the brownstock storage tower prior to the chlorination stage. The enzyme is supplied to the mill as an aqueous solution of protein and

additives such as stabilizers and preservatives. The commercially used xylanases have primarily been of *Trichoderma* and *Bacillus* origin. Because of the competitive nature of commercial processes, there have been few published reports of the ongoing industrial use of xylanase enzymes since the initial uses of xylanase in 1992.

The use of xylanase enzymes to enhance the bleaching of the pulp was first reported in 1986 (*4*). The first mill trials of xylanase in North America were carried out at Port Alberni in 1991 and ongoing usage in some mills started in 1992. Mill applications of xylanase have been widely reported (*5-10*).

Mill Survey. Given the increasing usage of xylanase for enhancing bleaching, a survey of mills regarding their xylanase usage was originated as a project by the Bleaching Committee of the Canadian Pulp and Paper Association. The objectives of the survey were to determine: 1. The major areas of effort of bleaching plants in Canada; 2. The extent to which mills are trialing and using the prominent additives that enhance bleaching: hydrogen peroxide (a common bleaching chemical), anthraquinone (a pulping additive that can decrease the Kappa number and hence bleaching chemical usage), and xylanase; 3. The reasons that motivated mills to trial, use, or discontinue xylanase; 4. The benefits and problems that have been observed with enzyme treatment; 5. The improvements that are desired in xylanase technology for bleaching.

The survey was carried out by telephone in September 1994 by co-authors R. Dines and D. Olson. All 40 Canadian bleached kraft mills and two bleached sulfite mills were contacted and responded to the survey. About two-thirds of the respondents were the mill representatives to the Canadian Pulp and Paper Association Bleaching Committee, with the remainder from mill technical and production departments. The results are reported in the following sections.

Mill Effort. For the past two years more of the effort in the bleach plants has been focused on decreasing chlorine usage than on other objectives. Of the nine bleaching objectives surveyed, the mills have spent the most effort decreasing AOX (by decreasing chlorine usage), followed closely by meeting customer demands (which in many cases was decreasing chlorine usage) and eliminating chlorine gas (Figure 1). These objectives were followed in effort by decreasing off-grade pulp, decreasing BOD, and cutting costs. The least effort was devoted to increasing throughput (not surprising in a recession), eliminating dioxin (which was largely completed more than two years ago) and converting to TCF, which has not occurred to any great extent yet.

Over the next two years, the bleach plant personnel anticipate a reprieve from the many demands and changes that they have been faced with. Effort on all of the objectives is expected to decrease, except cutting costs. Perhaps the mills will be trying to optimize the existing processes.

Usage of Additives to Enhance Bleaching. Mills have many bleaching technologies available to meet the objectives described. Some of these technologies involve additives of relatively low capital cost (usually less than $500,000). The survey considered four such additives: 1) xylanase, 2) anthraquinone, a digester additive that can act indirectly to improve the bleaching, 3) hydrogen peroxide at low levels of less than 5 kg/t, and 4) hydrogen peroxide at high levels of greater than 5 kg/t. The results are shown in Table I.

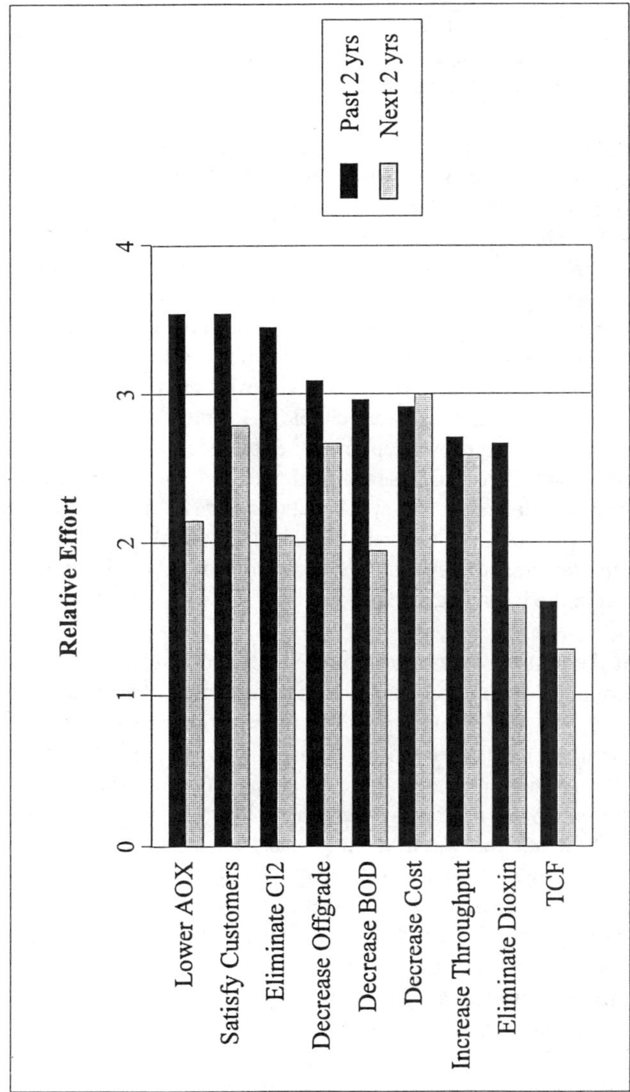

Figure 1. Relative bleach plant effort in several areas, on a scale of zero (no effort) to four (extreme effort).

Table I. Mill Trials and Usage of Four Additives

Additive	*Number of Mills Trialed*	*Number of Mills Using*
Low Peroxide	36	32
High Peroxide	19	11
Anthraquinone	19	7
Xylanase	18	6

Low peroxide is extremely widely used, with regular usage in 80% of Canada's mills. The most common usage is 2 to 3 kg/t of peroxide in the first extraction stage.

Perhaps surprisingly, the other three additives have similar trial and usage experience, with around 50% of the mills having trialed high peroxide, anthraquinone, and xylanase, and about one-third of the trials resulting in usage on a regular basis. Clearly the mills do not have a consensus on the additives (beyond low levels of peroxide). This variety of mill practices is no doubt a reflection of the effects of pulp properties, bleaching sequence, and bleaching objectives on the relative benefits of these additives. These results might bode well from the point of view of future xylanase usage, in that xylanase is the most recent of these bleaching technologies and thus might be the most likely to improve with mill and supplier experience.

Xylanase Enzyme Usage. Of the 42 bleaching mills surveyed, 18 have run xylanase trials, and six are regular users of the enzyme on at least 20% of their pulp. This accounts for 750,000 tonnes of enzyme treated pulp produced per year in Canada, which is 8% of the total bleached pulp production.

All of the enzyme usage is in kraft mills, as xylanase treatment is not generally effective on sulfite pulp. The bleaching sequences of the 10 grades which use enzyme treatment are listed in Table II. All six of the mills bleach softwood, and one also treats hardwood. Two of the mills have oxygen delignification, which is interesting because only three mills with oxygen delignification have run xylanase trials. All six of the mills have five stage bleach plants. The bleached brightness targets for most of the grades are 88 to 89.5 ISO Brightness, which is 1-2 points lower than typical for Canadian mills. This is somewhat surprising in that the bleaching benefit of xylanase is generally thought to increase with increasing brightness target.

Motivation for Xylanase Trials and Usage. For regular users of xylanase, the leading motivations for xylanase use are increasing pulp throughput (such as when the ClO_2 generator is the mill's bottleneck) and obtaining a marketing advantage (Table III). Cost savings, pulp quality, and environmental benefit were cited in decreasing order. The mills who have tried xylanase and decided not to use it regularly have different motivations. For these mills, the leading reasons xylanase trials were run was to meet environmental targets and to obtain a cost savings in bleaching chemicals.

Table II. Profile of Regular Xylanase Users

Mill	*Grade*	*Furnish*	*Bleaching Sequence*	*Brightness (ISO)*
1	1	SWD	D_{100}(EOP)DED	89.0
2	2	SWD	($C_{50}D_{50}$)(EO)DED	91.5
	3	SWD	D_{100}(EO)DED	91.5
	4	HWD	D_{100}(EO)DED	89.0
3	5	SWD	OD_{100}(EO)D(EP)D	89.5
4	6	SWD	D_{100}(EOP)DED	89.5
	7	SWD	($C_{45}D_{55}$)(COP)DED	89.5
5	8	SWD	D_{100}(EOP)DED	89.5
6	9	SWD	OD_{100}(EOP)DED	89.0
	10	SWD	O($C_{50}D_{50}$)(EOP)DED	89.0

Table III. Motivation for Xylanase Usage (Scale: 0=low to 4=high)

Factor	*Mills Using Xylanase*	*Mills Tried, Not Using*
Increase Throughput	3.6	2.7
Marketing Advantage	3.3	2.9
Cost Savings	3.2	3.4
Pulp Quality	3.1	2.2
Environmental Advantage	1.8	3.5

Enzyme Treatment Operations. The mills have added the enzyme to the pulp in a wide variety of locations between the brownstock decker and the brownstock storage tower, with no one location emerging as a clear preference (Table IV). The variety of treatment locations results from the differences in mill equipment and layout and also, perhaps, to the newness of the technology.

Sulfuric acid was used to adjust the pH of the alkali brownstock in all but one of the mills, with the other using chlorination filtrate. Several mills tried sulfurous acid and carbon dioxide for pH adjustment but abandoned these in favour of sulfuric acid. The typical range of pH for xylanase treatment is pH 5.0 to 8.5 .

There is significant concern that acidification of the brownstock changes the demand for bleaching chemicals. Of the sixteen mills that addressed this issue, six felt

Table IV. Enzyme Treatment Locations

Location	*Number of Mills*
Decker	4
Repulper	3
Chute	4
Stock Pump	2
Hi-Density Storage	3

that adjusting the pH decreased chemical usage, two felt it increased it, and eight detected no effect. The mills that reported a decrease in chemical usage with acidification observed that adding acid to the brownstock decreased the pH to the D100 stage and improved its performance. This is particularly true with low Kappa factor (<0.15) bleaching of low Kappa number (<22) pulp. In this case, the acid content of the bleached filtrate alone is insufficient to adjust the pH of the pulp to less than pH 3 for optimum D100 performance. The acid added for enzyme treatment then can have a beneficial effect on the subsequent bleaching. Mills should be aware of the effect of acid on the D100 pH before running an enzyme trial.

Bleaching Benefits and Problems with Enzyme Treatment. The most widely reported benefit of enzyme treatment is a savings in bleaching chemicals (Table V). The chemical savings was 8% to 15%, with an average of 11% of the total chemical across the bleach plant. The other widespread benefits were in improved effluent, including decreases in AOX of 12% to 25%, decreases in effluent colour, and other improvements to the effluent. Other benefits of enzyme treatment reported included increased bleached brightness (1 point gain), tear strength (5% gain), and pulp throughput (10% increase).

Table V. Benefits of Xylanase Treatment

Benefit	*Number of Mills*
Save Chemicals	8
Decrease AOX	4
Improve Effluent	4
Increase Throughput	2
Higher Brightness	2
Increase Tear Strength	2

The most common problems with enzyme treatment cited were corrosion of equipment and maintaining the brownstock residence time (Table VI). Sulfuric acid corrosion of mild steel has been encountered in several mills. The brownstock residence time must be maintained for as long as possible, but usually at least 1 to 2 hours, to obtain the maximum benefits of xylanase treatment. This sometimes means that the mills must maintain the storage tower nearly full, which curtails its ability to act as a buffer between the pulping mill and the bleach plant.

Table VI. Problems with Xylanase Treatment

Problem	*Number of Mills*
Corrosion	8
Maintain Residence Time	7
General Hassle	4
Bleach Plant Control	3
Decrease Tear Strength	2
Safety (acid)	1
Pitch	1

Other problems reported with enzyme treatment included difficulties in application and in bleach plant control. These relate to the subtle action of the enzymes, which is not easily observed on-line or in rapid testing. A decreased tear strength and pitch formation were also reported in some mills. One mill expressed a concern over the use of acid in an area where the mill has not previously had acid lines. This mill reported no safety problems with their trials, but had a long term concern.

Reasons For Not Using Xylanase. For the mills that have had enzyme trials, the clear leading reason for not continuing to use xylanase was an insufficient benefit (Figure 2). This was followed by difficulties in measuring performance, a feeling that resources are better used elsewhere, and inadequate mixing of enzyme into the pulp. The corrosion and residence time problems observed during the trials are viewed as solvable, but the other issues are perceived as more difficult to deal with. The mills had few complaints from customers or about the level of supplier preparation.

The mills that have not had enzyme trials had the same reasons for deciding not to use xylanase, and in addition loss of pulp yield was widely cited. This was a concern primarily of hardwood mills, because yield loss from xylanase treatment is often greater on hardwood pulp than on softwood.

Desired Improvements in Xylanase Technology. The most often cited improvements that are desired in the technology are higher pH operation and lower acid usage, which go hand in hand and would help decrease the corrosion problems (Figure 3). The next desired improvements are in increasing the benefit, followed by the capability for the

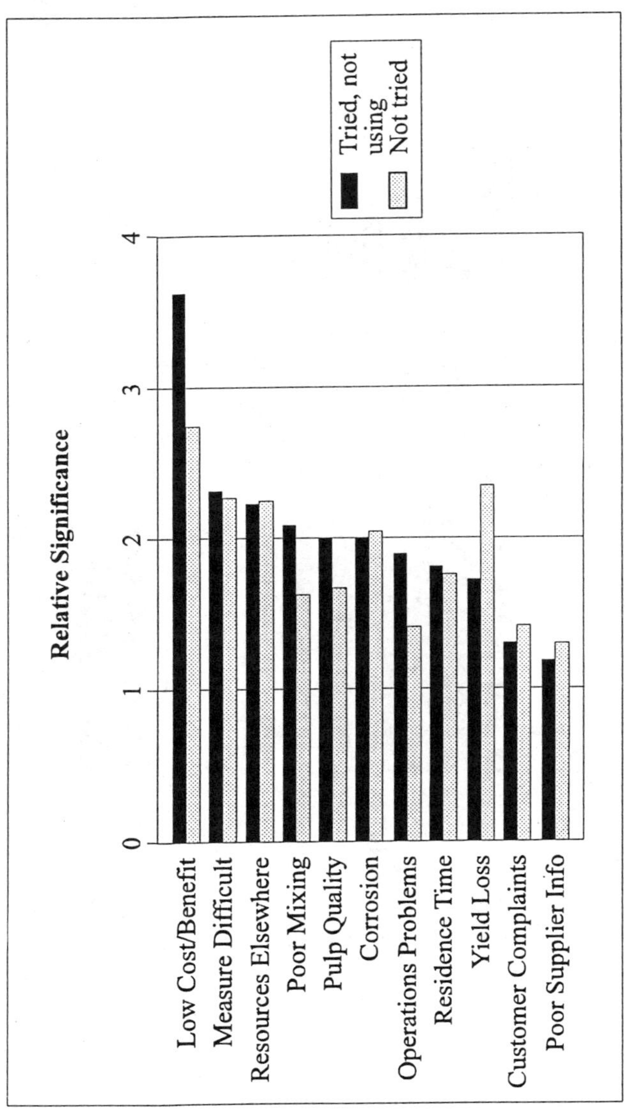

Figure 2. Factors that cause mills not to use enzymes.

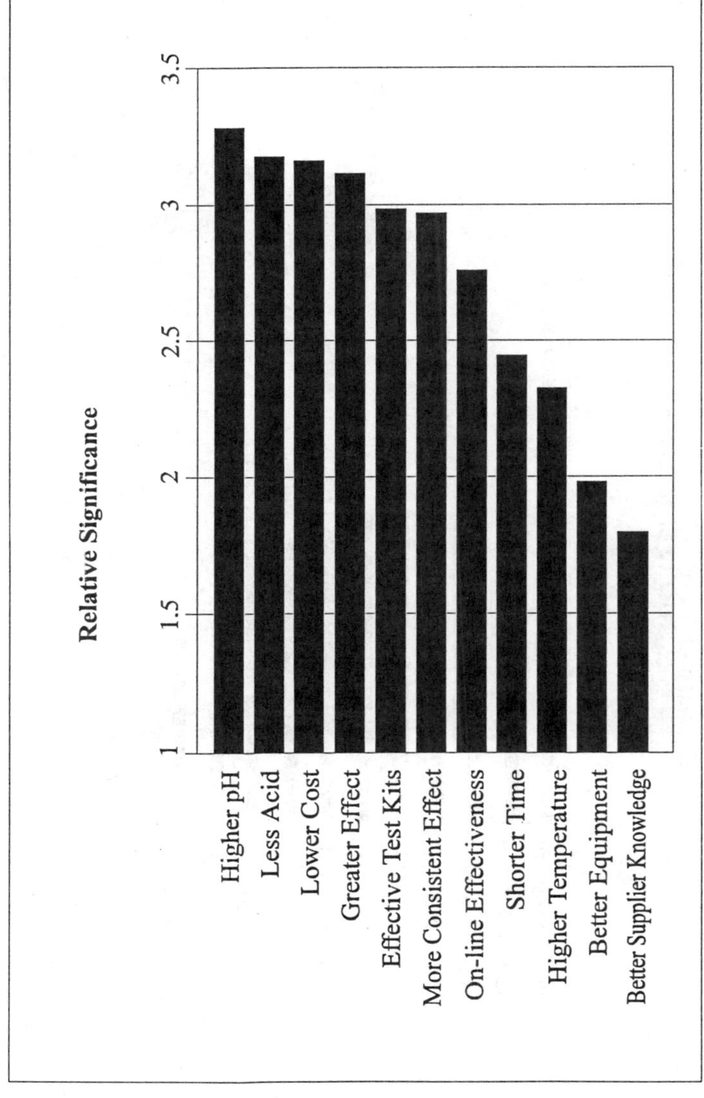

Figure 3. Improvements desired in xylanase bleaching technology.

off-line and on-line monitoring of the enzyme effect, which is not currently available and would improve the bleach plant control and decrease the general difficulties of running enzymes, as well as improve performance. There appears to be less of a desire for shorter time or higher temperature operation, or improvements in equipment or supplier knowledge.

The relative lack of interest in higher temperature operation is somewhat surprising, in that much discussion and research has focused on increasing the operating temperatures of xylanase, which are currently 35°C to 60°C. Apparently, the majority of the mills are able to maintain the brownstock in this temperature range.

Conclusions.

Xylanase enzymes were used to treat 750,000 tonnes per year of pulp in six mills in 1994. This represents 8% of Canada's bleached kraft pulp production.

Acknowledgments.

The authors thank the staff at each of the mills for their thoughtful, timely, and open responses to the survey. The help of Catherine Seelye and the late Maryse Guenette of Iogen Corporation with the manuscript is appreciated.

Literature Cited.

(*1*) Britt, K.; *Handbook of Pulp and Paper Technology*. Van Nostrand Reinhold Company, (1970).
(*2*) Rydholm, S.; *Pulping Processes*. Krieger Publishing Company, Malabar, Florida, (1965).
(*3*) Singh, R.; *The Bleaching of Pulp*. TAPPI Publishing, Atlanta, Georgia, (1979).
(*4*) Viikari, L. Ranua, M., Kantelinen, A., Sundquist, J., Linko, M.; *Bleaching With Enzymes*, in Proc. Int. Conf. Biotechnol. Pulp and Paper Ind. Stockholm, (1986) p. 67-69.
(*5*) Luers, M.; *Biobleaching at Skeena Cellulose*, in Proc. Canadian Pulp and Paper Association Spring Conference, Jasper, Alberta, (1992).
(*6*) Scott, B., Young, F., Paice, M.; *Biobleaching at Canadian Forest Products,* in Proc. CPPA Spring Conference, Jasper, Alberta, (1992).
(*7*) Hitzroth, A.; *Use Of Enzyme Treatment At Crestbrook Forest Industries*, in Proc. CPPA Spring Conference, Jasper, Alberta, (1992).
(*8*) Fasten, H.;*TCF Pulp at Aspa Mill: History and Future*, in Proc. 1993 Non-chlorine Bleaching Conference, Hilton Head, S.C., March 12-17 (1993).
(*9*) Reilama, I., Kukkonen, K.; *Experiences and Conclusions with TCF Bleaching at Metsa-Botnia*, in Proc. Non-Chlorine Bleaching Conference, Hilton Head, S.C. March 12-17, (1993).
(*10*) Turner, J.C., Skerker, P.S., Burns, B.J., Howard, J.C., Alonso, M.A., Andres, J.L.; *Bleaching With Enzymes Instead Of Chlorine: Mill Trials*, (1992), TAPPI 75(12):83-89.

Substrate Characterization

Chapter 4

Enzymatic Characterization of Pulps

J. Buchert, A. Suurnäkki, M. Tenkanen, and Liisa Viikari

VTT Biotechnology and Food Research, P.O. Box 1501, FIN–02044 VTT, Finland

The specificity of enzymes acting on pulp carbohydrates can be exploited in structural and chemical analysis of pulp fibres. Using the selective enzymatic solubilization, pulp components can be qualitatively and quantitatively analyzed, without attacking other components in pulps. Furthermore, the role of pulp components, such as xylan or glucomannan, in the technical properties of pulp can be evaluated after their enzymatic removal from pulp. This paper reviews the use of enzymes as analytical tools in the characterization of chemical pulps.

A clear breakthrough of biotechnical methods in the pulp and paper industry is a reality. In addition to methods already used in full industrial scale, several new methods are being developed. In all these applications a very limited enzymatic hydrolysis results in improved processability of fibres.

In addition to using carbohydrate degrading enzymes as process aids, the specific hydrolysis obtained with these enzymes can also be exploited in the analysis of structures and structure-function relationships in fibres. The conditions used in enzymatic solubilization of pulp carbohydrates are very mild, generally the temperature is about 45-50°C and the pH 5 (*1*). Thus, no destruction of the components is occuring. By using specific hemicellulases in one-stage treatment, up to 30-40% of the corresponding hemicellulose in pulp can be selectively solubilized without affecting other components in pulp (*2*). The degree of solubilization can be adjusted by controlling enzyme dosage and hydrolysis time. The use of enzymes as analytical tools in fibre characterization requires high purity from the enzyme preparates. Hence, availability of commercial enzymes for fibre analysis is limited. Furthermore, when enzymes are used in structure-function characterization of pulps, pilot-scale protein purification is a necessity due to the high amount of pulp to be treated.

0097–6156/96/0655–0038$15.00/0
© 1996 American Chemical Society

In this paper the use of enzymes as analytical tools in the characterization of the carbohydrate composition and structure-function relationships in different types of pulps is reviewed. In Figure 1 the principle of enzymatic characterization of fibres is presented.

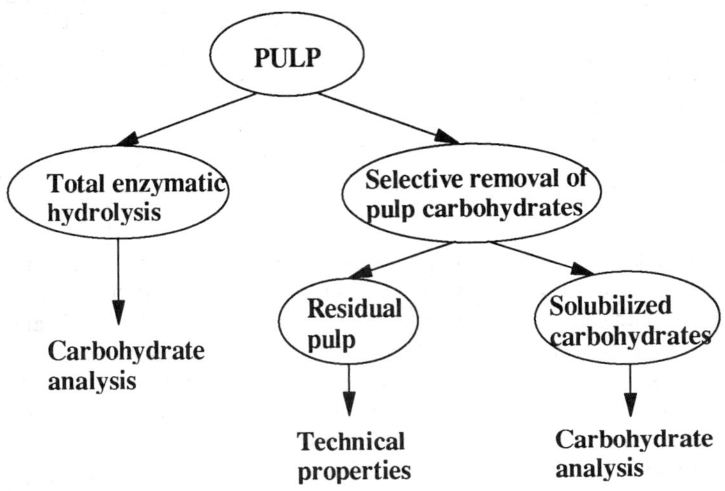

Figure 1. Use of carbohydrate degrading enzymes in fibre characterization.

Enzymes in Characterization of the Carbohydrates of Chemical Pulps

The analysis of pulp components by acid hydrolysis has proven to be difficult in some cases due to potential degradation of acid labile components and incomplete hydrolysis of certain linkages. Purified *Trichoderma reesei* xylanase has been used for selective solubilization of pulp xylan from different types of pulps, whereafter the chemical composition of the solubilized fraction has been determined either by proton NMR (*3, 4*) or by HPLC after a secondary enzymatic hydrolysis to monomers (*1*). With this selective enzymatic solubilization method a novel type of uronic acid side group, i.e. hexenuronic acid, has been identified and quantified in kraft pulp xylan (*3, 4*). Hexenuronic acid (HexA) is formed by ß-elimination from 4-O-methylglucuronic acid during kraft pulping and also the intermediate isomerization product, i.e. methylidoronic acid (MeIdoA) has been detected in enzymatically solubilized kraft xylan (Teleman, A. *et al. Carbohydr. Res.*, in press). Hexenuronic acid has not been detected after traditional acid hydrolysis of pulps due to its degradation to furanderivatives at acidic conditions (*5*). Hexenuronic acid is the major uronic acid component in both softwood and hardwood kraft xylan, whereas sulphite pulps cooked at acidic conditions contained only MeGlcA. Small amounts of HexA could, however, be detected in alkaline sulphite pulp, although also in this case MeGlcA was the major uronic acid (Table 1).

Table 1. Structure of enzymatically solubilized xylan in different unbleached pulps. The extensive xylanase treatment and subsequent NMR analysis was carried out as described previously (*4*). Sulphite pulps were cooked as described previously (*6*). nd= not detected

Pulp	*Sidegroups mol/100 mol xylose*		
	MeGlcA	HexA	Ara
Spruce sulphite (Acid Mg)	10.6	nd	nd
Spruce sulphite (ASAQ)	11.3	1.1	9.2
Birch kraft	2.2	4.1	-
Pine kraft	1.1	4.7	8.2

For the total characterization of pulp components, and especially the uronic acids in pulp xylan, a cellulolytic and hemicellulolytic enzyme mixture for complete hydrolysis of pulp carbohydrates to monosaccharides has been developed (*7*). Enzymatic total hydrolysis is a suitable for the hydrolysis of chemical pulps in which the accessibility of the carbohydrates is sufficient for the enzymatic attack. Compared with acid hydrolysis the yield of neutral sugars are generally slightly lower after enzymatic hydrolysis (*7*). However, in addition to the presence of HexA in the enzymatically hydrolyzed sample, also the amount of MeGlcA detected was higher than obtained in acid hydrolyzed sample. Thus, enzyme-aided analysis is especially suitable for the quantification of uronic acids in chemical pulps.

Table 2. Comparison of acid hydrolysis and enzymatic hydrolysis combined to HPLC for the analysis of uronic acids in unbleached birch kraft pulp. n.d.= not detected

Method of hydrolysis	*HexA, % of d.w.*	*MeGlcA, % of d.w.*	*Total, % of d.w.*
Acid	n.d.	0.30	0.30
Enzymatic	1.41	0.88	2.29

The relative amounts of different types of carboxyl groups in kraft pulps as well as the distribution of acidic groups in different xylans has been determined by measuring the carboxyl groups in the residual pulps by high precision potentiometric titration (*8*). Approximately 80% and 90% of the carboxyl groups in unbleached pine and birch brownstock kraft pulps were uronic acids (*8*).

Enzymes in Analysis of the Organization of Pulp Components in Fibres

The organization of xylan and lignin on the outermost surface of kraft fibres has been studied by combining enzymatic peeling and ESCA analysis (Electron Spectroscopy for Chemical Analysis) (*9*). After extensive xylanase treatment (about 30 % of the pulp xylan removed) more lignin could be detected on the surface of pine kraft fibres. Thus,

within the sampling depth of ESCA (about 10 nm), xylan was apparently partly covering lignin in pine kraft fibres. The selective removal of glucomannan did not uncover lignin in pine kraft fibres. Unlike in pine kraft pulp, the removal of as much as about 40 % of the xylan in birch kraft pulp did not expose lignin on the outermost surface of fibres. This was probably due to the relatively high content of xylan on the surface of birch kraft fibres (*10*).

By measuring the effect of extensive xylan or glucomannan removal on the pore size distribution of unbleached kraft pulps, further information on the location of hemicelluloses has been obtained (Suurnäkki, A. *et al. Holzforschung*, in press). Even with low hydrolysis levels the pore size distributions were increased indicating the presence of hemicelluloses on the surface of pores. This result is consistent with the previous observation of increased leachability of lignin after xylanase or mannanase treatment (*11*).

Enzymatic Characterization of the Structure-Function Relationships of Pulps

The papermaking properties of kraft pulps are generally considered to be dependent on the composition and properties of individual fibres (*12*). Also for the studies of the structure-function dependancy selective enzymes solubilizing pulp carbohydrates are indispensable tools. The role of hemicelluloses in the properties of kraft pulps has been investigated by removing different amounts of xylan from pulps by *T. reesei* xylanase. The effect of xylan on the beatability of kraft pulps was found to depend on the pulp origin. In brownstock birch kraft pulp the enzymatic removal of xylan decreased the beatability of fibres whereas after ECF-bleaching the effect was less pronounced indicating that the chemical structure, i.e the uronic acid profile, might have a role in the beatability (Buchert, J. *et al.*, manuscript in preparation).

The brightness stability of birch and pine kraft pulps has also been found to be affected by the uronic acid profile. During extensive removal of xylan by *T. reesei* xylanase, the uronic acid content was significantly decreased due to solubilization of uronic acid substituted xylan and this resulted in improved brightness stability of different kraft pulps. Enzymatic glucomannan removal had no effect on ageing (Buchert, J. *et al. Tappi J.*, accepted for publication).

Basic phenomena occuring during recycling of chemical pulps, i.e. hornification, has been investigated using selective removal of xylan and glucomannan combined to drying and analysis of the pulp properties. By removing the accessible hemicelluloses from the pulp the hornification was increased, which was observed in decreased swelling, strength and flexibility of the pulp (Oksanen *et al.* , submitted).

Conclusions

A deep understanding of the role of different pulp components on the final pulp properties requires the combination of novel and traditional techniques. Enzymes have proved to be powerful tools in fibre analysis. Combined to traditional and modern analysis methods enzymatic treatments provide valuable information of the structure and composition of fibre surfaces as well as the overall pulp matrix (Figure 2). The non-destructive conditions used in enzymatic treatments render the enzyme-aided analysis especially suitable for the analysis of labile components in pulp fibres. The

specificity of enzymes can also be exploited in the characterisation of the relationships between pulp components and papermaking properties in an unique manner. Recent reports on the use of purified enzymes in the characterization of kraft pulps are summarized in Table 3.

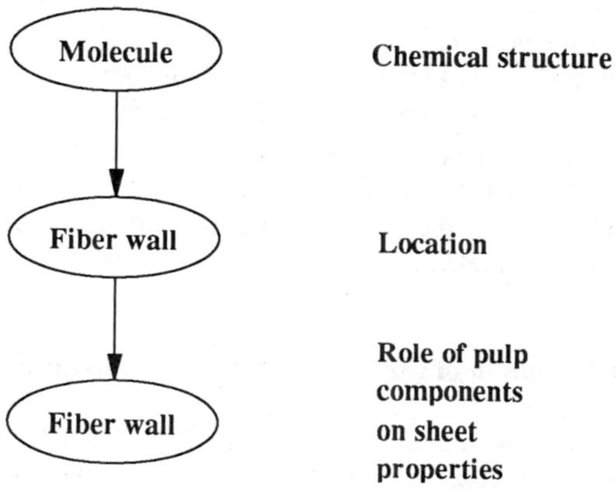

Figure 2. Information obtained from pulp and fibres by enzyme-aided analysis

Table 3. Recent reports on enzymatic characterization of pulps

Combined to	*Information obtained*	*Reference*
^{1}H NMR	Structure of solubilized pulp xylan	*3, 4*
High precision potentiometric titration, polyelectrolyte adsoption	Distribution of carboxylic groups in pulp xylan	*8*
^{1}H NMR, solute exclusion	Effect of xylan and glucomannan on pore size distribution	Suurnäkki *et al.*, *Holzforschung*, in press
Thermal ageing	Role of xylan and glucomannan in brightness stability	Buchert *et al.* *Tappi J.*, accepted
Drying, fibre characterization	Role of xylan and glucomannan in hornification	Oksanen *et al.*, submitted
PFI-refining, sheet properties	Role of hemicelluloses on technical properties of pulps	Buchert *et al.*, in preparation
ESCA	Location of lignin and hemicelluloses in pulp surface	*9*
HPLC	Total carbohydrate composition of pulps	*7*

Literature Cited

1. Buchert, J.; Siika-aho, M.; Bailey, M.; Puls, J.; Valkeajärvi, A.; Pere, J.; Viikari, L. *Biotechnol. Tech.* **1993**, *7*, pp 785-790.
2. Suurnäkki, A. *Hemicellulases in the bleaching and characterization of kraft pulps.* VTT Publications 267; Technical Research Centre of Finland: Espoo, 1996.
3. Teleman, A.; Harjunpää, V.; Tenkanen, M.; Buchert, J.; Hausalo, T.; Drakenberg, T.; Vuorinen, T. *Carbohydr. Res.* **1995,** *272,* pp 55-71.
4. Buchert, J.; Teleman, A.; Harjunpää, V.; Tenkanen, M.; Viikari, L.; Vuorinen, L. *Tappi J.* **1995,** *78,* pp 125-130.
5. Teleman, A.; Hausalo, T.; Tenkanen, M.; Vuorinen, T. *Carbohydr. Res.* **1996**, *280*, pp 197-208.
6. Buchert, J.; Kantelinen, A.; Suurnäkki, A.; Viikari, L.; Janson, J. *Holzforschung* **1995**, *49*, pp 439-444.
7. Tenkanen, M.; Hausalo, T.; Siika-aho, M.; Buchert, J.; Viikari, L. *Proc. 8th Int. Symp. Wood and Pulping Chemistry*, Helsinki, 6-9 June 1995. Vol. III, pp 189-194.
8. Laine, J.; Buchert, J.; Viikari, L.; Stenius, P. *Holzforschung,* **1996**, *50*, pp 208-214.
9. Buchert, J.; Carlsson, G.; Viikari, L.; Ström, G. *Holzforschung*, **1996**, *50*, pp 69-74.
10. Suurnäkki, A.; Heijnesson, A.; Buchert, J.; Viikari, L.; Westermark, U. *J. Pulp Paper Sci.* **1996**, *22*, pp J43-J47.
11. Hortling, B.; Korhonen, M.; Buchert, J.; Sundquist, J.; Viikari, L. *Holzforschung*, **1994**, *48*, pp 441-446.
12. Clark, J. D. A. *Pulp Technology and Treatment for Paper*, Miller Freeman Publications Inc.: San Fransisco, CA, 1978, 751 p.

Chapter 5

Molecular Mass Distribution of Materials Solubilized by Xylanase Treatment of Douglas-Fir Kraft Pulp

Ed de Jong, Ken K. Y. Wong[1], Lori A. Martin, Shawn D. Mansfield, F. M. Gama[2], and John N. Saddler

Chair of Forest Products Biotechnology, Department of Wood Science, University of British Columbia, 270–2357 Main Mall, Vancouver, British Columbia V6T 1Z4, Canada

Irgazyme, a commercial xylanase preparation from *Trichoderma longibrachiatum*, and xylanase D a purified enzyme from *Trichoderma harzianum* E58 were tested for their ability to enhance peroxide bleaching of Douglas-fir (*Pseudotsuga menziesii*) kraft pulp. A treatment with Irgazyme caused a much larger increase in brightness than did xylanase D. A double xylanase treatment with Irgazyme, before and after peroxide bleaching, resulted in the highest final brightness. Alkaline extraction increased the brightness of Douglas-fir brownstock. Treatment with Irgazyme released more lignin and carbohydrates than did xylanase D. The molecular mass of the lignin extracted from Irgazyme-treated brownstock was much larger than that from the control pulp. The lignin-like macromolecules directly solubilized from peroxide bleached pulps were substantially larger than those solubilized from the brownstock, irrespective of whether they were produced during xylanase or control treatments. This indicates that different kinds of materials were solubilized when a xylanase treatment was applied at different points in the bleaching sequence and raises concerns about the role of lignin entrapment in the mechanism by which xylanase enhances peroxide bleaching.

During the last few years the pulp and paper industry has rapidly moved away from using elemental chlorine as a bleaching agent. In a recent survey of Canadian pulp mills, 41 of the 51 bleach plants produced elemental chlorine free (ECF) grades in 1995, while prior to 1987 all of them used chlorine (*1*). However, to achieve the

[1]Current address: PAPRO, NZ FRI Ltd., Private Bag 3020, Rotorua, New Zealand
[2]Current address: Department of Biological Engineering, University of Minho, Braga, Portugal

0097–6156/96/0655–0044$15.00/0
© 1996 American Chemical Society

desired brightness levels without elemental chlorine several changes have had to be implemented. The major changes included prolonged cooking regimes during the pulping stage to reach lower kappa values, as occurs in extended delignification, superbatch and extended modified continuous cooking. In the bleach plant, the benefits of chlorine dioxide, oxygen, ozone, peroxide and enzymes have also been shown.

With oxygen based chemicals, desired brightness levels are often only reached at the expense of relatively high cost and/or low strength qualities, making the use of enzymes an interesting alternative. Two classes of enzymes have been shown to be useful in the bleaching of kraft pulp. Oxidative enzymes such as laccases and manganese peroxidases have been shown to bleach brownstocks to high brightnesses. However, this has only been recently demonstrated at the laboratory scale (*2*). In 1986 Viikari and co-workers (*3*) showed that xylanases can be used as a prebleaching step in conventional bleaching sequences. Xylanase treatment is effective on both softwood and hardwood kraft pulps (*4, 5*) and has been implemented in several kraft mills in North America and Europe (*5-8*). In 1994, six kraft mills in Canada treated at least 20% of their pulp with xylanase, totalling 7.5 million tonnes/year and representing 8% of the Canadian kraft pulp production (*9*). Xylanase pretreatment has a strong boosting effect on pulps being bleached with elemental chlorine and permits substantial reductions in both the chlorine charge required and the subsequent amounts of organochlorine generated (*4, 5*). Xylanase treatment is also effective on pulps produced with modern pulping processes such as superbatch and modified continuous cooking (*10-13*) and when it is incorporated into modern bleaching processes, including those that involve elemental chlorine-free (ECF) and totally chlorine-free (TCF) bleaching sequences (*5, 12, 14-16*). Xylanase treatment not only reduces the chemical loadings required to reach target brightness, certain reports also suggest that a xylanase treatment can improve the brightness ceiling of pulp bleached with peroxide (*17, 18*).

As the enzyme appeared to be most effective at the beginning of the bleaching sequence where its application enhances subsequent chemical bleaching, the phenomenon has been coined 'bleach boosting' (*19*). Several mechanisms have been suggested to explain the bleach boosting effect of xylanase prebleaching. Kantelinen and co-workers (*20*) have proposed that the enzyme attacks xylan redeposited on the surface of the pulp at the end of the kraft cook. However, pulps from which the surface xylans had been removed with dimethyl sulfoxide still responded to xylanase prebleaching (*19, 21*). Similarly, xylanase prebleaching remained effective on pulps containing low amounts of redeposited xylan, such as those produced with flow-through and extended cooking methods (*11, 12, 22*). More recent work using mechanical peeling indicated that the outer surface of pine fibers was not enriched in xylan deposits and that its removal did not inhibit xylanase-aided bleaching (*23, 24*). A comparison of the outer and inner fractions of pulp fiber walls showed a surprisingly low bleachability of the outer layer during xylanase-aided peroxide bleaching (*25*). These authors suggested that bleachability depended on the composition of the lignin-carbohydrate complex (LCC) as well as the extent to which the outer layers were maintained during the pulping process. Other studies have indicated that xylanase treatments led to an extensive depolymerization of xylan

without solubilization (*26*) and the liberation of lignin-carbohydrate complexes (*27*). Other evidence suggests that the release of chromophores may play an important role (*28, 29*). The exact mechanism of xylanase prebleaching thus remains unclear, although several hypotheses have been made concerning the nature of the target xylan (*30*). Possible substrates include: xylan in lignin-carbohydrate complexes; xylan that physically entraps residual lignin; xylan that modifies fiber porosity; and xylan that contains chromophoric carbohydrate units.

Our research group has investigated the validity of these hypotheses. Our work with softwood and hardwood kraft pulps and model pulps, has shown that a xylanase treatment can cause both a bleach boosting as a direct brightening effect in peroxide bleaching sequences (*31-34*). Although xylan chromophores were formed under pulping conditions, they appeared to be of minor importance in xylanase bleaching (*31, 35*). Alternatively, LCC's could be isolated from kraft pulps and were shown to be hydrolysed by xylanases (*36*). Finally, a difference in the susceptibility of distinct fiber length fractions to xylanase bleaching was observed, with the smallest fibers showing the highest response (*37*). The research presented in this paper has focused on several aspects of xylanase bleaching of Douglas-fir kraft pulp. Size exclusion chromatography was used to assess the molecular mass distribution of the material solubilized by xylanase and that which becomes extractable in alkaline solutions. In addition, the bleaching activity of a purified 22 kDa β-1,4-xylanase from *Trichoderma harzianum* E58 was compared with a commercial xylanase preparation from *Trichoderma longibrachiatum*.

Materials and Methods

Enzymes. The commercial xylanase studied, Irgazyme 40S-4X (Genencor, San Fransisco, CA, USA), was an enzyme preparation from a selected strain of *Trichoderma longibrachiatum*. It contained a mixture of several extracellular enzymes but was essentially free of cellulases. The purified enzyme studied, xylanase D, was the 22 kDa, pI 9.4 xylanase from *Trichoderma harzianum* E58 (*38*). The assay described by Bailey et al. (*39*) was used with birchwood xylan as substrate to measure xylanase activity at pH 7 for Irgazyme and pH 5 for xylanase D. Each nkat of activity represents the release of 1 nmol xylose equivalent per second.

Kraft Pulp. Brownstock (unbleached, never dried kraft pulp) derived from Douglas-fir (*Pseudotsuga menziesii*) was prepared by Mr. Wai Gee in the pilot plant of the Pulp and Paper Research Institute of Canada (Vancouver, B.C.). All pulps were washed extensively by screening with a 200 mesh screen in the Bauer-McNett fiber length classifier. The composition of the pulp was determined as: glucose 83.1%; xylose 4.75%; mannose 3.99%; galactose 0.48%; arabinose 0.36%; acid insoluble lignin 4.00% and acid soluble lignin 0.42%.

Bleaching Conditions. The bleaching sequence used was as described before (*33*). All bleaching and extraction stages were done in polyester bags, while enzyme treatments of partially bleached pulps were done in polyethylene (Ziploc) bags. The

pulps were mixed vigorously by hand at the start of each stage, and before filtrate samples were taken after all enzyme (X) and alkaline extraction (E) stages. The pulps were washed at 1% consistency with de-ionized water after each stage. Each time a pulp was drained the filtrate was passed through the pulp cake three times to collect the fines.

Pulps were treated with a high loading of 400 nkat xylanase g $pulp^{-1}$ or with the same amount of heat inactivated enzyme at 10% consistency. The start pH was 7.0 for Irgazyme and 4.8 for xylanase D. The reaction was carried out for 60 min at 50 °C. The filtrate was analyzed for total sugars, and approximately 40 ml was brought to pH 7 and subsequently freeze-dried with a Speed Vac plus SC210A centrifuge (Savant, Farmindale, NY, USA). The E-stage conditions were 10% pulp consistency, 2% NaOH for 1 hour at 80 °C. For chelation (Q-stage) the pulp was adjusted to pH 5.5 with sulphuric acid and brought to a 3% consistency. The pulp was then incubated with $Na_2 \cdot EDTA \cdot 2H_2O$ (1% loading based on dry weight of pulp) for 30 min at 50 °C. For the peroxide bleaching (P-stage) the pulp was brought to 10 % consistency and bleached with 4% H_2O_2, 2% NaOH and 0.05% $MgSO_4 \cdot 7H_2O$ for 3 hours at 80 °C. The bleached pulp was adjusted to pH 4.5, thoroughly washed with de-ionized water, and used to prepare optical handsheets.

Brightness. Handsheets for measuring brightness were made from 4 g dry weight pulp (CPPA standard C.5). ISO-brightness of handsheets was measured on a Technibrite TB-1C (Technidyne Corp., New Albany, IN, USA). Both sides of each handsheet were read five times, and the overall average values calculated.

Lignin Determination. The residual lignin in pulp samples was determined in duplicate using the microkappa method (TAPPI Useful Methods UM 246, 1991) with 0.5 g dry weight of pulp in 25 ml. Lignin and chromophore content in filtrate samples was analyzed by measuring absorbance at 280 and 457 nm, respectively. These samples were diluted with 50 mM Na-phosphate buffer, pH 7.0, to give readings between 0.2 and 0.7 AU.

Total Sugar in Pulp Filtrates. The total sugars in the liquid samples were determined using the phenol-sulphuric acid method (*40*). Samples were also spiked with a 0.2 mM xylose standard in order to verify the assay, and absorbance of the reaction mixture was measured at 490 nm.

Size Exclusion Chromatography. The molecular mass distribution of lignins and xylans was analyzed with Toyopearl HW-55S and HW-50S resins (TosoHaas, Montgomeryville, PA, USA), packed in series in two 0.5 x 20 cm HR columns (Pharmacia, Uppsala, Sweden). The resins have a separation range for dextrans from 0.34 to 40 kDa (*41*). Elution with 0.3 M NaOH was carried out at 3.6 ml $hour^{-1}$ using a Dionex DX 500 HPLC system, and monitored using a Dionex ED40 electrochemical detector and a Dionex AD20 absorbance detector (280 nm, pathlength 6 mm, set at low) in series (*41*). The Pulsed Ampherometric Detector (PAD, electrochemical detector) is a sensitive detector for carbohydrates, although other oxidizable compounds also register (*41*). Immediately prior to injection, the

sample of freeze-dried filtrate was diluted to 5% of its original volume with 0.3 M NaOH and a 900 μl aliquot was mixed with 100 μl of an internal standard (vanillin, 5 mg/ml). The sample was then filtered through a Millipore 0.45 μm HV filter (Millipore, Bedford, MA) and 20 μl was injected for chromatographic analysis.

Results

Pulp Brightness. To assess the effectiveness of the two enzymes with respect to their ability to enhance peroxide bleaching, the ISO-brightness of optical handsheets was measured (Table I). Handsheets were made after all xylanase and control treatments (X, C, XQPX, XQPC, CQPX, CQPC, XQPPX, XQPPC, CQPPX, CQPPC), peroxide bleaching steps (XQP, CQP, XQPP, CQPP) and alkaline extraction steps (XE, CE, XQPXE etc). The results clearly show that a treatment with Irgazyme, a commercial xylanase preparation from *Trichoderma longibrachiatum*, caused a much larger increase in brightness than the purified xylanase D from *Trichoderma harzianum* E58. It was also found that a double xylanase treatment with Irgazyme, before and after peroxide bleaching, resulted in the highest final brightness. This was not the case with xylanase D, which was only marginally beneficial to the bleaching of Douglas-fir kraft pulps when it was applied after peroxide bleaching. Alkaline extraction increased the brightness of Douglas-fir brownstock and of pulp bleached by one peroxide stage. An Irgazyme treatment of the brownstock led to a relatively large increase in brightness after alkaline extraction. However, alkaline extraction of xylanase treated pulps bleached by one peroxide step resulted in a marginal brightness increase compared to the pulps treated with heat inactivated enzyme. Although alkaline extraction decreased the brightness of pulps bleached by two peroxide stages, it did not decrease the brightness gain of the xylanase treatments relative to the control treatments (Table I).

Carbohydrate solubilization. All control treatments at pH 7 and pH 5 released sugars from pulp with 0.63 mg/g pulp and 0.42 mg/g pulp, respectively. Size exclusion chromatography of these samples showed that the molecular mass of apparent carbohydrates had a binodal distribution (Figures 1A, 2 and 4). The smaller component had a molecular mass of 0.4 kDa while the larger component had a molecular mass of 1 kDa. In all cases, an enzyme treatment increased the total amount of sugars solubilized. The Irgazyme treatment released much more sugars (2-4.4 mg/g) than did the xylanase D treatment (0.4-1 mg/g pulp). On average this was four times greater at each corresponding bleaching stage.

Chromophore solubilization. The filtrates obtained after the enzyme treatment and alkaline extraction were analyzed for compounds absorbing in the UV (280 nm) and visible (457 nm) range of the spectrum (Table II). Irrespective of when and how often an Irgazyme treatment was applied this enzyme treatment released more UV-Vis absorbing material than did the xylanase D treatment. Another significant difference between the two enzymes was observed after subsequent alkaline extraction, where Irgazyme treatments led to increases in the total amount of

Table I

ISO-brightness values of xylanase bleached Douglas-fir kraft pulp handsheets.

		X[a]	XE	XQP		XQPX	XQPXE	XQPP	XQPPX	XQPPXE
Irgazyme	C[b]	28.8	29.7	52.0	CC	52.2	52.4	67.7	67.7	67.38
	X	-0.1	1.2	0.5	CX	0.9	1.2	1.0	0.9	0.79
					XC	0.5	0.5		1.2	1.18
					XX	1.1	1.2		1.7	1.79
Xylanase D	C[b]	29.1	30.0	51.2	CC	50.9	51.3	68.1	67.5	67.48
	X	-0.5	0.2	0.1	CX	0.2	0.3	0.3	0.4	0.39
					XC	-0.4	-0.4		-0.1	0.05
					XX	-0.2	-0.1		0.1	0.24

[a] The handsheets were made directly after xylanase (X) and control treatments (C), after subsequent chelation (Q) and peroxide bleaching (P) and after an alkaline extraction (E) step. The averages of replicate experiments are given.

[b] The ISO-brightness values for all the handsheets which had received one or two control (C and CC) treatments are given. Absolute increases (or decreases) are given for all the handsheets which had received one (X, XC and CX) or two (XX) xylanase treatments.

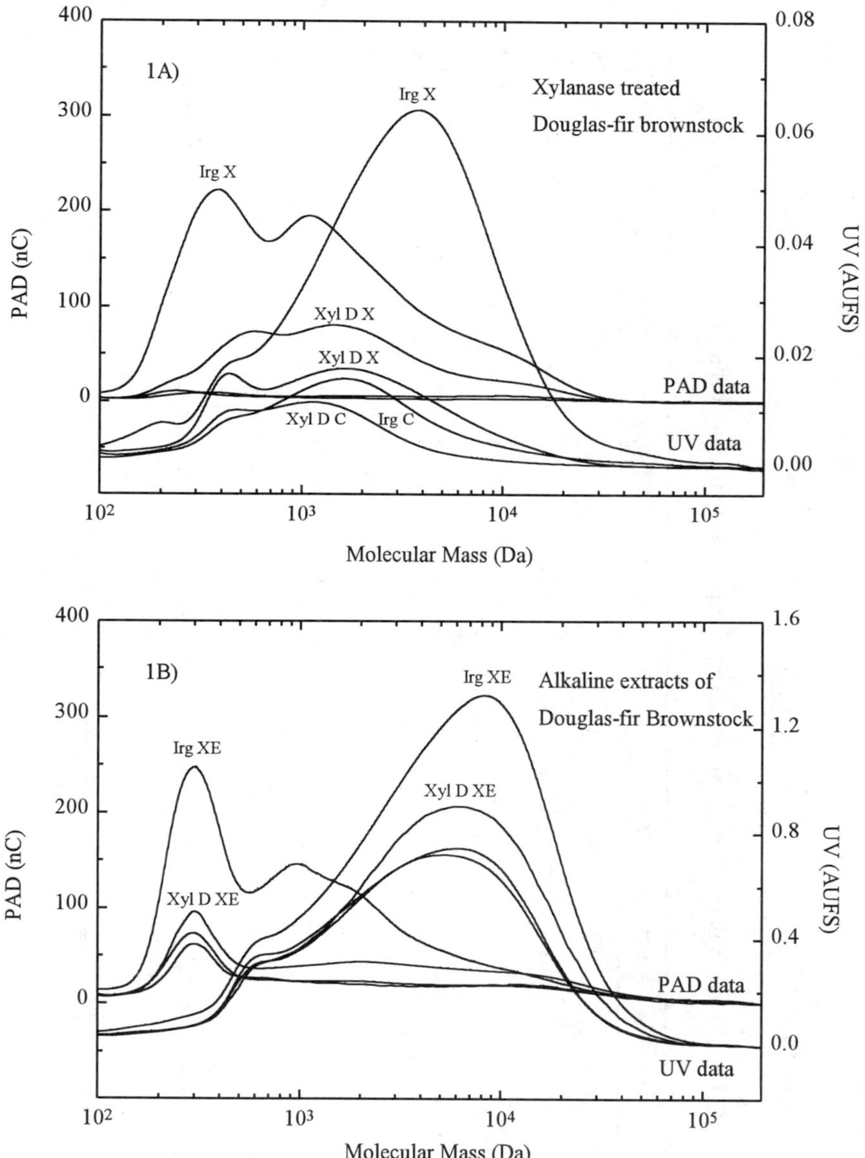

Figure 1. 1A) Size exclusion chromatograms showing the molecular mass distribution of material solubilized from Douglas-fir kraft pulps after xylanase and control treatments. Irgazyme was applied at pH 7 and xylanase D at pH 5. PAD = pulsed amperometric detector, UV = absorbance at 280 nm. 1B) Size exclusion chromatograms showing the molecular mass distribution of alkaline extracts from pulp after xylanase or control treatment. For analytical details see the materials and methods section.

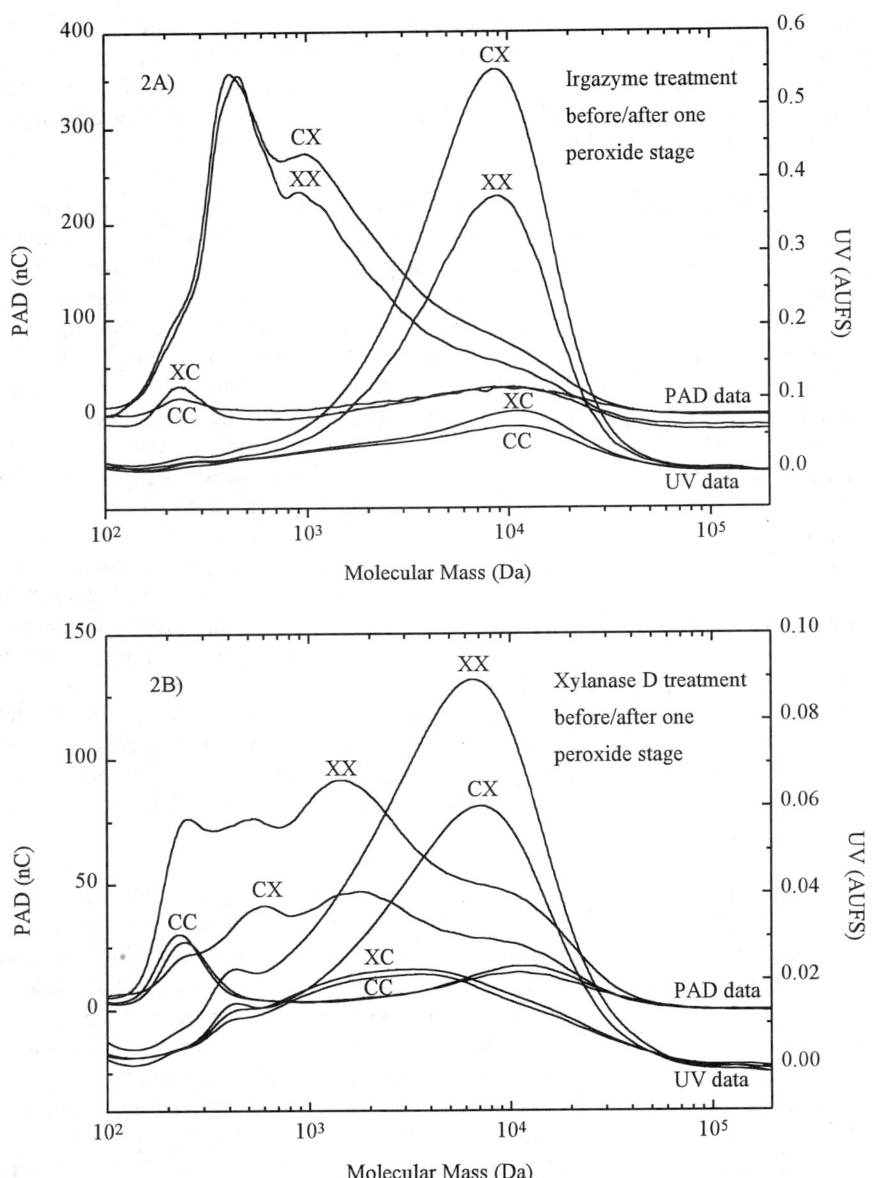

Figure 2. Size exclusion chromatograms showing the molecular mass distribution of material solubilized from Douglas-fir kraft pulps by xylanase (CX, XX) or control (CC, XC) treatments. The pulps were treated with two xylanase or control treatments that were separated by one peroxide bleaching stage. 2A) Pulp treated with Irgazyme 2B) Pulp treated with xylanase D.

material solubilized while xylanase D often led to a decrease (Table II). The ratio of absorbances at 280 and 457 nm for enzyme filtrates was basically the same as that for their corresponding alkaline extracts (Table II). However, this ratio was generally three times higher for the material solubilized from peroxide bleached pulp as compared to unbleached pulp.

Molecular Mass Distribution of the Solubilized Material. Xylanase treatment of Douglas-fir brownstock and peroxide bleached pulps released UV-absorbing material of around 3-10 kDa, while there were as many as three distinct PAD-positive peaks (Figures 1A, 2 and 4). The UV-detector registers lignin and lignin-like material, while the PAD-detector is a sensitive detector for carbohydrates. The majority of the PAD-positive material was always of a much lower molecular mass, i.e. smaller than 2 kDa, than the UV-absorbing material. There were substantial differences in the amount of material solubilized by xylanase D and Irgazyme. The latter enzyme always released more UV-absorbing and more PAD-positive material. It was noticeable that both Irgazyme and xylanase D released substantially more UV-absorbing material after the pulp had been bleached by one or two peroxide steps. In all cases, there was not much difference between the amount released after one or two peroxide stages (Figures 2 and 4). All of the chromatographic data were in agreement with the amounts of UV-absorbing material measured in the reaction filtrates (Table II). With both enzymes, a xylanase treatment after one or two peroxide bleaching steps released material with higher average molecular mass than was obtained with a xylanase treatment of the brownstock (Figures 1A, 2 and 4). A xylanase treatment before one or two peroxide bleaching stages did not have much influence on the amount or molecular mass of the material solubilized during the control treatment of bleached pulps (CC versus XC, Figures 2 and 4). However, it is clear that one Irgazyme treatment (CX) released more material than the second of two xylanase treatments (XX).

An alkaline extraction of a xylanase treated pulp solubilized more UV-absorbing and PAD-positive material than an extraction of a control treated pulp (Figures 1B, 3, 5). There was a considerable difference between Irgazyme and xylanase D in the total amount of material and the molecular mass of PAD-positive material extracted after the enzyme treatment. Irgazyme released substantial amounts of higher molecular mass compounds (0.6-3 kDa) which were almost completely absent in the xylanase D filtrates. Similarly, after one and two peroxide bleaching steps much higher molecular mass, PAD-positive material was extracted from the Irgazyme samples with NaOH, although, the differences between Irgazyme and xylanase D were much smaller. If the molecular mass distribution of material solubilized directly by the enzyme treatment was compared with the material subsequently extracted with NaOH, the latter material always had a higher molecular mass, irrespective of how many times the pulp had been bleached with peroxide (Figures 1, 2, 3, 4 and 5).

After two peroxide bleaching stages there was no significant difference between the amount of the high molecular mass (5-20 kDa) UV-absorbing material solubilized by alkaline extraction after an Irgazyme or xylanase D treatment (Figure 5). However, there were still differences in the amounts of UV-absorbing and PAD-

Table II

UV/Vis absorbing materials solubilized after control and xylanase treatments of Douglas-fir kraft pulps.

		X[a]	XE		XQPX	XQPXE	XQPPX	XQPPXE
Irgazyme	C[b]	0.13/*8*	7.21/*678*	CC	0.56/*18*	1.99/*64*	0.48/*17*	1.76/*32*
	X	0.37/*39*	2.70/*120*	CX	2.20/*56*	0.86/*39*	1.35/*7*	0.59/*35*
				XC	0.01/*-2*	0.16/*3*	-0.12/*-9*	-0.27/*-3*
				XX	1.52/*29*	0.48/*24*	0.82/*7*	0.32/*23*
Xylanase D	C	0.12/*3*	7.45/*663*	CC	0.28/*6*	2.88/*94*	0.50/*17*	2.75/*94*
	X	0.04/*2*	1.36/*117*	CX	0.67/*15*	-0.54/*-11*	0.06/*-3*	-0.88/*-45*
				XC	0.22/*9*	-0.10/*17*	-0.15/*-12*	-1.95/*-56*
				XX	0.55/*15*	0.08/*-44*	-0.01/*-11*	-1.82/*-56*

[a] The amounts of UV (280 nm, AUFS) and visible (*457 nm, mAUFS, in italic*) absorbing materials were measured directly after xylanase (X) and control treatments (C), and after a subsequent alkaline extraction (E) step. The averages of replicate experiments are given.

[b] The absolute amounts of solubilized materials are given for all the pulps which had only received one or two control (C and CC) treatments. Absolute increases (or decreases) are given for all the pulps which had received one (X, XC and CX) or two (XX) xylanase treatments.

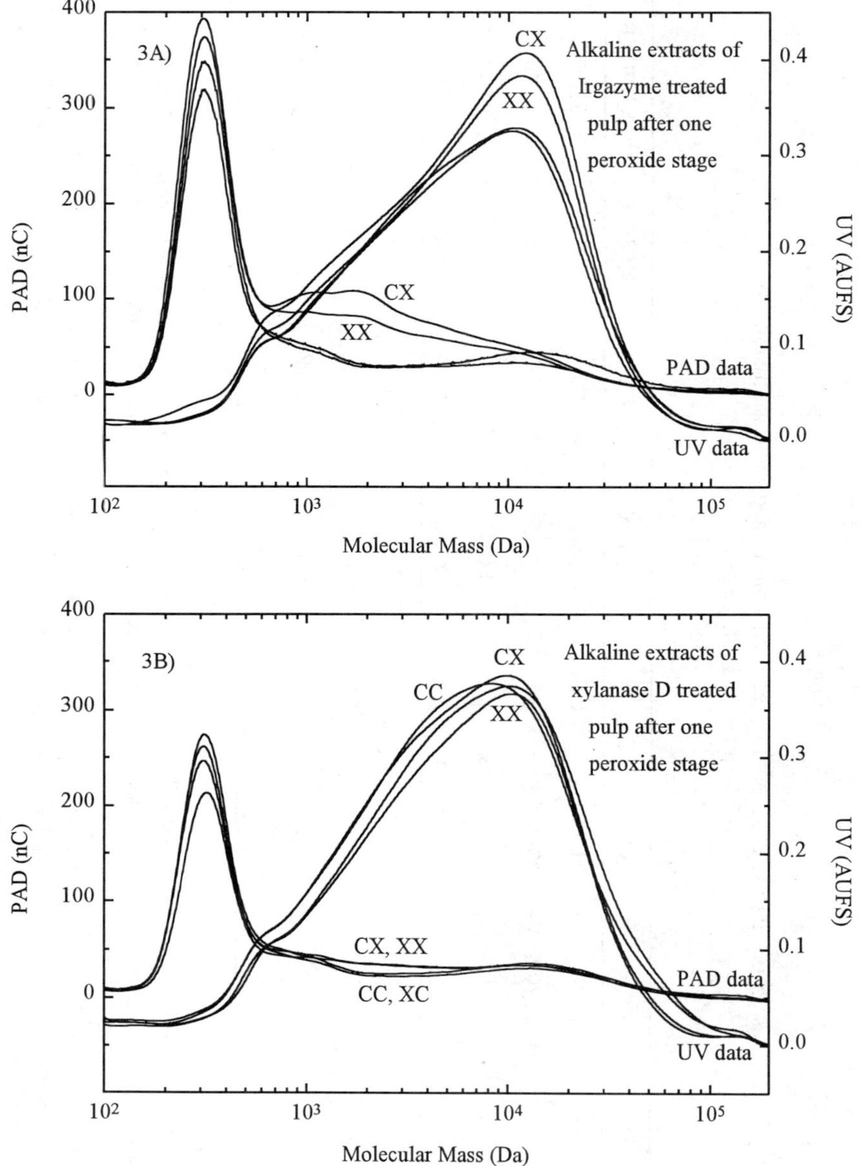

Figure 3. Size exclusion chromatograms showing the molecular mass distribution of material solubilized from Douglas-fir kraft pulps with an alkaline solution. The pulps were treated with two xylanase and/or control treatments that were separated by one peroxide bleaching stage. 3A) Pulp treated with Irgazyme 3B) Pulp treated with xylanase D.

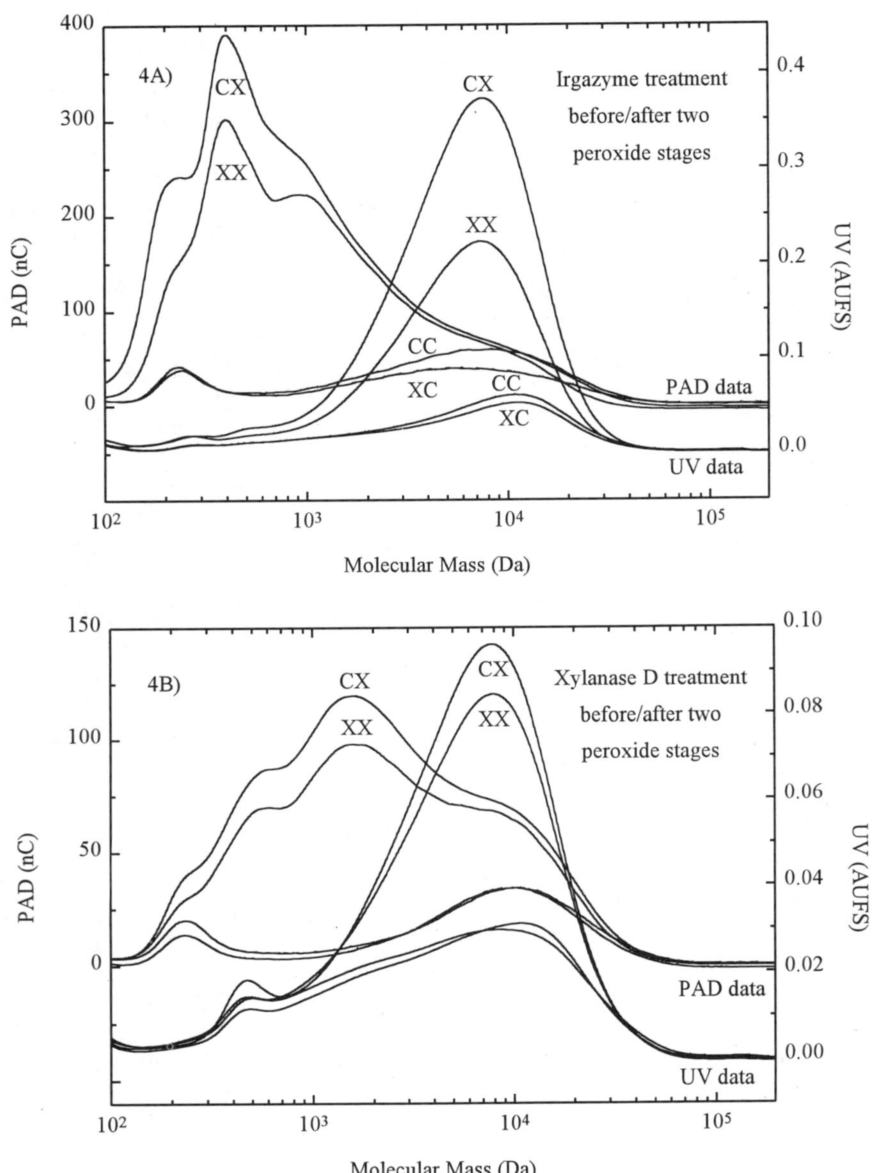

Figure 4. Size exclusion chromatograms showing the molecular mass distribution of material solubilized from Douglas-fir kraft pulps by xylanase (CX, XX) or control (CC, XC) treatments. The pulps were treated with two xylanase or control treatments that were separated by two peroxide bleaching stages. 4A) Pulp treated with Irgazyme 4B) Pulp treated with xylanase D.

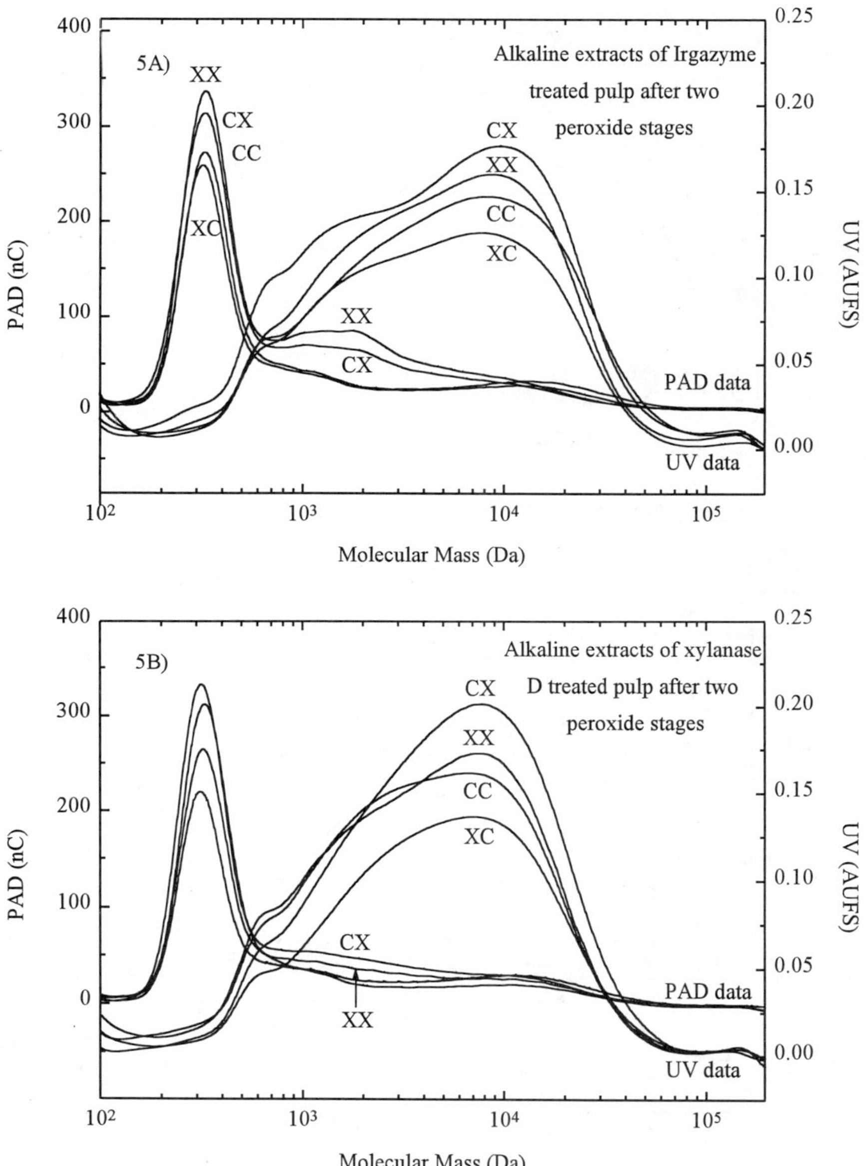

Figure 5. Size exclusion chromatograms showing the molecular mass distribution of material solubilized from Douglas-fir kraft pulps after alkaline extraction. The pulps were treated with two xylanase and/or control treatments separated by two peroxide bleaching stages. 5A) Pulp treated with Irgazyme 5B) Pulp treated with xylanase D.

positive material of intermediate molecular mass (0.6-3 kDa). In contrast, an alkaline extraction of brownstock treated with Irgazyme showed a substantial increase in material solubilized over the whole molecular mass range (Figure 1B).

Discussion

Irgazyme showed both a much stronger bleach boosting effect on unbleached pulp and a stronger direct brightening effect on peroxide bleached pulp when compared to xylanase D (Table I). Indeed, xylanase D hardly showed any bleach boosting effect, and in fact, led to a loss in brightness when the bleached pulp was subjected to another enzyme control treatment (compare CQP(P) with CQP(P)C and XQP(P)C). Although direct brightening did occur with xylanase D, it was marginal and, contrary to the results obtained with Irgazyme, a second xylanase treatment was not beneficial. It is well documented that not every xylanase is as effective in biobleaching of kraft pulps (*29, 42, 43*).

The activity of a xylanase preparation is usually measured by the hydrolysis of an isolated xylan in a short-term assay, such as that used in the present study (*39*). However, this study has shown that even when the two different xylanases were applied on pulp with the same amounts based on their determined activity, the total amount of sugars released differed considerably. A similar result was also observed using two xylanases purified from *T. reesei* (*42*). It would appear that conventional activity assays do not correlate well with hydrolysis rates occurring in pulps, nor with an enzyme's effectiveness in enhancing pulp bleaching.

Two families of xylanases have now been distinguished according to their amino acid sequence (*44*). Most of the pulp bleaching work carried out with purified xylanases has used enzymes from family 11. In particular, two family 11 xylanases from *Trichoderma reesei* have been examined extensively (*16, 42, 45, 46*). The enzyme with the lower isoelectric point (pI 5.5) was much more effective in solubilizing sugars from pulp and in enhancing pulp brightness than was the one with a pI of 9. Interestingly, the optimal pH for the activity of these two enzymes on soluble xylan differed from that for their optimal activity on pulp (*42*). The two enzymes also differed in their response to the chelation stage, with the pI 9 enzyme being less tolerant (*16*). Part of this difference may be attributed to charge interactions between pulp and enzyme (*16, 45*). Two xylanases from *Streptomyces* sp. have also been compared for their bleaching effect on both softwood and hardwood kraft pulps. Xyl 1a is a low molecular mass, high pI xylanase and xyl 3 is a high molecular mass, low pI xylanase. At pH 9.0 the action patterns of the two enzymes were similar. However, xyl 3 released 12% more chromophores from hardwood and 30% more from softwood pulp than did xyl 1a. Both enzymes enhanced brightness while reducing chemical demand, although xyl 3 was more effective (*29*).

Commercial xylanase preparations also vary in their effectiveness in pulp bleaching (*13, 32, 34, 47-52*) Although, this may be due to differences in their pH and temperature optima, the formulation of the enzyme preparations and contamination with other proteins would likely contribute to the observed differences. From our comparison of Irgazyme with Pulpzymes HB and HC at equal

enzyme dosages (Novo Nordisk, Denmark) (*32, 34*), we hypothesized that the higher brightnesses obtained with Irgazyme may be explained by a synergistic action of different enzymes in this commercial preparation because the Pulpzymes were highly purified products of individual xylanases (*32, 34*). Synergism has been noted among distinct xylanases from *Streptomyces* sp. in the prebleaching of softwood but not of hardwood pulp (*29*). However, the differences between treatments with Irgazyme and xylanase D might also depend on whether Irgazyme is more tolerant to the reaction conditions, such as the requirements for metal ions (*16*) or the much lower pH used with the xylanase D enzyme.

It was apparent that after one or two Irgazyme treatments, there was always an increase in the amount of UV-Vis absorbable material that could be extracted by alkali. This increase in lignin extractability has often been detected as an increase in absorbance of the extract at 280 nm and by a drop in the kappa number of the pulp (*12, 33, 53-55*). It has also been reported that the molecular mass of the lignin extracted after a xylanase treatment tends to be larger (*12, 53, 54, 56*). In our work the molecular mass of lignin extracted from Irgazyme-treated brownstock was much larger than that from the control pulp (Figure 1B). However, the differences in molecular mass of the alkaline extracts of the control and enzyme treated pulps were already marginal after one peroxide bleaching stage (Figure 3) and were completely absent after two peroxide bleaching stages (Figure 5).

The release of higher molecular mass, lignin-like material would apparently suggest that xylanase treatments affect the porosity of brownstock pulp. However, xylanases have been reported to decrease the water retention value of pulp (*57, 58*), with a corresponding increase in pulp freeness (*58*). These results therefore suggest there is a decrease in pulp swelling. However, since xylan is relatively hydrophillic, it is possible that its removal reduces the capacity of wood fibers to hold water. Therefore, the solute exclusion method (*59*) seems a more appropriate method for measuring fiber porosity because this method does not apply force on the fibers and relies on the diffusion of probes in the aqueous phase. To-date we have not been able to resolve any changes in the microporosity of pulp fibers as a result of xylanase treatment using this method (unpublished results).

Another interesting point was that the lignin-like macromolecules directly solubilized from peroxide bleached pulps (Figures 3 and 5) were substantially larger than those solubilized from the brownstock (Figure 1B), irrespective of whether they were produced during xylanase or control treatments. Since a xylanase treatment released higher molecular mass lignin from peroxide bleached pulps than from the brownstock, this is an indication that different kinds of materials were solubilized when a xylanase treatment was applied at different points in the bleaching sequence. However, this observation is probably not completely conclusive as the molecules could be subjected to all kinds of association/dissociation interactions and other modifications under alkaline conditions (*41, 60*). Nevertheless, it seems that the increased extractability of high molecular mass, lignin-like material after xylanase treatment of brownstock can also be achieved by peroxide bleaching. This result raises doubts concerning the role of lignin entrapment in the mechanism with which xylanase enhances peroxide bleaching.

Thus it seems unlikely that the bleach boosting effect on peroxide bleaching

is solely the result of high molecular mass compounds being made extractable by xylanase. Indeed our previous results have suggested that the role of high molecular mass lignin might be related to its abundance in shorter pulp fibers than to its entrapment underneath fiber surfaces (*37*). Other results suggested a role for lignin-carbohydrate complexes (LCC's) (*31, 36, 61*). In the present work some PAD-positive material did co-elute with UV-absorbing material during size exclusion chromatography, although the majority of the UV-absorbing material solubilized from Douglas-fir kraft pulp was larger than the PAD-positive material. This result would suggest that fragments of LCC's are present in the material solubilized during xylanase treatment and subsequent alkaline extraction. Other chromatographic and spectroscopic techniques should be used to verify this apparently close association between part of the lignin and carbohydrate components.

Conclusion

Although the mechanism of xylanase–aided bleaching is still unclear, a prevalent thought is that the enzymes act by enhancing pulp bleachability without directly brightening the pulp. This mechanism is generally accepted both because a xylanase treatment is not expected to directly modify lignin, which is the major contributor to pulp colour and because it permits chemical savings during subsequent pulp bleaching. Immediately after the enzyme stage there is nevertheless a drop in the kappa number of the pulp and an increase in the amounts of lignin that are solubilized and become extractable with alkali. The extractable lignin also tends to have a larger molecular mass after xylanase pretreatment of kraft pulp.

However, the present work has shown that even larger fragments could be extracted from peroxide bleached pulp, with or without xylanase treatment. Therefore, it seems unlikely that a xylanase treatment would yield a bleach boosting effect only by increasing the extractability of high molecular mass compounds in a peroxide bleaching sequence. The presented work again showed that distinct xylanase preparations have quite different capacities to enhance pulp bleaching and that certain xylanases can be applied later in the bleaching sequence. Indeed, the final brightness achieved by xylanase post–treatment can be as high or higher than that achieved with xylanase prebleaching. Although bleach boosting could still be observed in the xylanase prebleached pulps where direct brightening is small or absent, its importance to pulp bleaching is now open to question.

Acknowledgments

We like to thank the Pulp and Paper Research Institute of Canada (Vancouver) for the preparation of the kraft pulp and the use of their handsheet making and testing laboratories. The work was partly financed by grants from the Science Council of British Columbia and the Natural Sciences and Engineering Research Council of Canada.

Literature Cited

1. Pryke, D.C.; McKenzie, D.J. *Pulp Pap. Can.* **1996**, *97*(1), 27-29.
2. Paice, M.G.; Bourbonnais, R.; Reid, I.D. *Tappi J.* **1995**, *78*(9), 161–169.
3. Viikari, L.; Ranua, M.; Kantelinen, A.; Sundquist, J.; Linko, M. *Proc. 3rd Int. Conf. Biotechnol. Pulp Pap. Ind.* **1986**, pp. 67–69.
4. Senior, D.J.; Hamilton, J.; Bernier, R.L.; du Manoir, J.R. *Tappi J.* **1992**, *75*(11), 125–130.
5. Viikari, L.; Kantelinen, A.; Sundquist, J.; Linko, M. *FEMS Microbiol. Rev.* **1994**, *13*, 335–350.
6. Jurasek, L.; Paice, M.G. *Proc. Int. Symp. Poll. Prev. Manufac. Pulp Pap.*, **1992**, pp. 105-107.
7. Fastén, H. *Proc. Non-Chlorine Bleach. Conf.* **1993**, paper 18.
8. Kukkonen, K.; Reilama, I. *Proc. Non-Chlorine Bleach. Conf.* **1993**, paper 20.
9. Tolan, J.S., Olson, D. Dines, R.E. *Pulp Pap. Can.* **1995**, *96*(12), 107-110.
10. Yang, J.L.; Cates, D.H.; Law, S.E.; Eriksson, K–E.L. *Tappi J.* **1994**, *77*(3), 243–250.
11. Suurnäkki, A.; Buchert, J.; Viikari, L. *Proc. 48th Appita Annu Gen. Conf.*, **1994**, pp. 169–172
12. Suurnäkki, A.; Kantelinen, A.; Buchert, J.; Viikari, L. *Tappi J.* **1994**, *77*(11), 111–116.
13. Allison, R.W.; Clark, T.A.; Suurnäkki, A. *Appita* **1996**, *49*, 18–22.
14. Scott, B.P.; Young, F.; Paice, M.G. *Pulp Pap. Can.* **1993**, *94*(3), 57–61.
15. Lundgren, K.R.; Bergkvist, L.; Högman, S.; Jöves, H.; Eriksson, G.; Bartfai, T.; van der Laan, J.; Rosenberg, E.; Shoham, Y. *FEMS Microbiol. Rev.* **1994**, *13*, 365–368.
16. Buchert, J.; Viikari, L. *Paperi ja Puu* **1995**, *77*(9), 582–587.
17. Tolan, J.S. *Proc. Non-Chlorine Bleach. Conf.* **1993**, paper 40.
18. Pham, P.L.; Alric, I.; Delmas, M. *Appita* **1995**, *48*, 213–217.
19. Pedersen, L.S.; Kihlgren, P.; Nissen, A.M.; Munk, N.; Holm, H.C.; Choma, P.P. *Tappi Proc. Pulp. Conf.* **1992**, pp. 31–37.
20. Kantelinen, A.; Hortling, B.; Ranua, M.; Viikari, L. *Holzforschung* **1993**, *47*, 29–35.
21. Holm, H.C.; Skjold–Jørgensen, S.; Munk, N.; Pedersen, L.S. *Proc. 1992 Pan–Pac Pulp Pap. Technol. Conf.* **1992**, *Vol A*, 53–57.
22. Hortling, B.; Turunen, E.; Sundquist, J. *Nord. Pulp Pap. Res. J.* **1992**, *7*, 144-151.
23. Suurnäkki, A.; Heijnesson, A.; Buchert, J.; Westermark, U.; Viikari, L. *Proc. 8th Int. Symp. Wood Pulp. Chem.* **1995**, *Vol. III*, pp. 237–242.
24. Suurnäkki, A.; Heijnesson, A.; Buchert, J.; Viikari, L.; Westermark, U. *J. Pulp Paper Sci.* **1996**, *22*, J43-J47.
25. Leite, M.; Eremeeva, T.; Treimanis, A.; Egle, V.; Bykova, T.; Purina, L.; Viesturs, U. *Acta Biotechnol.* **1995**, *15*, 199-209.
26. Paice, M.G.; Gurnagul, N.; Page, D.H.; Jurasek, L. *Enzyme Microb. Technol.* **1992**, *14*, 272–276.
27. Yang, J.L.; Eriksson, K–E.L. *Holzforschung* **1992**, *46*, 481–488.

28. Patel, R.N.; Grabski, A.C.; Jeffries, T.W. *Appl. Microbiol. Biotechnol.* **1993**, *39*, 405-412.
29. Elegir, G.; Sykes, M.; Jeffries, T.W. *Enzyme Microb. Technol.* **1995**, *17*, 955–959.
30. Wong, K.K.Y.; Saddler, J.N. *Crit. Rev. Biotechnol.* **1992**, *12*, 413-435.
31. de Jong, E.; Wong, K.K.Y.; Saddler, J.N. *Holzforschung* **1996**, in press.
32. Nelson, S.L.; Wong, K.K.Y.; Saddler, J.N.; Beatson, R.P. *Pulp Pap. Can.* **1995**, *96*(7), 42–45.
33. Wong, K.K.Y.; Martin, L.A.; Gama, F.M.; Saddler, J.N.; de Jong, E. **1996**, Submitted.
34. Wong, K.K.Y.; Nelson, S.L.; Saddler, J.N. *J. Biotechnol.* **1996**, in press.
35. Wong, K.K.Y.; Clarke, P.; Nelson, S.L. *ACS Symp. Ser.* **1995**, *618*, 352-362.
36. Wong, K.K.Y.; Yokota, S.; Saddler, J.N.; de Jong, E. *J. Wood Chem. Technol.* **1996**, in press.
37. Mansfield, S.D.; Wong, K.K.Y.; de Jong, E.; Saddler, J.N. *Appl. Microbiol. Biotechnol.* **1996**, in press.
38. Maringer, U.; Wong, K.K.Y.; Saddler, J.N.; Kubicek, C.P. *Biotechnol. Appl. Biochem.* **1995**, *21*, 49-65.
39. Bailey, M.J.; Biely, P.; Poutanen, K. *J. Biotechnol.* **1992**, *23*, 257-270.
40. Dubois, M.; Gilles, K.A.; Hamilton, J.K.; Rebers, P.A.; Smith, F. *Anal. Chem.* **1956**, *28*, 350-356.
41. Wong, K.K.Y.; de Jong, E. *J. Chromatogr. A* **1996**, *737*, 193-204.
42. Buchert, J.; Ranua, M.; Kantelinen, A.; Viikari, L. *Appl. Microbiol. Biotechnol.* **1992**, *37*, 825–829.
43. Da Silva, R.; Yim, D.K.; Park, Y.K. *Rev. Microbiol.* **1994**, *25*, 112–118.
44. Henrissat, B.; Bairoch, A. *Biochem. J.* **1993**, *293*, 781-788.
45. Buchert, J.; Tenkanen, M.; Pitkänen, M.; Viikari, L. *Tappi J.* **1993**, *76*(11), 131–135.
46. Tenkanen, M.; Puls, J.; Poutanen, K. *Enzyme Microb. Technol.* **1992**, *14*, 566–574.
47. Allison, R.W.; Clark, T.A.; Ellis, M.J. *Appita* **1995**, *48*, 201–206.
48. Lavielle, P. *Proc. EU CE PA Int. Environ. Symp.* **1993**, *Vol. 1*, 151–160.
49. Bajpai, P.; Bhardwaj, N.K.; Bajpai, P.K.; Jauhari, M.B. *J. Biotechnol.* **1994**, *38*, 1-6.
50. Brown, J.; Cheek, M.C.; Jameel, H.; Joyce, T.W. *Tappi J.* **1994**, *77*(11), 105–109.
51. Leduc, C.; Daneault, C.; Delauois, P.; Jaspers, C.; Pennickx, M. *Appita* **1995**, *48*, 435–439, 448.
52. Vicuña, R.; Oyarzún, E.; Osses, M. *J. Biotechnol.* **1995**, *40*, 163–168.
53. Buchert, J.; Siika-aho, M.; Ranua, M.; Kantelinen. A.; Viikari, L.; Linko, A. *Proc. 7th Int. Symp. Wood Pulp. Chem.* **1993**, *Vol 2*, 664–632.
54. Hortling, B.; Korhonen, M.; Buchert, J.; Sundquist, J.; Viikari, L. *Holzforschung* **1994**, *48*, 441–446.
55. Allison, R.W.; Clark, T.A. *Tappi J.* **1994**, *77*(7), 127–134.
56. Kantelinen, A.; Suurnäkki, A.; Buchert, J.; Hortling, B.; Viikari, L. *Proc. 7th Int. Symp. Wood Pulp. Chem.* **1993**, *Vol 2*, 626–632.

57. Saake, B.; Clark, T.; Puls, J. *Holzforschung* **1995**, *49*, 60–68.
58. Siles, J.; Vidal, T.; Colom, J.F.; Torres, A. *Proc. 8th Int. Symp. Wood Pulp. Chem.* **1995**, *Vol. II*, 391–396.
59. Stone, J.E.; Scallan, A.M. *Cellul. Chem. Technol.* **1968**, *2*, 343–358.
60. Dutta, S.; Garver Jr, T.M.; Sarkanen, S. *ACS Symp. Ser.* **1989**, *397*, 155-176.
61. Yokota, S.; Wong, K.K.Y.; Saddler, J.N.; Reid, I.D. *Pulp Pap. Can.* **1995**, *96*(4), 39-41.

Cellulases and Xylanases

Chapter 6

β-1,4-Glycanases of *Cellulomonas fimi*: Families, Mechanisms, and Kinetics

M. R. Bray[1], A. L. Creagh[3], H. G. Damude[1], N. R. Gilkes[1], C. A. Haynes[3], E. Jervis[3], D. G. Kilburn[1], A. M. MacLeod[1], A. Meinke[1], R. C. Miller, Jr.[1], D. R. Rose[4], H. Shen[1], P. Tomme[1], D. Tull[2], A. White[4], S. G. Withers[2], and R. A. J. Warren[1]

[1]Department of Microbiology and Immunology, University of British Columbia, 300–6174 University Boulevard, Vancouver, British Columbia V6T 1Z3, Canada
[2]Department of Chemistry, University of British Columbia, 2036 Main Mall, Vancouver, British Columbia V6T 1Z1, Canada
[3]Department of Chemical Engineering, University of British Columbia, 2216 Main Mall, Vancouver, British Columbia V6T 1Z4, Canada
[4]Department of Medical Biophysics, University of Toronto, 500 Sherbourne Street, Toronto, Ontario M4X 1K9, Canada

Four endoglucanases, two cellobiohydrolases and a mixed function exoglucanase-xylanase from *Cellulomonas fimi* are modular proteins comprising from two to six domains. All of them contain a catalytic domain (CD) and at least one cellulose-binding domain (CBD). The CDs come from five of the families of glycoside hydrolases; the CBDs from three of the families of CBDs, although all but one of the enzymes has a CBD from family II. The two cellobiohydrolases attack cellulose molecules from opposite ends. The CDs and the CBDs function independently of each other when separated by proteolysis or genetic engineering. The enzymes interact with cellulose in two ways. The CDs have weak affinity for substrate, relative to the CBDs, and catalyze hydrolysis of glycosidic bonds with inversion or retention of anomeric configuration, depending on the CD. The CBDs have much greater affinities for cellulose, with K_a values of the order of 0.5-1.0 μM for the family II CBDs. The family II CBDs adsorb to both crystalline and amorphous cellulose; the family IV CBD from endoglucanase CenC adsorbs to amorphous but not to crystalline cellulose. CBD_{Cex} from the exoglucanase-xylanase Cex, is a β-barrel in solution, with extensive β-sheet structure; three tryptophans, which participate in binding to cellulose, are adjacent in space and exposed on the surface of the β-barrel. Adsorption of CBD_{Cex} to crystalline cellulose is entropically driven. Although CBD_{Cex} appears to bind irreversibly, the binding is dynamic and the polypeptide is mobile on the cellulose surface.

0097–6156/96/0655–0064$15.25/0
© 1996 American Chemical Society

The basic properties of β-1,4-glycanases are (i) hydrolysis of glycosidic bonds with either retention or inversion of configuration at the anomeric centre; (ii) a degree of specificity for a particular polysaccharide; and (iii) an endo- or an exoglycanolytic mode of action (1). Until quite recently, except for lysozyme, little was understood of the details of these characteristics. The use of low molecular weight, soluble substrates, of various inhibitors, and of various modifying reagents, had given some indications of enzyme mechanisms and of the amino acid residues involved (2,3). Now, amino acid sequence alignment, structure determination by crystallography and NMR spectroscopy, the development and application of mechanism-based inhibitors, and site-directed mutation are revealing much about the structures, mechanisms and specificities of these important enzymes. The cellulase system produced by the aerobic, mesophilic bacterium *Cellulomonas fimi* will serve as a model to outline these developments.

Modules and families

Analysis of the amino acid sequences and of the catalytic and substrate-binding properties of several cellulases showed them to be modular proteins. Subsequently, the modular nature of many other β-1,4-glycanases was deduced simply from sequence alignments (1).

All of the enzymes have at least a catalytic domain (CD) or module. The commonest ancillary module is a cellulose-binding domain (CBD). The enzymes are grouped into families of related amino acid sequences, based on the sequences of the CDs only (4,5). All of the CDs in a family have similar three-dimensional structures and the same stereospecificity of hydrolysis (1). The enzymes characterized from *C. fimi* come from five families (Fig. 1). Endoglucanases CenA, CenB and CenC, and cellobiohydrolases CbhA and CbhB, are inverting enzymes; endoglucanase CenD and the xylanase/glucanase Cex, are retaining enzymes.

CBDs are also classified into families of related amino acid sequences (6). The CBDs of the *C. fimi* enzymes come from three families (Fig. 1). All of the enzymes except CenC have a CBD from family II. Two of the enzymes have two CBDs: CenC has tandem CBDs from family IV; CenB has CBDs from families II and III (Fig. 1).

Structures of catalytic domains

CD_{Cex} is the only CD from the *C. fimi* enzymes for which the structure has been solved (7). It is a 4/7 $(\beta/\alpha)_8$ barrel, in which the catalytic carboxyls are at the C-termini of β-strands four and seven (8). The distance of 5.5 Å between the catalytic carboxyl groups, which were identified previously (9,10), is similar to that in other retaining β-glycanases (11). The ionization states of these carboxyls, Glu127, the acid/base catalyst, and Glu233, the nucleophile, are stabilised by an extensive hydrogen-bonding network. Three other CDs from family 10 are also 4/7 $(\beta/\alpha)_8$ barrels with structures very similar to that of Cex (12,13,14). In a covalent enzyme-substrate intermediate of Cex, carbon C-1 of the proximal glucose in the attached cellobiose is bound by an α–linkage of 1.43 Å to O ε1 of the nucleophilic carboxyl of Glu 233, *syn* to the ester group. This is a preferential location for a nucleophilic attack by a carboxyl group. The carboxyl group of Glu127 is 4.5 Å

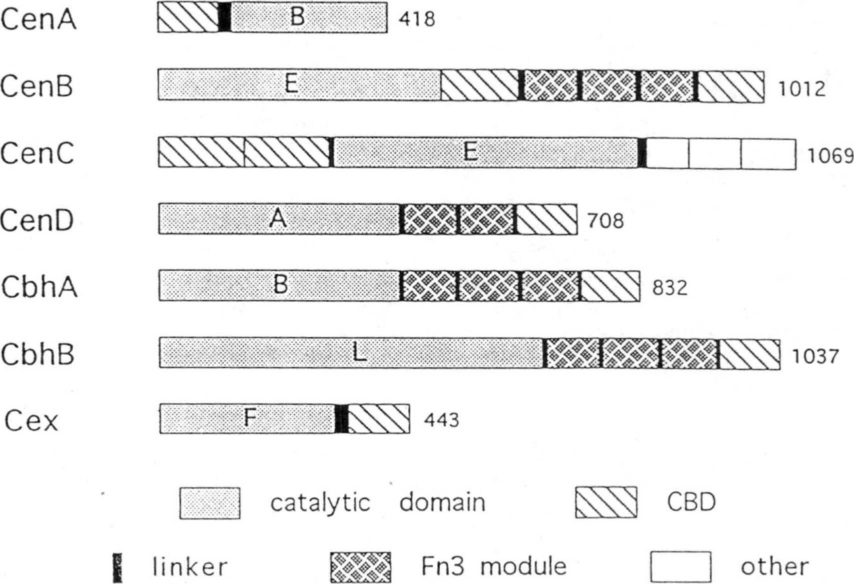

Figure 1. Structural and functional organization of *C. fimi* β-1,4-glycanases. The mature enzymes are represented as linear maps drawn approximately to scale. CenA, CenB, CenC and CenD are endoglucanases. CbhA and CbhB are cellobiohydrolases. Cex is an exoglucanase and xylanase.

from C-1 of the covalently attached cellobiose, consistent with its roles of protonating the glycosidic bond during formation of the glycosyl enzyme intermediate and then deprotonating an incoming water molecule during its hydrolysis (15).

CenA and CbhA belong to family 6 of glycoside hydrolases. The crystal structures of the CDs of two members of this family of inverting enzymes, cellobiohydrolase CBHII from *Trichoderma reesei* (16) and endoglucanase E2 from *Thermomonospora fusca* (17), show that the CDs in this family are $(\beta/\alpha)_8$ barrels. The E2 and CBHII structures and alignment of their sequences with that of the CD of CenA were used to target conserved aspartates in CenA for site-directed mutation, thereby allowing identification of its acid and base catalysts (18). The corresponding aspartates in E2 and CBHII are in the active sites of the enzymes, with their carboxyls 9 Å apart, as is usual in inverting enzymes (11,16,17).

CenD is in family 5 of glycoside hydrolases. The crystal structures of the CDs of two members of this family of retaining enzymes, endoglucanase CelCCA from *Clostridium cellulolyticum* (19) and endoglucanase CelC from *Clostridium thermocellum* (14), show that the CDs in this family are $4/7(\beta/\alpha)_8$ barrels, like the CDs in family 10. The structure of CD_{CelC} confirmed the earlier identification of Glu280 as the nucleophile (20). The carboxyl of Glu140 is ~5.0 Å from that of Glu280. Site-directed mutation of the glutamates corresponding to Glu140 in endoglucanases from *Bacillus* spp. showed them to be essential for catalytic activity (21), and site-directed mutation of Glu140 in CelC inactivates the enzyme (22). It is very likely that Glu140 is the acid/base catalyst in CelC (14). Glutamates corresponding to Glu280 and Glu140 of CelC are strictly conserved throughout family 5. In CenD these are Glu280 and Glu140, respectively (23).

It should be pointed out that families 5 and 10 are in a superfamily of $4/7(\beta/\alpha)_8$ barrel structures comprising at least nine families of β-glycoside hydrolases (24,25).

Endoglucanase CelD from *C. thermocellum* is the only CD of the inverting enzymes in family 9 for which the crystal structure is known (26). It is an α_{12} barrel in which the carboxyl groups of Glu555 and Asp201 are 10 Å apart in the substrate binding cleft on the enzyme. Site-directed mutation shows these residues to be essential for catalytic activity, and it is suggested that they are the acid and base catalysts, respectively (26,27). These residues are strictly conserved throughout family 9. The corresponding residues in CenB and CenC from *C. fimi* are Glu425 and Asp44, and Glu859 and Asp474, respectively (28,29). More detailed characterization of appropriate mutants is necessary before it can be concluded with any certainty that these amino acids are the acid and base catalysts (18).

The CD of CbhB is the only one from the enzymes of *C. fimi* characterised to date for which there is no structural information. All current members of this family of inverting enzymes are cellobiohydrolases. The family is too small at present to target with confidence conserved glutamates and/or aspartates for site-directed mutation.

Structures of cellulose-binding domains

CBDs vary widely in size: those in families I, II, III and IV are ~36, ~100, ~150 and ~150 amino acids long, respectively (6). At present, the structures of only two CBDs are known. CBD_{CBHI} from cellobiohydrolase CBHI of *T. reesei* is in family I; it is a 3-stranded, antiparallel β-barrel (30). CBD_{Cex} from Cex of *C. fimi* is in family II; it is a 9-stranded, antiparallel β-barrel (31).

CBD_{Sca}, the family III CBD of scaffoldin from the cellulosome of *C. thermocellum*, has been crystallised (32) but the structure has not been reported.

Three tryptophans are aligned and exposed on one face of CBD_{Cex} (Fig. 2). Two of them correspond to tryptophans which are strictly conserved in family II; the third corresponds to a strictly conserved aromatic residue (6). Site-directed mutation of the corresponding residues in CBD_{XynA} of xylanase A from *Pseudomonas fluorescens* subsp. *cellulosa* (33), and in CBD_{CenA} of endoglucanase CenA from *C. fimi* (34), shows them to be involved in binding to cellulose. Aromatic residues are also involved in the binding to cellulose of CBDs from family I (35).

Stereochemistry of glycoside hydrolysis

Following the suggestion that all members of a family of glycoside hydrolases would exhibit the same stereochemistry of glycoside hydrolysis (4), determination of the stereochemistry for a number of enzymes has yet to reveal an exception to this rule (1,36). Furthermore, the superfamily of $4/7(\beta/\alpha)_8$ barrels contains only families of retaining enzymes (24,25).

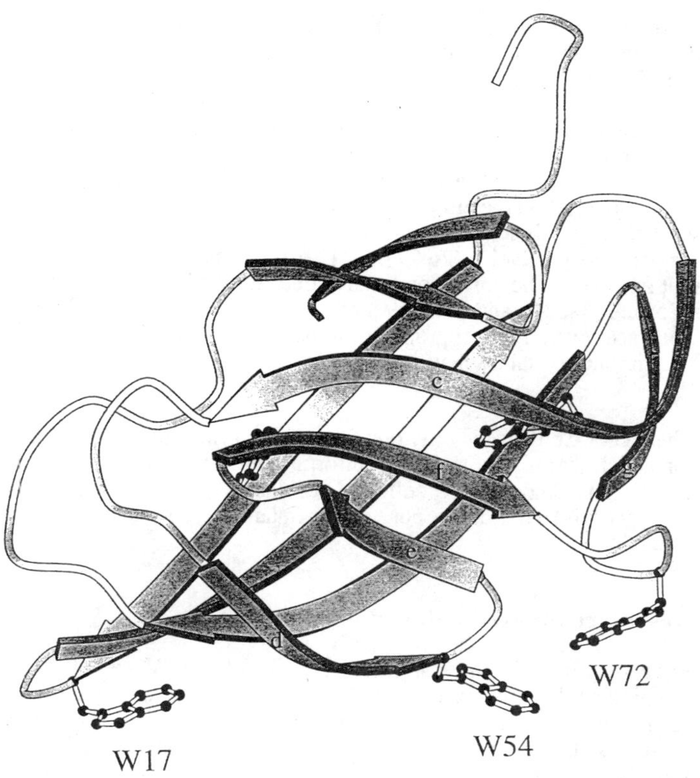

Figure 2. Nine-stranded, antiparallel β-barrel structure of CBD_{Cex}. The three exposed tryptophans are shown.

Identification of the catalytic carboxyl amino acids

In retaining enzymes, a carboxyl group acts as a nucleophile to displace the aglycone and form a covalent enzyme-substrate intermediate with the glycone; a second carboxyl group acts as an acid/base catalyst, protonating the aglycone as it leaves, then deprotonating a water molecule, with the hydroxyl ion being formed displacing the enzyme carboxyl group from the glycone. In inverting enzymes, a carboxyl group acts as a base catalyst to deprotonate a water molecule; the hydroxyl ion displaces the aglycone, which is protonated by a second carboxyl acting as an acid catalyst (Fig. 3; ref. 2,11). The carboxyl amino acids involved are strictly conserved in a family of CDs, so that their identification in one member of a family in effect identifies them in all members of the family.

A variety of methods is used to identify the crucial carboxyl amino acids, but with one exception, and only for retaining enzymes, none of them identifies such a residue unequivocally. A combination of methods is usually required.

Determination of the crystal structure of its CD will reveal which residues comprise the active site of an enzyme. Experience shows that in retaining enzymes, the carboxyls of the nucleophile and the acid/base catalyst are ~ 5.3 Å apart; in inverting enzymes, the carboxyls of the acid and the base catalysts are ~9.5 Å apart (11). The greater separation is required in inverting enzymes because a water

RETAINING MECHANISM

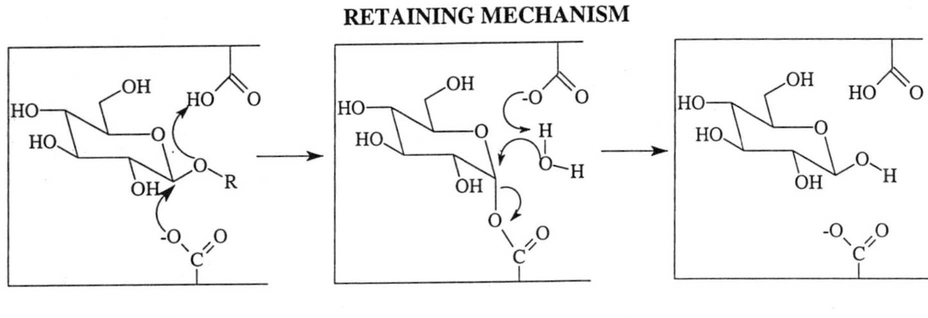

INVERTING MECHANISM

GENERIC TRANSITION STATE

Figure 3. Proposed mechanisms of inverting and retaining β-glucanases.

molecule must bind with the substrate between the carboxyls. In retaining enzymes, a water molecule enters the active site after displacement of the aglycone by the nucleophile. Experience also shows that the presence of carboxyl groups the requisite distance apart may be insufficient evidence for assigning them particular roles in the reaction because crystal structures may not represent catalytically active conformations. Mutation of the true catalytic residues should reduce the activity of the enzyme several orders of magnitude, but detailed analyses of the mutants is required to differentiate the acid and base catalysts in an inverting enzyme and the nucleophile and acid/base catalyst in a retaining enzyme.

Alignment of the amino acid sequences of the CDs in a family can be useful in identifying catalytic residues in the absence of other structural information for any member, especially if the family is large. In general, the number of strictly conserved amino acids diminishes as more members are assigned to a family. Conserved carboxylic amino acids are prime targets for mutation. Again, mutation of true catalytic residues should reduce activity several orders of magnitude.

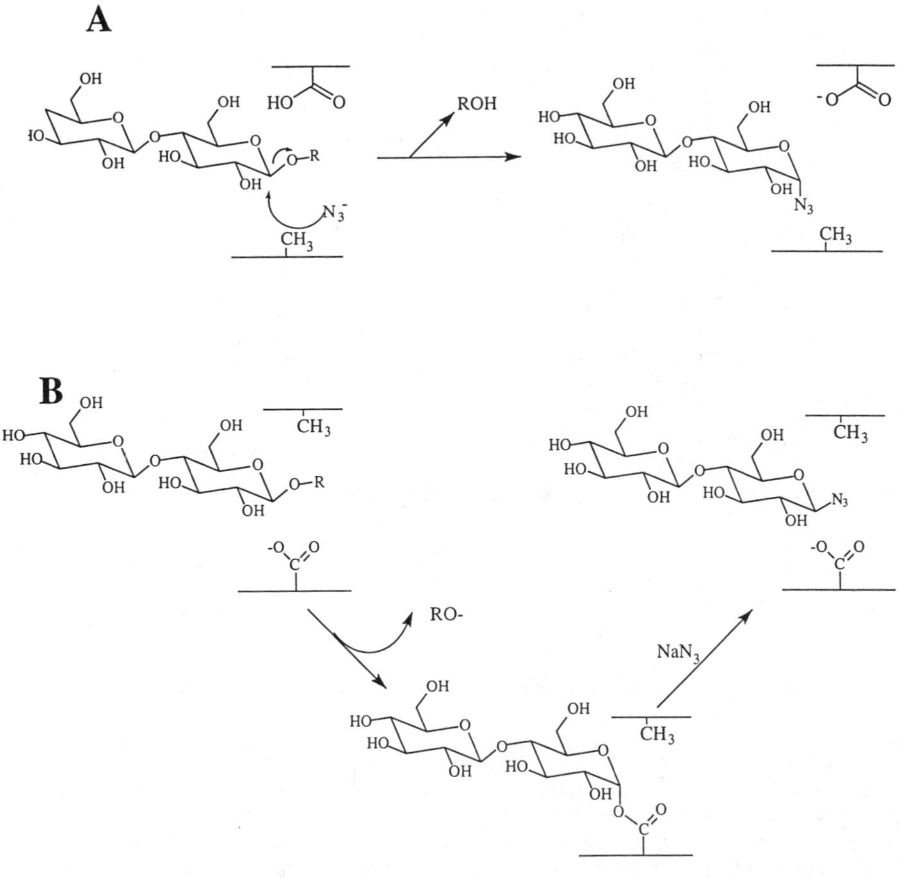

Figure 4. Proposed mechanism for the formation of (A) α-cellobiosyl azide by the nucleophile mutant and (B) β-cellobiosyl azide by the acid/base catalyst mutant of a retaining β-glucanase.

The only direct method for identifying a catalytic residue is applicable to retaining but not to inverting enzymes. It depends on retaining enzymes catalysing hydrolysis via formation of a covalent enzyme-substrate intermediate between the nucleophile and the glycone moiety of the bond being hydrolysed. A mechanism-based inhibitor is used to trap a covalent enzyme-inhibitor intermediate. The enzyme is not irreversibly inactivated, and the half-life of the intermediate is long enough to enable identification of the carboxyl amino acid to which the inhibitor is attached (11). Since its introduction (37), the method has been improved by the development of a tandem mass spectrometric method for identification of the labelled amino acid (38). It can be applied to any retaining enzyme for which an appropriate inhibitor can be synthesized. It is emphasized again that this needs to be done for only one enzyme in a family.

Identification of the catalytic carboxyl amino acids in CenA and Cex from *C. fimi* is described to illustrate the application of the various methods.

Catalytic carboxyl amino acids in Cex

The mechanism-based inhibitor 2',4'-dinitrophenyl 2-deoxy-2-fluoro-β-D-glucopyranoside labels Glu233 in Cex, identifying it as the nucleophile (9). The 2-fluoro substituent slows both the formation and the hydrolysis of the covalent enzyme-substrate intermediate, but the dinitrophenol is a reactive leaving group which enhances the rate of formation of the intermediate. The net effect is the formation of the 2-deoxy-2-fluoro-α-D-glucosyl enzyme, which has a half-life of >800h. The half-life is reduced to ~250h in the presence of cellobiose, with reactivation of the enzyme occurring by transglycosylation. The mutation E233A reduces k_{cat} for the hydrolysis of ρ-nitrophenyl-β-D-cellobioside (PNPC) and 2,4-dinitrophenyl-β-D-cellobioside (DNPC) about 10^7-fold, as would be expected for mutation of the nucleophile (39). The anionic nucleophiles formate and azide increase k_{cat} for the hydrolysis of DNPC by the mutant $4x10^4$ and $1.2x10^5$-fold, respectively, substituting for the missing carboxyl group on the enzyme. Furthermore, the product formed from DNPC by the mutant in the presence of azide is α-cellobiosyl azide, an analogue of the α-cellobiosyl-enzyme intermediate formed transiently by the wild-type enzyme (39). The stereochemistry of the product is further proof that azide substitutes for the nucleophilic carboxyl in the mutant (Fig. 4).

The acid/base catalyst in Cex was identified indirectly. Glu127 corresponds to a residue that is strictly conserved in family 10 enzymes. PNPC and DNPC, which do not require protonic assistance for initial bond cleavage, exhibit k_{cat}/K_m values with the E127A mutant which are similar to those for the wild-type enzyme. ρ-Bromophenyl-β-D-cellobioside (PBrPC), which does require protonic assistance, exhibits a k_{cat}/K_m value with the mutant which is about $6x10^3$-fold lower than that for the wild-type enzyme. This is strong evidence for Glu127 being the acid/base catalyst (10). Furthermore, azide increases k_{cat} for hydrolysis of DNPC and PNPC by the mutant about 200- and 8-fold, respectively, without changing k_{cat}/K_m significantly. Azide does not affect the hydrolysis of PBrPC by the mutant nor does it affect the hydrolysis of any substrate by the wild-type enzyme. The product formed from DNPC in the presence of azide is β-cellobiosyl azide (10). It seems that azide occupies a vacant anionic site, normally occupied by the side-chain of the missing acid/base catalyst, and reacts with the glycosyl enzyme intermediate to form the β-glycosyl product (Fig. 4).

As mentioned earlier, x-ray crystallography confirmed Glu233 and Glu127 as the nucleophile and acid/base catalyst in Cex, respectively (7,15). The methods are generally applicable to retaining glycosidases with or without crystal structures being available (20,37,40-43).

The kinetics of hydrolysis by Cex are consistent with the double-displacement mechanism via a covalent enzyme-substrate intermediate, with the formation of the intermediate being rate-limiting (44). The pK_as of the nucleophile and the acid/base catalyst are 4.1 and 7.7, respectively (44).

Catalytic carboxyl amino acids in CenA

Four aspartates are conserved in the active sites of the inverting enzymes of family 6 (16,17). In CenA, they are Asp216, Asp252, Asp287 and Asp392. Mutants at these positions were analyzed with DNPC, a substrate which does not require acid assistance for its hydrolysis, and with carboxymethylcellulose (CM-cellulose) and phosphoric acid-swollen cellulose (PAS-cellulose), substrates which do need assistance for their hydrolysis (18). Removal of the acid catalyst in an inverting enzyme should have much less effect on the hydrolysis of DNPC than on the hydrolysis of CM-cellulose and PAS-cellulose. Removal of the base catalyst should affect significantly the hydrolysis of all three substrates because deprotonation of a water molecule is essential in the single-displacement reaction mechanism.

The mutation D252A reduces k_{cat} for the hydrolysis of CM-cellulose 2×10^5-fold; it eliminates activity on PAS-cellulose; but it has no effect on k_{cat} for hydrolysis of DNPC. This is consistent with Asp252 being the acid catalyst in CenA. The mutation D392A reduces k_{cat} for the hydrolysis of CM-cellulose and PAS-cellulose about 3×10^4-fold and eliminates activity on DNPC. This is consistent with Asp392 being the base catalyst in CenA. The mutation D287A reduces k_{cat} for the hydrolysis of both CM-cellulose and PAS-cellulose about 10^4-fold, but that for DNPC only 37-fold. This is consistent with Asp287 playing a role in acid catalysis by being positioned to raise the pK_a of Asp252, the acid catalyst. The mutation D216A reduces k_{cat} for the hydrolysis of DNPC, CM-cellulose and PAS-cellulose by 18-, 132- and 1380-fold, respectively, indicating that Asp216 is not directly involved in catalysis.

The spatial distribution in the crystal structures of CbhII and E2 of the aspartates corresponding to Asp252, Asp287 and Asp392 of CenA are consistent with the roles assigned to these three residues in CenA (16-18). Furthermore, the pH profiles for hydrolysis of DNPC by wild-type CenA show that k_{cat} is strongly dependent on the presence in the enzyme of both a protonated group and a deprotonated group, consistent with the presence of an acid and a base catalyst in the active site. The pH profile for the D252A mutant shows a dependence only on a base catalytic group, confirming the role of Asp252 as the acid catalyst (45).

Values of k_{cat} for hydrolysis of DNPC, PNPC, cellotriose and cellotetraose by CenA are 0.37, 0.00078, 0.27 and 220 s^{-1}, respectively. The pK_as for the aglycones in these substrates are 4.0, 7.2, ~16.0 and ~16.0, respectively. The Bronsted plot (log k_{cat}/K_m vs. aglycone pK_a) for these substrates has a slope of $\beta_{1g} = -0.9$, indicative of substantial charge development on the leaving oxygen at the transition state, consistent with complete bond cleavage and little proton donation. The enzyme, via specific binding interactions with sugar aglycones, appears to

make glucose as good a leaving group as DNP, and cellobiose 1000-fold better than DNP. The effective pK_a values of the departing sugar aglycones are therefore reduced from ~16.0 to ~0 (45).

Exoglycanases of *C. fimi*

CbhA (46) and CbhB (47,48) are inverting cellobiohydrolases. Both release cellobiose as the major soluble product from cellulose. The action of both enzymes on CM-cellulose is indicative of an exoglucanolytic mode of action: plots of specific fluidity vs. reducing sugar released have the shallow slopes characteristic of exo-acting enzymes. The ratio of the α- and β- anomers of cellotetraose at mutarotational equilibrium is 1:2. The ratio for the cellotetraose released from cellohexaose by CbhB is 1.0:0.9 before mutarotation, 1:2 after allowing time for mutarotation, showing that the enzyme removes cellobiose from the reducing end of the hexamer. The ratio for the cellotetraose released from cellohexaose by CbhA is 1:2 before and after allowing time for mutarotation, showing that the enzyme removes cellobiose from the non-reducing end of the hexamer (46,48,49).

Cellobiohydrolases are relatively inefficient degraders of CM-cellulose. The enzymes remove cellobiose units processively from the ends of cellulose molecules, but processivity is blocked by the first carboxymethyl substituent an enzyme encounters. As expected, CbhA releases a burst of reducing sugar from CM-cellulose. The subsequent addition of CbhB but not of more CbhA releases a second burst of reducing sugars. Similarly, two bursts of reducing sugar are released by CbhB followed by CbhA, but only one by two additions of CbhB. This can happen only if the enzymes act on different ends of the cellulose molecules (Fig. 5).

Cellobiose is the only soluble sugar released from cellulose by CenC; it is the major product of the hydrolysis of soluble cellodextrins by this enzyme. This suggests that CenC is another cellobiohydrolase. It cleaves cellopentaose to cellobiose and cellotriose. The ratio of the α- and β-anomers of the cellotriose before allowing time for mutarotation is 2:1, showing that CenC is an inverting enzyme which removes cellobiose units from the reducing ends of cellulose molecules. However, CenC also has significant endoglucanase activity. The slope for the plot of specific fluidity versus reducing sugar release for the hydrolysis of CM-cellulose by CenC lies between those for CbhA and a true endoglucanase, such as CenA. Furthermore, CenC is significantly more active than CbhA in causing clearing on CM-cellulose-Congo red plates, considered to be diagnostic of endoglucanase activity. Thus, CenC behaves like a semi-processive enzyme with both endo- and exoglucanase activities (50).

Cex is in family 10 of retaining xylanases. Although 50-fold more active on xylosides, Cex, like other enzymes in family 10, has weak but detectable activity on glucosides (51). The plot of specific fluidity versus reducing sugar release for the hydrolysis of CM-cellulose by Cex has the shallow slope typical of exoglycanolytic activity (52). Thus, Cex appears to be a fourth exo-acting enzyme from the plant cell wall-hydrolyzing system of *C. fimi*.

CbhA, CbhB and Cex have endoglycanase activity

The production of zones of clearing on CM-cellulose-Congo red plates is considered to be diagnostic of endoglucanase activity. Cellobiohydrolases and

other exo-acting enzymes apparently do not cause clearing, and this was one of the criteria for classifying CbhA and Cex as exoglucanases (46,52). CbhB was classified initially as an endoglucanase because it did cause clearing on CM-cellulose-Congo red plates (47). It is clear from a more detailed analysis, however, that CbhB is a cellobiohydrolase, albeit with endoglucanase activity (47). When 2000 pmol of CbhA, CbhB or Cex are spotted on CM-cellulose-Congo red plates, clearing is seen, although it is less than that caused by 20 pmol of the endoglucanase CenA (48). Thus, these three enzymes are predominantly exo-acting but they do have low but detectable endo-activity. CenC is also an exo-acting enzyme but with significantly more endo-activity than CbhA, CbhB and Cex (49). The shallow but obvious slopes for the plots of specific fluidity versus reducing sugar release for the hydrolysis of CM-cellulose by each of the four "exoglycanases" are also indicative of endoglucanase activity (46,48,50,52).

What determines exo- versus endo-activity? Endoglucanase E2 from *T. fusca* and cellobiohydrolase CbhII from *T. reesei* are both in family 6 of inverting enzymes. Both have $(b/a)_8$ CDs. The active site of E2 is an open cleft, that of CBHII is tunnel-shaped. The tunnel is formed by two loops covering the active site cleft in CBHII. The C-terminus proximal loop is formed by a sequence of amino acids present in CBHII but not in E2. The N-terminal proximal loop corresponds to a sequence which is present in both E2 and CBHII, but in E2 it is pulled back from the active-site cleft by the deletion of a short sequence adjacent to the loop in CBHII (16,17). It is proposed that CBHII is an exoglycanase because cellulose molecules can enter its active site only by being threaded in from one end (16).

CbhA is also in family 6, and it also has extra amino acids, relative to E2, corresponding to those forming the loops in CBHII (46). Deletion from CbhA of a sequence of amino acids corresponding to those forming the C-proximal loop in CBHII enhances the endoglucanase activity of CbhA (53). This supports the idea that exoglucanase activity in CBHII and CbhA is a consequence of tunnel-shaped active sites. Interestingly, cellobiohydrolase CBHI from *T. reesei*, a family 7 enzyme, also has a tunnel-shaped active site (54). The endoglucanase activity of CbhA suggests that the loops occluding its active site are flexible enough to allow a cellulose chain occasionally to enter through the roof of the tunnel rather than from one end. A tunnel-shaped active site is not a pre-requisite for exoglycanase activity. Cex, which exhibits exoglucanase activity on CM-cellulose, has an open active site cleft (7).

CBHI and CBHII from *T. reesei* also appear to have endoglucanase activity (55), blurring further the distinction between exo- and endo-acting enzymes. It is perhaps more accurate to say that some enzymes but not others can act processively from the ends of cellulose molecules. Occasionally, the processive enzymes will initiate hydrolysis from sites within cellulose molecules rather than at the ends by making an initial endoglycanolytic cut. The relative frequencies of the two modes of hydrolysis will depend on the accessibility of the active site.

Specificities and activities of β-1,4-glycanases

The enzymes are relatively specific for the glycone, less specific for the aglycone, of the β-1,4 bonds they hydrolyse. Although an enzyme may lack absolute specificity, it is usually most active on a particular glycone. Most, if not all, of the enzymes hydrolyse substrates with non-carbohydrate aglycones. The chromophoric nitrophenyl and fluorogenic methylumbelliferyl-β-D-glycosides are

used routinely for screening and quantifying enzyme activities, and in probing enzyme mechanisms. The specificities of 15 β-1,4-glycanases for a series of chromophoric or fluorophoric glycosides, derived from glucose, cellobiose, soluble cellooligosaccharides, lactose, xylose and xylobiose, differentiate the enzymes into six groups, each with a characteristic specificity pattern. The groups correspond to families of CDs. The substrates do not differentiate endo- and exoglycanases in the same family (56). The use of nitrophenyl glycosides in identifying the catalytic carboxyls in glycoside hydrolases is outlined above. Such substrates, however, reveal little about the hydrolysis of the insoluble, polymeric substrates of the enzymes. Nor do soluble derivatives of β-1,4-glycans, despite the usefulness of a substrate such as CM-cellulose for differentiating endo- and exoglycanases.

The activity of an enzyme may be very different not only on different β-1,4-glycans, but also on different forms of a particular glycan. The rate of hydrolysis of crystalline cellulose may differ markedly from that of amorphous cellulose. Activity on xylan can be influenced by the level of decoration of the xylan backbone.

The activity of an enzyme on an insoluble substrate is determined routinely by measuring the increase in reducing sugar, usually in the soluble products, sometimes in both the insoluble and soluble products. The enzymes from *C. fimi* differ significantly in their activities on several β-1,4-glycans and on different physical forms of cellulose (Table I). The activities of an enzyme on different β-1,4-glycans may indicate the relative specificities for particular glycones, but it must be borne in mind that the rates will also depend on the ease with which an enzyme molecule can sequester a single molecule of substrate in its active site. The latter factor will influence the activity of an enzyme on different forms of a glycan, such as amorphous and crystalline cellulose. Such measurements, however, do not reveal the modes of attack of different enzymes on a particular substrate, although the release of cellobiose as the only soluble product from cellulose is an indicator of cellobiohydrolase activity.

Determination of the sizes of the insoluble products of the action of an enzyme on an insoluble substrate does give clues as to the mode of attack. Derivatizing the insoluble products to make them soluble in tetrahydrofuran allows determination of their average degree of polymerization (DP, number of glucosyl residues per cellulose chain) by size exclusion chromatography (57). The procedure is non-degradative (58). The kinetics of the change in DP relates to the mode of action of an enzyme on a substrate. The reduction in weight of the underivatized, insoluble material indicates the relative efficiency of an enzyme in solubilizing a substrate. The preference of an enzyme for amorphous or crystalline cellulose can be evaluated with substrates of different degrees of crystallinity. The relative crystallinities of cotton, bacterial cellulose (BC) and bacterial microcrystalline cellulose, for example, are cotton <BC<BMCC.

C. fimi enzymes exhibit different modes of attack on these substrates (59). CenA, CenB and CenD have similar activities on cotton and all three enzymes decrease the DP of the residual insoluble substrate progressively. CenA differs from CenB and CenD because a low molecular weight product (DP~11) accumulates in the later stages of digestion. The modes of action of the three enzymes on BC are similar to their actions on cotton, with CenB and CenD releasing monodisperse products, and CenA releasing bidisperse products. However, CenB and CenD are ~2-fold better than CenA at solubilizing cotton. CenA is ~6-fold less active than CenB and CenD on BMCC, but again it produces

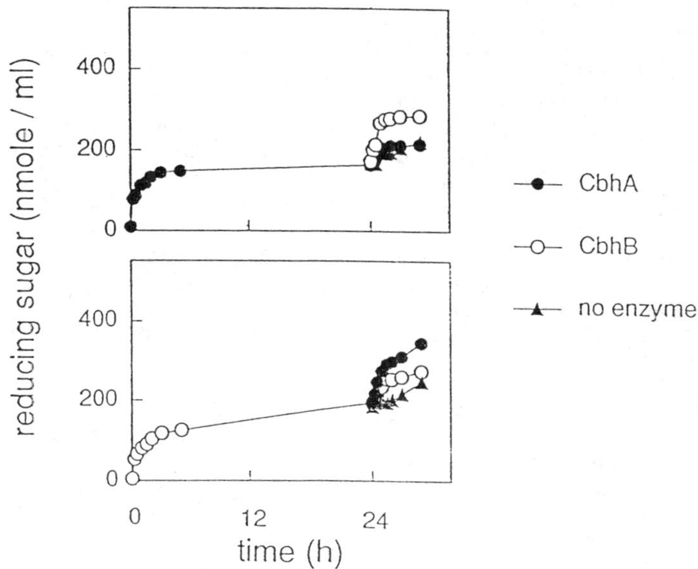

Figure 5. Release of reducing sugars from CM-cellulose by the sequential action of cellobiohydrolases CbhA and CbhB. The enzymes act on opposite ends of the CM-cellulose molecules.

Table I. Activities of enzymes from *C. fimi* on various β-glucans

Enzyme	Avicel	BMCC	CM-cellulose	PAS-cellulose	β-glucan	Lichenan	glucoman
CenA	2.18	0.21	760	244	2180	3940	N.D.
CenB	2.22	10.87	928	66	5700	3240	912.5
CenC	0.99	1.55	1016	114	3900	3220	722.0
CenD	2.42	9.66	47	81	940	1220	N.D.
Cex	0.16	N.D.	10	46	N.D.	N.D.	N.D.

Activities (37°C, pH 7.0) expressed as μmoles Glc. min^{-1}. $(\mu mole\ enzyme)^{-1}$.
CMCase and β (1-3, 1-4) glucanase (lichenase) activities measured at 30°C
N.D. : not detected

bidisperse products. CenD produces monodisperse products from BMCC, as it does from cotton and BC. CenB, however, produces bidisperse products from BMCC, similar to those produced by CenA. This contrasts with its action on cotton and BC.

The patterns of products released from cotton, BC and BMCC by CenA suggest that it preferentially attacks amorphous regions in the cellulose microfibrils. The amorphous regions are digested completely to fragments of low DP, leaving crystalline fragments of the microfibrils. The surfaces of the crystalline fragments are attacked only slowly by CenA (Fig. 6).

Although CenB and CenD attack both the amorphous and the crystalline regions of the microfibrils, they are more active on the latter, as evidenced by their extensive solubilization of BMCC. Molecules of high DP persist throughout the digestion as the average DP is decreasing. The enzymes appear to degrade the surfaces of the microfibrils progressively (Fig. 6).

CbhA solubilizes BMCC more effectively than cotton and BC but it does not reduce the DP of any of the substrates significantly, as is expected for a cellobiohydrolase. The more effective solubilization of BMCC suggests that the ends of the cellulose molecules are more accessible to CbhA in crystalline than in amorphous regions of the microfibrils. The smaller microfibrils of BMCC have a greater surface area to volume ratio than cotton and BC.
Cex has insignificant activity on cotton, BC and BMCC when assayed in this way, emphasizing its predominantly xylanolytic activity.

The properties of the enzymes from *C. fimi* suggest a model for the degradation of cellulose microfibrils by this organism (Fig. 7). The model is incomplete because CenC and CbhB have yet to be characterized in detail, and there may be as yet unidentified components of the system.

Binding of CBDs to cellulose

All of the β-1,4-glycanases from *C. fimi* that have been characterized to date have at least one CBD. The exact role(s) of the CBDs in the hydrolysis of cellulose are unclear. Removal of its CBD reduces the activity of CD_{CenA} on insoluble but not on soluble cellulose (60), suggesting that one role of the CBD is to maintain a high local concentration of an enzyme on the surface of cellulose. The CBD may play a more direct role in the hydrolysis of cellulose, however, because CBD_{CenA}, which is devoid of hydrolytic activity, disrupts the surfaces of cellulose and releases small particles from cotton fibres (61). Furthermore, CBD_{CenA} synergizes with CD_{CenA} in the hydrolysis of cotton fibres (62).

The K_a for the binding of CBD_{CenA} to BMCC is ~4 x 10^7 M^{-1}. The Scatchard plot of the binding isotherm is non-linear, reflecting multiple interactions between the CBD and BMCC (63). The binding of CBD_{CenA} is strongly dependent on at least two of the tryptophans that are conserved in CBDs in family II (34).

An understanding of the interactions between a CBD and cellulose requires determination of the structure of the CBD, detailed analysis of its interactions with a defined matrix, and identification of the amino acid residues which interact directly with the cellulose. This in turn requires relatively large quantities of the CBD and an appropriate cellulosic matrix. BMCC and *Valonia* cellulose are both highly

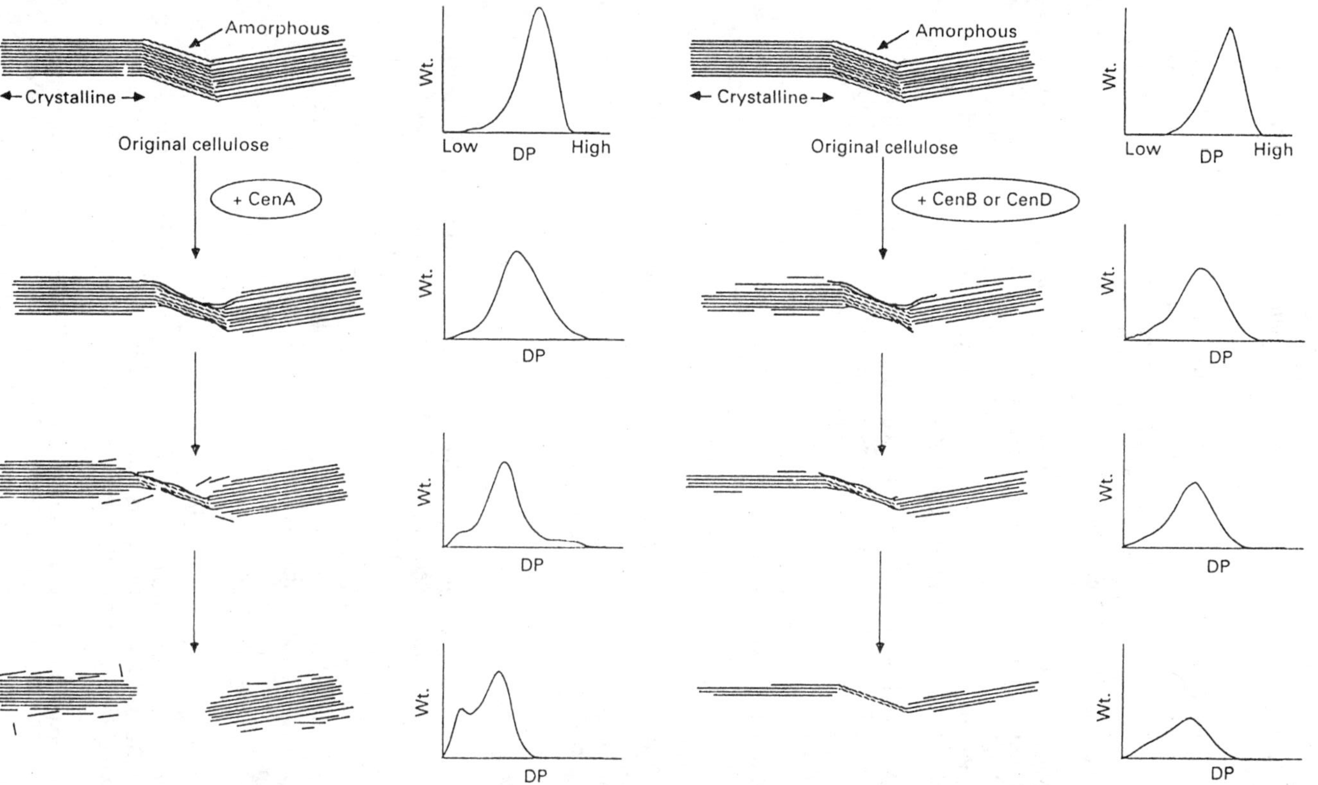

Figure 6. Proposed modes of hydrolysis of cellulose by endoglucanase CenA and by endoglucanases CenB and CenD. DP is the degree of polymerization of the cellulose, Wt. is the weight of the insoluble substrate (Reproduced with permission of reference 59. Copyright 1994 Biochemical Society and Portland Press.)

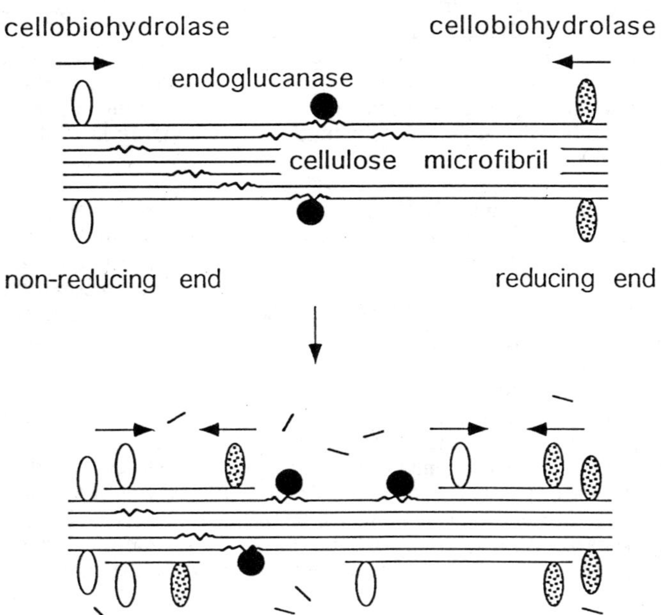

Figure 7. Cartoon of the hydrolysis of cellulose by the enzymes from *C. fimi*.

crystalline materials, use of which largely obviates having to distinguish between binding to amorphous and binding to crystalline regions of the cellulose. CBD_{Cex}, also in family II, is readily produced and purified (64). The relative affinity of CBD_{Cex} is 10-fold greater for BMCC than for amorphous cellulose (phosphoric acid swollen cellulose) (65), emphasizing the need to analyse binding using a matrix of reproducible crystallinity. CBD_{Cex} binds to cellulose irreversibly under the appropriate conditions: a β-glucosidase-CBD_{Cex} fusion protein does not desorb from cellulose during prolonged washing of the cellulose with buffer (65,66).

The BMCC surface presents two independent classes of binding sites to CBD_{Cex}. ~70 per cent are high affinity sites, the others are low affinity sites. The K_a for the high-affinity site is $\sim 6 \times 10^7 M^{-1}$, that for the low affinity site is $\sim 1 \times 10^6 M^{-1}$, as determined by isothermal titration microcalorimetry. The binding of CBD_{Cex} to either site is exothermic, but driven mostly by a large positive change in entropy, a consequence of both polypeptide and cellulose dehydration (67). CBD_{Cex} is a nine-stranded antiparallel β-barrel, with three conserved tryptophans aligned and exposed on one surface (31). The thermodynamic data are consistent with these tryptophan residues making enthalpically weak but sufficient contacts with the largely inflexible BMCC surface to dehydrate both the binding face of CBD_{Cex} and the cellulose (67).

N-Bromosuccinimide (NBS) selectively oxidizes the indole sidechains of tryptophan residues in proteins to oxindole derivatives, changing the planarity, aromaticity and hydrophobiticy of the sidechains. In general, only tryptophans exposed on the surface of a protein are oxidized. If any of the exposed tryptophans are involved in substrate or ligand binding, the substrate or ligand will usually protect them from oxidation by NBS. CBD_{Cex} is unusually sensitive to NBS. There are five tryptophans in CBD_{Cex}, two inside the β-barrel of the polypeptide, three exposed on the surface (31). NBS oxidizes all five tryptophans in CBD_{Cex}, but the exposed tryptophans are oxidized before the buried ones. Oxidation of the exposed tryptophans does not affect the conformation of the backbone of CBD_{Cex}, but oxidation of all five tryptophans denatures the polypeptide. Oxidation of the equivalent of two tryptophans per molecule of CBD reduces the binding of the polypeptide to cellulose by 90 per cent. Surprisingly, binding of CBD_{Cex} to BMCC does not prevent oxidation of CBD_{Cex}, but it does afford some protection because a higher concentration of NBS is required for oxidation of the bound polypeptide (68).

Despite the apparent irreversibility of binding, the failure of cellulose to protect CBD_{Cex} against NBS shows that binding is dynamic. In fact, CBD_{Cex} is mobile on the surface of crystalline cellulose. This is demonstrated very elegantly by confocal microscopic analysis of fluoresceinated CBD_{Cex} adsorbed to sheets of *Valonia* cellulose. When spots of the adsorbed polypeptide are photobleached irreversibly, the fluorescence of the spots recovers with time as unbleached molecules of fluoresceinated CBD_{Cex} migrate into the bleached areas. The rate of recovery of fluorescence is a function of the rate of migration of the polypeptide on the cellulose. Only 35 per cent of the adsorbed polypeptide is mobile, possibly because that is the proportion bound to the low affinity sites on the cellulose (69). The cellulose surface can be envisioned as a complex potential energy surface with numerous local minima which trap the CBD with high affinity. CBDs bound away from the minima search dynamically for an unoccupied minimum.

Family II CBDs bind to both amorphous and crystalline cellulose. Their affinities generally are greater for crystalline than amorphous cellulose. N1 and

N2, the CBDs in CenC, are from family IV. They bind to amorphous but not to crystalline cellulose (70), and to soluble oligosaccharides. N1 and N2 may serve to target CenC to amorphous rather than crystalline cellulose (71).

Conclusions

The cellulase system of *C. fimi* comprises enzymes with different mechanisms, different modes of attack on cellulose, and different affinities for cellulose. Retaining, but not inverting, enzymes can transglycosylate. It is not clear if transglycosylation plays a role in cellulose degradation, perhaps by producing soluble inducers from cellulose. The combination of multiple endoglucanases, specific for amorphous or crystalline cellulose, and two cellobiohydrolases that digest cellulose molecules from opposite ends, all of them with individual CBDs, is seen in a number of aerobic microorganisms, both bacteria and fungi, that degrade cellulose. Exactly how these enzymes interact to form an effective system is not yet clear.

Acknowledgments

We thank Carrie Hirsch and Laura Minato for invaluable help in the preparation of the manuscript. We also thank Karen Kleman-Leyer and T. Kent Kirk for help with the DP experiments. Our work is supported by grants from the Natural Sciences and Engineering Research Council of Canada and the Protein Engineering Network of Centres of Excellence of Canada

Literature Cited

1. Tomme, P.; Warren, R.A.J.; Gilkes, N.R. *Adv. Microb. Physiol.* **1995**, *37*, 1-81.
2. Sinnott, M.L. *Chem. Rev.* **1990**, *90*, 1171-1202.
3. Legler, G. *Adv. Carbohydr. Chem. Biochem.* **1990**, *48*, 317-384.
4. Henrissat, B. *Biochem. J.* **1991**, *280*, 309-316.
5. Henrissat, B.; Bairoch, A. *Biochem. J.* **1993**, *293*, 781-788.
6. Tomme, P.; Warren, R.A.J.; Miller, R.C. Jr.; Kilburn, D.G.; Gilkes, N.R. In *Enzymatic Degradation of Insoluble Carbohydrates*; J.N. Saddler, M.H. Penner Eds.; ACS Symposium Series No. 618; Americal Chemical Society: Washington, DC, **1995**; pp 142-163.
7. White, A.; Withers, S.G.; Gilkes, N.R.; Rose, D.R. *Biochemistry* **1994**, *33*, 12546-12552.
8. Jenkins, J.; Leggio, L.L.; Harris, G.; Pickersgill, R. *FEBS Lett.* **1995**, *362*, 281-285.
9. Tull, D.; Withers, S.G.; Gilkes, N.R.; Kilburn, D.G.; Warren, R.A.J.; Aebersold, R. *J. Biol. Chem.* **1991**, *266*, 15621-15625.
10. MacLeod, A.M.; Lindhorst, T.; Withers, S.G.; Warren, R.A.J. *Biochemistry* **1994**, *33*, 6371-6376.
11. McCarter, J.D.; Withers, S.G. *Curr. Opin. Struct. Biol.* **1994**, *4*, 885-892.
12. Derewenda, U.; Swenson, L.; Green, R.; Wei, Y.; Morosoli, R.; Shareck, F.; Kluepfel, D.; Derewenda, Z.S. *J. Biol. Chem.* **1994**, *269*, 20811-20814.

13. Harris, G.W.; Jenkins, J.A.; Connerton, I.; Cummings, N.; Leggio, L.L.; Scott, M.; Hazlewood, G.P.; Laurie, J.I.; Gilbert, H.J.; Pickersgill, R.W. *Structure* **1994** *2,* 1107-1116.
14. Dominguez, R.; Souchon, H.; Spinelli, S.; Dauter, Z.; Wilson, K.S.; Chauvaux, S.; Béguin, P.; Alzari, P.M. *Nature Struct. Biol.* **1995,** *2,* 569-576.
15. White, A.; Tull, D.; Johns, K.; Withers, S.G.; Rose, D.R. *Nature Struct. Biol.,* **1996,** *3,* 149-154.
16. Rouvinen, J.; Bergfors, T.; Teeri, T.; Knowles, J.K.C.; Jones, T.A. *Science* **1990,** *249,*380-386.
17. Spezio, M.; Wilson, D.B.; Karplus, P.A. *Biochemistry* **1993,** *32,* 9906-9916.
18. Damude, H.G.; Withers, S.G.; Kilburn, D.G.; Miller, R.C. Jr.; Warren, R.A.J. *Biochemistry* **1995,** *34,* 2220-2224.
19. Ducros, V.; Czjzek, M.; Belaich, A.; Gaudin, C.; Fierobe, H.-P.; Belaich, J.-P.; Davies, G.J.; Haser, R. *Structure* **1995,** *3,*939-949.
20. Wang, Q.; Tull,D.; Meinke, A.; Gilkes, N.R.; Warren, R.A.J.; Aebersold, R.; Withers, S.G. *J. Biol. Chem.* **1993,** *268,* 14096-14102.
21. Baird, S.D.; Hefford, M.A.; Johnson, D.A.; Sung, W.L.; Seligy, V.L. *Biochem. Biophys. Res. Commun.* **1990,** *169,* 1035-1039.
22. Navas, J.; Béguin, P. *Biochem. Biophys. Res. Commun.* **1992,** *189,* 807-812.
23. Meinke, A.; Gilkes, N.R.; Kilburn, D.G.; Miller, R.C. Jr.; Warren, R.A.J. *J. Bacteriol.* **1993,** *175,* 1910-1918.
24. Henrissat, B.; Callebaut, I.; Fabrega, S.; Lehn, P.; Mornon, J.-P.; Davies, G. *Proc. Natl. Acad. Sci.* USA, **1995,** *92,* 7090-7094.
25. Jenkins, J.; Leggio, L.L.; Harris, G.; Pickersgill, R. *FEBS Lett.* **1995,** *362,* 281-285.
26. Juy, M.; Amit, A.G.; Alzari, P.M.; Poljak, R.J.; Claeyssens, M.; Béguin, P.; Aubert, J.-P. *Nature* **1992** *357,* 89-91.
27. Chavaux, S.; Béguin, P.; Aubert, J.-P *J. Biol. Chem.* **1992,** *267,* 4472-4478.
28. Meinke, A.; Braun, C.; Gilkes, N.R.; Kilburn, D.G.; Miller, R.C. Jr.; Warren, R.A.J. *J. Bacteriol.* **1991,** *173,* 308-314.
29. Coutinho, J.B.; Moser, B.; Kilburn, D.G.; Warren, R.A.J.; Miller, R.C. Jr. *Mol. Microbiol.* **1991,** *5,* 1221-1233.
30. Kraulis, P.J.; Clore, G.M.; Nilges, M.; Jones, T.A.; Petterson, G.; Knowles, J.; Gronenborn, A.M. *Biochemistry* **1989,** *28,* 7241-7257.
31. Xu, G.-Y.; Ong, E.; Gilkes, N.R.; Kilburn, D.G.; Muhandiram, D.R.; Harris-Brandts, M.; Carver, J.P.; Kay, L.E.; Harvey, T.S. *Biochemistry* **1995,** *34,* 6993-7009.
32. Lamed, R.; Tormo, J.; Chirino, A.J.; Morag, E.; Bayer, E.A. *J. Mol. Biol.* **1994,** *244,* 236-237.
33. Poole, D.B.; Hazlewood, G.P.; Huskisson, N.S.; Virden, R.; Gilbert, H.J. *FEMS Microbiol. Lett.* **1993,** *106,* 77-84.
34. Din, N.; Forsythe, I.J.; Burtnick, L.D.; Gilkes, N.R.; Miller, R.C. Jr.; Warren, R.A.J.; Kilburn, D.G. *Mol. Microbiol.* **1994,** *11,* 747-755.
35. Reinikainen, T; Ruohonen, L.; Nevanen, T.; Laaksonen, L.; Kraulis, P.; Jones, T.A.; Knowles, J.K.C.; Teeri, T.T. *Proteins* **1992**, *14,* 475-482.
36. Warren, R.A.J. *Annu. Rev. Microbiol.* **1996**, *50,* 183-212.
37. Withers, S.G.; Warren, R.A.J.; Street, I.P.; Rupitz, K.; Kempton, J.B.; Aebersold, R. *J. Amer. Chem. Soc.* **1990,** *112,* 5887-5889.
38. Withers, S.G.; Aebersold, R. *Protein Sci.* **1995,** *4,* 361-372.

39. MacLeod, A.M.; Tull, D.; Warren, R.A.J.; Withers, S.G. Submitted to *Biochemistry.*
40. Wang, Q.; Trimbur, D.; Graham, R.; Warren, R.A.J.; Withers, S.G. *Biochemistry* **1995,** *34,* 14554-14562.
41. Miao, S.; Ziser, L.; Aebersold, R.; Withers, S.G. *Biochemistry* **1994,** *33,* 7027-7032.
42. Miao, S.; McCarter, J.D.; Grace, M.; Grabowski, G.; Aebersold, R.; Withers, S.G., *J. Biol. Chem.* **1994,** *269,* 10975-10978.
43. Gebler, J.C.; Aebersold, R.; Withers, S.G. *J. Biol. Chem.* **1992,** *167,* 11126-11130.
44. Tull, D.; Withers, S.G. *Biochemistry* **1994,** *33,* 6363-6370.
45. Damude; H.G., Ferro, V.; Withers, S.G.; Warren, R.A.J. *Biochem. J.* **1996,** *315,* 467-472.
46. Meinke, A.; Gilkes, N.R.; Kwan, E.; Kilburn, D.G.; Warren, R.A.J.; Miller, R.C. Jr.; *Mol. Microbiol.* **1994,** *12,* 413-422.
47. Shen, H.; Tomme, P.; Meinke, A.; Gilkes, N.R.; Kilburn, D.G.; Warren, R.A.J.; Miller, R.C. Jr. *Biochem. Biophys. Res. Commun.* **1994,** *199,* 1223-1228.
48. Shen, H.; Gilkes, N.R.; Kilburn, D.G.; Miller, R.C. Jr.; Warren, R.A.J. *Biochem. J.* **1995,** *311,* 67-74.
49. Shen, H.; Meinke, A.; Tomme, P.; Damude, H.G.; Kwan, E.; Kilburn, D.G.; Miller, R.C. Jr.; Warren, R.A.J.; Gilkes, N.R. In *Enzymatic Degradation of Insoluble Carbohydrates;* J.N. Saddler, M.H. Penner, Eds., ACS Symposium Series No. 618; American Chemical Society: Washington, DC, **1995,** pp 174-196.
50. Tomme, P.; Kwan, E.; Gilkes, N.R.; Kilburn, D.G.; Warren, R.A.J. *J. Bacteriol.* **1996,** *178,* in press.
51. Gilkes, N.R.; Claeyssens, M.; Aebersold, R.; Henrissat, B.; Meinke, A.; Morrison, H.D.; Kilburn, D.G.; Warren, R.A.J.; Miller, R.C. Jr. *Eur. J. Biochem.* **1991,** *202,* 367-377.
52. Gilkes, N.R.; Langsford, M.L.; Kilburn, D.G.; Miller, R.C. Jr.; Waren, R.A.J. *J. Biol. Chem.* **1984,** *259,* 10455-10459.
53. Meinke, A.; Damude, H.G.; Tomme, P.; Kwan, E.; Kilburn, D.G.; Miller, R.C. Jr.; Warren, R.A.J.; Gilkes, N.R. *J. Biol. Chem.* **1995,** *270,* 4383-4386.
54. Divne, C.; Ståhlberg, J.; Reinikainen, T.; Ruohonen, L.; Petterson, G.; Knowles, J.K.C.; Teeri, T.T.; Jones, T.A. *Science* **1994,** *265,* 524-528.
55. Ståhlberg, J.; Johansson, G.; Pettersson, G. *Biochim. Biophys. Acta* **1993,** *1157,* 107-113.
56. Claeyssens, M.; Henrissat, B. *Protein Sci.* **1992,** *1,* 1293-1297.
57. Wood, B.F.; Conner, A.H.; Hill, C.G., Jr. *J. Appl. Polymer Sci.* **1986,** *32,* 3703-3712.
58. Dañhelka, J.; Kössler, I.; Bohácková, B. *J. Polym. Sci. Polym. Chem. Ed.* **1976,** *14,* 287-298.
59. Kleman-Leyer, K.M.; Gilkes, N.R.; Miller, R.C., Jr.; Kirk, T.K. *Biochem. J.* **1994,** *302,* 463-469.
60. Gilkes, N.R.; Warren, R.A.J.; Miller, R.C. Jr.; Kilburn, D.G. *J. Biol. Chem.* **1988,** *263,* 10401-10407.
61. Din, N.; Gilkes, N.R.; Tekant, B.; Miller, R.C. Jr.; Warren, R.A.J.; Kilburn, D.G. *Bio/Technology* **1991,** *9,* 1096-1099.
62. Din, N.; Damude, H.G.; Gilkes, N.R.; Miller, R.C., Jr.; Warren, R.A.J.; Kilburn, D.G. *Proc. Natl. Acad. Sci. USA* **1994,** *91,* 11383-11387.
63. Gilkes, N.R.; Jervis, E.; Henrissat, B.; Tekant, B.; Miller, R.C. Jr.; Warren, R.A.J.; Kilburn, D.G. *J. Biol. Chem.* **1992,** *267,* 6743-6749.

64. Ong, E.; Gilkes, N.R.; Miller, R.C. Jr.; Warren, R.A.J.; Kilburn, D.G. *Biotechnol. Bioeng.* **1993,** *42,* 401-409.
65. Ong, E.; Gilkes, N.R.; Warren, R.A.J.; Miller, R.C. Jr.; Kilburn, D.G. *Bio/Technology* **1989,** *7,* 604-607.
66. Ong, E.; Gilkes, N.R.; Miller, R.C. Jr.; Warren, R.A.J.; Kilburn, D.G. *Enzyme Microb. Technol.* **1991,** *13,* 59-65.
67. Creagh, A.L.; Ong, E.; Jervis, E.; Kilburn, D.G.; Haynes, C.A. *Proc. Natl. Acad. Sci. USA* **1996,** *93,* in press.
68. Bray, M.R.; Johnson, P.E.; Gilkes, N.R.; Mcintosh, L.P.; Kilburn, D.G.; Warren, R.A.J. *Protein Sci.,* submitted.
69. Jervis, E.; Kilburn, D.G. Manuscript in preparation.
70. Coutinho, J.B.; Gilkes, N.R.; Warren, R.A.J.; Kilburn, D.G.; Miller, R.C., Jr. Mol. *Microbiol.* **1992,** *6,* 1243-1252.
71. Coutinho, J.B.; Gilkes, N.R.; Kilburn, D.G.; Warren, R.A.J.; Miller, R.C., Jr. *FEMS Microbiol. Lett.* **1993,** 211-218.

Chapter 7

Families and Functions of Novel Thermophilic Xylanases in the Facilitated Bleaching of Pulp

P. L. Bergquist[1,2], M. D. Gibbs[1,2], D. J. Saul[2], R. A Reeves[1,2], D. S. Morris[2], and V. S. J. Te'o[1,2]

[1]School of Biological Sciences, Macquarie University, North Ryde, Sydney, New South Wales 2109, Australia
[2]Centre for Gene Technology, University of Auckland, Private Bag 92019, Auckland, New Zealand

Hemicellulases have shown some promise as bleaching aids in the pulp and paper industry. Novel genetic techniques have been developed to isolate xylanase and mannanase genes from both culturable organisms isolated after enrichment and non-culturable organisms that cannot be maintained in pure culture. Genomic DNA is subjected to the polymerase chain reaction (PCR) using consensus primers. The products are cloned and sequenced and genes for novel enzymes are identified. The full length sequences are obtained by genomic walking PCR, allowing the amplification of the entire gene in a form suitable for direct cloning into an expression vector. Using this technique, we have revealed the presence of a number of novel genes for xylanases of family F and family G in culturable bacteria that were believed to express only a single enzyme and from an unculturable consortium. Selected candidate enzymes have been tested for their abilities to enhance the bleaching of kraft pulps and several function as well as commercially-available enzyme preparations but at alkaline pH and at temperatures up to 50°C higher than mesophilic xylanases.

Thermophilic enzymes may have significant advantages in some industrial processes due to their intrinsic stability at high temperature, in organic solvents and during immobilisation. We have cloned a number of thermophilic genes into mesophilic bacteria and fungi, realizing a cost advantage in that the mesophilic proteins can be removed by a simple heating step, which gives a facile 20-fold purification. Novel genetic techniques have been developed to isolate xylanase and mannanase genes from both culturable organisms isolated after enrichment and non-culturable organisms that cannot be maintained in pure culture.

Biotechnology has received increasing attention by the pulp and paper industry because of its potential for use in primary pulp manufacture. Interest has been focussed on a number of areas within the industry including pulp modification, waste treatment and by-product conversion. kraft pulping, a process widely used in paper manufacture, removes about 95% of the lignin by alkaline sulphate cooking. The remaining lignin gives the pulp a brown colour which is removed in a multi-stage

0097–6156/96/0655–0085$15.00/0
© 1996 American Chemical Society

bleaching process using a variety of agents. Currently, there is concern about the environmental impact of some of the compounds used in the process. This is particularly the case with chlorine and chlorine dioxide. Enzymes, including xylanases (endo-1,4-β-xylanase, EC 3.2.1.8), have been shown to reduce the amount of chlorine required to achieve comparable levels of paper brightness *(1-4)*. However, the mesophilic enzymes currently in use have limitations because high temperatures are used in bleaching.

We have cloned and over-expressed many of these enzymes from genomic libraries and genomic DNA preparations of *Caldicellulosiruptor saccharolyticus* and other anaerobic thermophiles (e.g. *Caldicellulosiruptor* strain Rt8B4, *Thermotoga* FSSJ3.B1, and *Dictyoglomus* sp. Rt46B.1). The temperature optima and the stability of the cellulolytic and hemicellulolytic enzymes produced by *Cs. saccharolyticus, Thermotoga* and *Dictyoglomus sp.* exceed those reported for other cellulases and xylanases.

Selective enrichment cultures from extreme natural environments have provided a large collection of micro-organisms in pure culture. Examination of these isolates has revealed information about the physiology and phylogeny of the inhabitants of extreme environments. However, microbiologists have speculated that pure-cultured strains represent less than 5% of the micro-organisms that exist in these habitats in nature. The enrichment procedures for isolating organisms from mixed populations have been shown to have limitations (refs *5-7*; Saul *et al.*, 1996, unpublished). Our research has focussed primarily on culturable strains but we have also designed specific PCR primers that allow us to amplify all genes for hemicellulolytic enzymes that have homology to known sequences. This technique can be used to isolate genes from a sample of DNA isolated directly from an enriched thermal site, or alternatively, we can amplify the genomic DNA from pure cultures of micro-organisms. In this way we have cloned fragments of hemicellulase genes in *Escherichia coli* and sequenced them, followed by PCR procedures that allow us to find the 5'- and 3'- ends of the genes. After cloning into an expression vector, we have tested their enzymatic characteristics (particularly, pH and temperature profiles) without having to embark on extensive and time-consuming library constructions. Simultaneously, we have amplified the SSU (16S) rRNA genes from the bacterial population DNA, allowing us to estimate the range of cultured and uncultured species in a given sample.

Our investigations of the enzymes from several thermophilic bacterial strains show that these organisms produce a number of xylanases which degrade fibre-bound substrates, each with differing temperature and pH optima. We have now isolated the individual genes for thermophilic xylanases and mannanases from culturable and non-culturable bacteria. The main theme of our current work is to over-express these genes in a mesophilic host that is easy and cheap to grow and which does not have the perceived disadvantages of *Escherichia coli*. Our preference is to find a simple and cheap host system that will allow excretion of the enzymes into the medium so that simple heat treatment would effectively eliminate contaminating mesophilic enzymes and would provide a low-cost source of enzyme for pulp bleaching. Accordingly, a major emphasis in our current work is centred on expression in the fungi *Kluveromyces lactis* and *Trichoderma reesei.*

Materials and Methods

DNA preparation and Library construction. All bacteria were from the collection of H.W. Morgan, Waikato University, New Zealand. Culturable bacteria had been isolated by enrichment and screening procedures on hemicelluloses. Unculturable bacteria were isolated *en masse* from enrichments on xylan or mannan, using an enclosed percolation system at 70-87°C. Bulk DNA was prepared from the resulting biomass by standard methods.

DNA preparation, manipulation and digestion with restriction enzymes were performed according to Sambrook *et al. (8)*. Gene libraries were constructed in λZAPII by the methods of Short *et al. (9)*. Xylanase-positive plaques were identified from replica plates that were overlaid with oat spelt xylan and stained with Congo red *(10)*. Genomic DNA from either culturable organisms or from biomass enriched *in situ* in the percolation system was prepared and subjected to PCR using consensus primers that span the active site sequences of family F or family G xylanases or two sets of mannanase primers that we have designed. The PCR products were cloned and sequenced and novel enzymes were identified. The full length sequences were obtained either by genomic walking PCR *(11)* or by using consensus amino- and carboxy- terminal primers that allow amplification of the entire gene in a form that allows direct cloning into the expression vector pJLA602 *(12)*.

Genomic-walking is achieved by a polymerase chain reaction using a specific walking primer based on known sequences. The second primer is the top- or bottom oligonucleotide of the linker; the choice depends on whether the 5' or 3' restriction site of the linker has been used for the library construction. The linkers are not phosphorylated to prevent the non-specific amplification of all fragments in the restriction fragment linker library. Consequently, only one strand becomes covalently attached to the genomic fragment and the linker primer binding site of the reciprocal strand is lost during the denaturation step of the PCR. The linker primer binding site can only be generated upon primer-extension from the specific walking primer. To help prevent the regeneration of restriction sites following the ligation of the linkers to the genomic restriction fragments, a G was selected as the first base after the 5'-overhang, and a T as the base before the 3'-overhang (very few 6 bp restriction sites generating 5'-overhangs end in a G, and very few 6 bp restriction sites generating 3'-overhangs begin with a T). The walking primer is designed to be relatively long to allow for high annealing temperatures as genomic-walking is a PCR with only one specific primer. The M13 universal reverse sequencing primer was incorporated into the upper-strands to allow for direct sequencing of the genomic-walking products, (Figure 1.).

Enzyme assays. Assays for enzymes involved in xylan degradation have been described elsewhere *(13)*. Quantitative assays were carried out by measuring the release of reducing sugar from oat spelt xylan *(14)*. In some cases, purified recombinant enzymes have been assayed for release of reducing sugar and lignin from kraft pulp.

Other Methods. Polymerase Chain Reactions (PCR) were performed in a Perkin-Elmer Cetus DNA Thermal Cycler. DNA sequencing was carried out using an Applied Biosystems 373A DNA sequencer and Catalyst 800 Robotic Workstation. Sequences were analysed on a Silicon Graphics IRIX workstation using the GCG Software *(15)*.

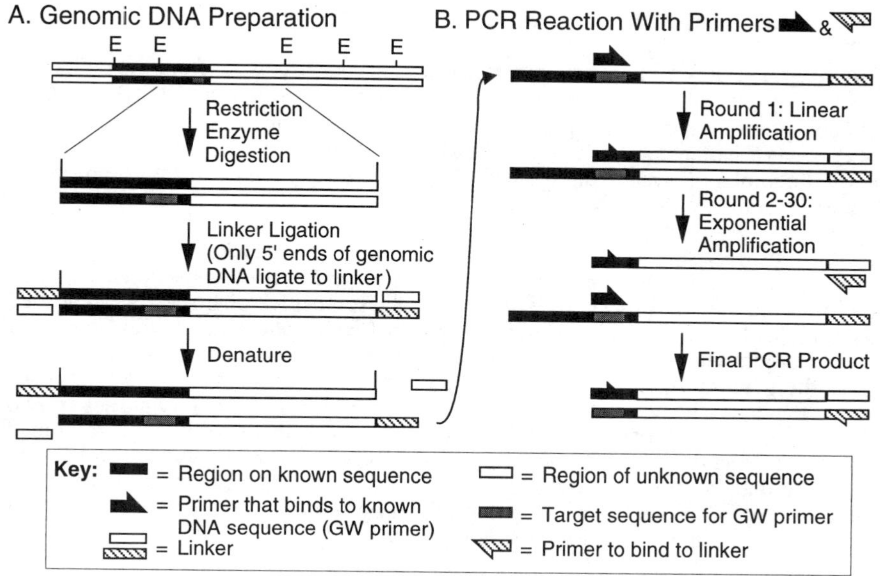

Figure 1. Diagrammatic representation of genomic walking PCR.

Results

Microbial diversity and enrichment cultures. The genus *Thermus* was used as a model system in these studies. Two New Zealand hot pools with different environments were selected and the diversity of *Thermus* strains growing within them was examined by isolating SSU rRNA genes and comparing their sequences. Although all of the sequences were similar, several variants were found in each pool. Standard methods for the enrichment of *Thermus* were carried out and the gene isolation and sequencing was repeated. The procedure resulted in a loss of heterogeneity. The the enrichments from the two pools yielded different strains but within each enrichment, no variants could be found. The results demonstrate the loss of diversity caused by enrichment methods and also indicate that minor differences in SSU rRNA sequence reflect physiological characteristics of the organism which effect success growth under selective conditions (Saul *et al.*, 1996, unpublished). Accordingly, in some of the work that is described below, we have used bulk genomic DNA from total biomass for gene isolation

Cloning and characterization of Xylanases. Currently, at least 80 true β-1,4-xylanase genes have been isolated from bacterial and fungal sources. Comparison of amino acid sequences shows that all fall into two distinct groups (families F and G, or 10 and 11, respectively) based on the homology of conserved amino acids *(16; 25)*.. Within each group are regions of highly conserved residues which presumably play some role in the active site of the enzyme. This conservation is also present at the DNA level in the homologous areas. We have designed primers based on these conserved regions to amplify internal portions of the different xylanase genes from genomic DNA extracted from culturable and non-culturable bacteria (Figure 2). Each amplified product is ligated directly into a standard plasmid vector and sequenced. Sequencing a number of recombinants from each PCR preparation allows a

determination of the number of xylanases within a single organism or within a mixture of DNA from unculturable organisms.

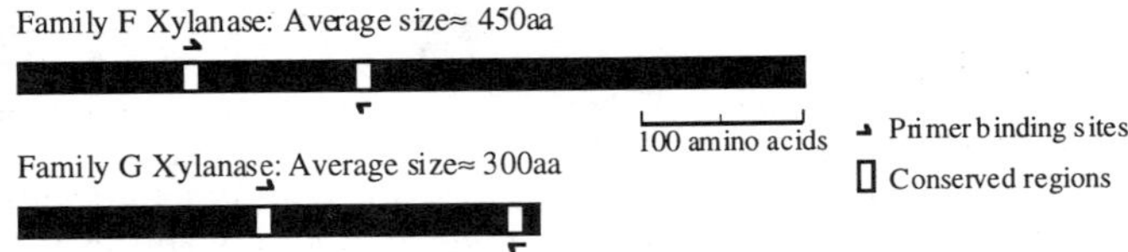

Figure 2. The two basic groups of xylanases based on regions of conserved amino acids. Arrows indicate primers for each group allowing PCR of internal portions of xylanase genes.

To isolate an entire xylanase gene identified by this method, two approaches are available, first, by employing standard Southern blot techniques, using the PCR product to probe back to restriction enzyme-digested genomic DNA, or secondly, by genomic walking PCR using new primers designed on the sequence of the PCR product *(11)*.

***Caldicellulosiruptor saccharolyticus* and *Caldicellulosiruptor* strain Rt8B.4.** *Cs saccharolyticus* has a high A+T DNA, in common with a number of other anaerobic extremely thermophilic bacteria in cluster B of Rainey *et al. (17)*. A *Bam*H1 partial digest gave an 19kb insert into the vector pBR322 (pNZ1084). The thermophilic DNA inserted into this recombinant plasmid has been completely sequenced and contains most of the genes coding for enzymes involved in xylan degradation in *Cs. saccharolyticus*. Three of the ORFs were identified with enzymes involved in xylan degradation by Luthi *et al. (13):* XynA, an endo-xylanase, XynB, a β-xylosidase and XynC, and an acetylxylan esterase.

A total of seven open reading frames were found clustered on the *Cs. saccharolyticus* genomic 12kb *Bam*HI fragment upstream of *xynA* (see Figure 3): 936 bp orf1; 920 bp orf10 (*xynG*); 1019 bp orf11 (*xynH*); 1785 bp orf2; 4043 bp orf9 (*xynF*); 2102 bp orf8 (*xynE*) and 1310 bp orf7 (*xynD*).

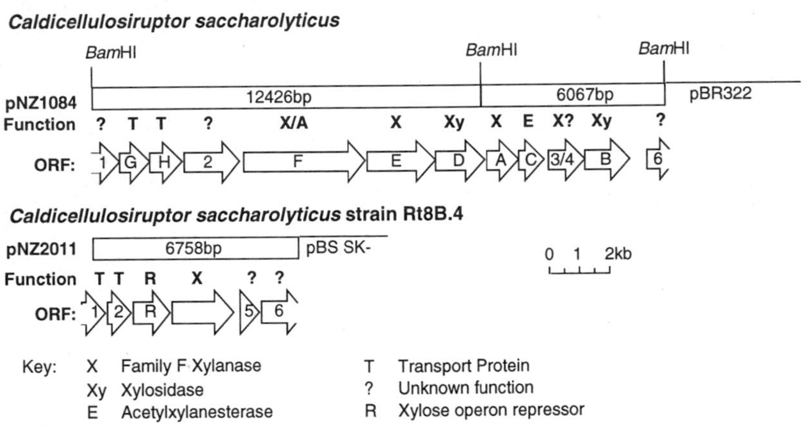

Figure 3. Multi-xylanase gene cluster of *Cs. saccharolyticus*. compared with the structure of a putative xylanase/xylose operon from*Caldicellulosiruptor* Rt8B.4.

Homology comparisons with sequences in the GenBank and EMBL databases showed both XynG and XynH to be putative membrane transport proteins and XynF to be a multi-domain enzyme with arabinosidase and possibly xylanase and xylosidase activities. XynE is another multi-domain enzyme with xylanase activity and XynD a single domain enzyme with xylosidase activity. XynF appears to have four putative domains which may have resulted from a gene fusion, with two arabinofuranosidase plus xylanase domains (domains 1 + 2 and domain 4) fused by a non-catalytic domain (domain 3). XynE has a two-domain structure with the catalytic domain located at the C-terminus and a non-catalytic domain at the N-terminus region. Surprisingly, the xylanase gene organization of *Cs. saccharolyticus* is quite different to that of its close relative *Caldicellulosiruptor* sp. Rt8B.4 (Figure 3). No multigene xylanase or cellulase/hemicellulase gene clusters were present on this portion of its genome, and other xylanases were not found in the expression gene library of this organism.

A xylanase domain has been identified as part of a mutidomain enzyme first described as part of a cellulase cluster of genes in *Cs. saccharolyticus* (Figure 4). This gene, *celB (11)*, is associated with a cellulose binding domain and an endoglucanase activity. The expression product from the *celB* catalytic domain has been shown to be a family F xylanase with a broad pH optimum and significant activity at alkaline pH. It is the only xylanase that we have found in *Cs. saccharolyticus* that is associated with proline-threonine linkers and a cellulose binding domain *(19)*.

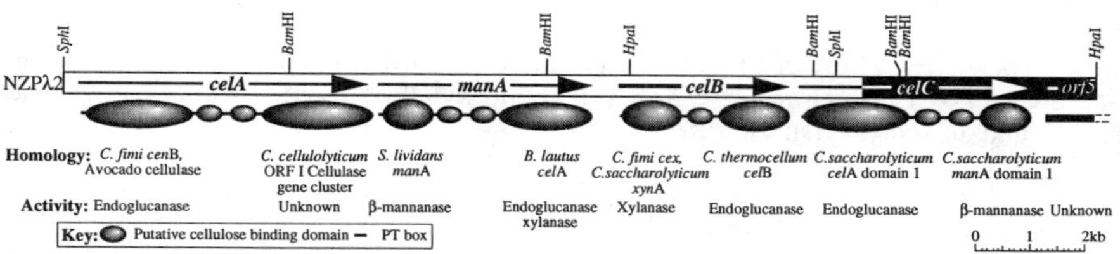

Figure 4. Diagrammatic representation of multidomain enzymes on a λ recombinant (white) λ2A from a genomic library of *Cs. saccharolyticus*. Other sequence (black) was obtained by genomic walking PCR (Modified from ref *11*).

Consensus family F and G primers were used to amplify internal portions (140-250 bp PCR products) of xylanases from both culturable and unculturable micro-organisms, including *Cs. saccharolyticus*. No PCR products belonging to the family G xylanases were amplified from the genome of *Cs. saccharolyticus*. Sequencing of the family F PCR products showed the presence of a new xylanase gene different from the xylanase cluster located on the 12kb *Bam*HI genomic DNA of *Cs. saccharolyticus*. This new gene (*xynI*) was found to be highly homologous to the xylanase gene *xynA* from *Caldicellulosiruptor* sp. Rt8B.4 (*18*). Primers used to amplify the catalytic domain consisting of a 1.1kb *xynI* DNA fragment which was later sequenced and found to be highly homologous to the second domain of *xynA* from Rt8B.4.

Genomic walking PCR was used to amplify upstream DNA fragments to obtain the complete *xynI* sequence, which was translated and found to contain two stop codons at the N-terminus region. At first glance, the presence of the stop codons were thought to have resulted from sequence misinterpretation, but detailed analysis of the single-stranded DNA templates covering the N-terminus region indicated that the

sequence was correct. Therefore, the N-terminal domain of XynI apparently was non-functional.

Thermotoga* FjSS3-B.1 *xynA, B and C. A gene expressing xylanase activity was isolated from a genomic expression library of *Thermotoga* sp. strain FjSS3-B.1. The gene was sequenced and shown to encode a single domain, family F xylanase (family 10 β-glycanase). This gene was designated *xynA*. The recombinant enzyme has extremely high thermal stability, activity over a relatively broad pH range and activity on *Pinus radiata* kraft pulp *(19)*. Amplification of genomic DNA using family F primers revealed the presence of two additional xylanase genes that were not apparent from a comprehensive analysis of an expression library. The sequence of products amplified by these primers showed that there were two further, undetected xylanase genes present in the organism. Genomic walking primers were designed from the sequences of the PCR products and flanking regions were obtained from genomic DNA. Two distinct genes were identified which were designated *xynB* and *C*. Unlike *xynA*, these genes code for multidomain enzymes with the xylanase domain flanked by two domains of unknown function but which are similar to domains found in *Thermoanaerobacter xynA* and *Cs. saccharolyticus xynE*. *xynB* produces an active xylanase, but *xynC* fails to give any activity. This may be the result of miss-folding of the truncated enzyme in *E. coli*. Alternatively, *xynC* may be a pseudogene analogous to other hemicellulase pseudogenes found in *Caldicellulosiruptor* (11). Winterhalter and Liebl *(20)* have also cloned the *xynB* gene from a related species and have shown that the apparently non-catalytic N-terminal domain enhances the thermostability of the expressed protein. We have confirmed this result with *Thermotoga* FjSS3-B.1.

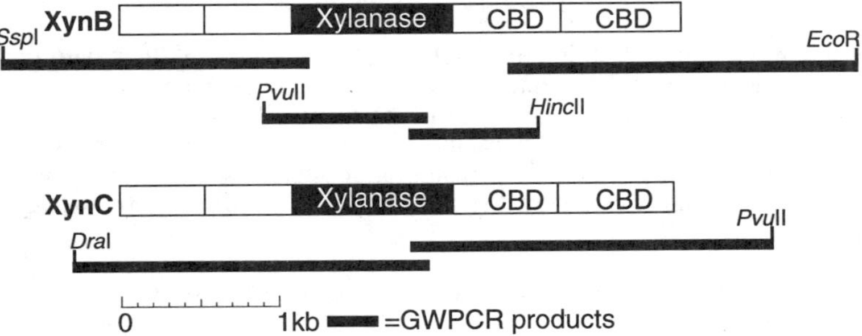

Figure 5. Genomic walking PCR products and the structure of *xynB* and *xynC* from *Thermotoga* strain FjSS3B.1.

Xylanases from *Dictyoglomus thermophilum* species Rt46B.1. The *xynA* gene of *Dictyoglomus thermophilum* is a family F xylanase (family 10 β-glycanase) which was cloned from *Dictyoglomus thermophilum* using standard genetic techniques (gene library construction and screening). The gene was sequenced and shown to contain a single complete open reading frame coding for a single domain xylanase, XynA, with a putative length of 352 amino acids. No other family F xylanase genes could be identified in the expression library and none were amplified using the F consensus primers. XynA exhibits optimal endoxylanase activity at 85°C, pH 7.0, and shows 50% activity between pH5.5-9.5 and between 65°C to 90°C (21). An identical gene has been cloned from DNA prepared after a percolation experiment using xylan as a substrate at high temperature and pH.

The *xynB* gene of *Dictyoglomus thermophilum* encodes a family G xylanase (family 11 β-glycanase), and was identified through the use of family G internal

consensus primers. These primers amplify an internal consensus fragment (ICF) of 300 bp specifically from family G xylanase genes. Sequencing of a representative sample of the family G ICFs amplified from *Dictyoglomus thermophilum* genomic DNA revealed only one sequence species, indicating that there was a single family G xylanase gene in the *Dictyoglomus thermophilum* genome. Genomic-walking PCR was then used to amplify fragments from the *Dictyoglomus thermophilum* genome up- and downstream of the family G internal consensus fragment. Sequencing of the appropriate GWPCR fragments allowed the full-length *xynB* sequence to be obtained. Oligonucleotide primers were then designed to amplify the *xynB* gene for cloning into an expression vector (Figure 6).

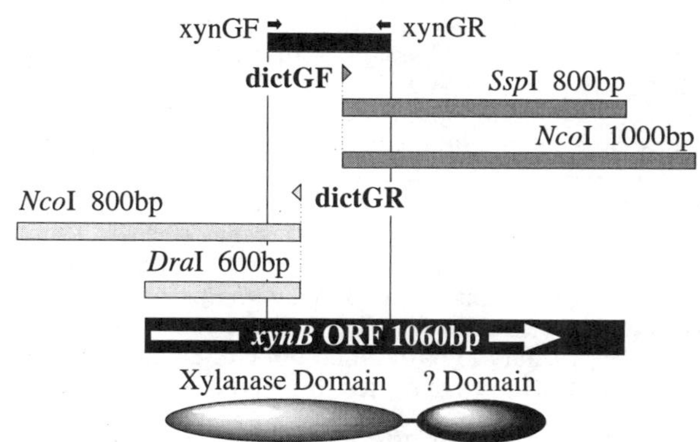

Figure 6. Genomic Walking of the *Dictyoglomus thermophilum xynB* gene. Boxes indicate GWPCR fragments; Arrows indicate the length and direction of nucleotide sequences obtained from the GWPCR fragments. The black box (g1) is the *xynB* internal consensus fragment amplified by the family G internal consensus primers. The resulting consensus sequence is shown at the bottom of the figure with the putative domain structure of the XynB xylanase.

Enzymatic activities of recombinant hemicellulases and activity on kraft pulp. The enzymatic characteristics of the expressed products of the *Cs. saccharolyticus xynA* (xylanase), *xynB* (β-xylosidase) and *xynC* (xylan acetyl esterase) genes have been described previously *(14, 22)*. It should be noted that ORF3/4 is almost certainly a pseudogene since we have demonstrated the presence of a frameshift that introduces a stop codon (Gibbs, unpublished, 1995). The other genes upstream have also been characterized. The N-terminus of the XynF 43kDa protein band was sequenced and found to map to the beginning of the putative domain four of XynF, which suggested that degradation of the full length XynF protein was occurring. No protein band was identified for the full length XynF with its expected molecular size of 152 kDa. Similarly, no protein band was identified for XynE, although the expected molecular size for its full length protein was 80.8 kDa. Although no protein band could be identified for the full length XynE, release of reducing sugars (small amounts of xylobiose and possibly xylobiose with a side-chain group but no xylose) from oat spelts xylan and arabinoxylan were detected. The same results were found for the catalytic domain of XynE. As expected, XynF was found to release predominantly arabinose from the xylan-based substrates with small amounts of xylobiose and possibly xylobiose with a side-chain group, but no xylose. Like a true endo-xylanase, the catalytic domain of XynI was found to release predominantly xylose and xylobiose from the xylan-based substrates with low amounts of xylotriose and

xylotetrose. The above results indicated that although XynF was an arabinosidase, the release of xylobiose and possibly xylobiose with a side-chain group indicated that XynF was an exo-acting type of xylanase as well. Similarly, XynE could be classified as an exo-acting type of xylanase based on the hydrolysis products released from the xylan-based substrates (also produced only xylobiose and possibly xylobiose with a side-chain group).

All three enzymes (full length XynE, XynF and XynI domain 2) were characterized to determine their temperature stabilities, pH and temperature optima. As a result, the temperature optima of XynE and XynF were found to be 75°C with 50% activities at 82°C. However, the catalytic domain of XynI had a temperature optimum of only 60 to 65°C with 50% activity at 76°C. The pH optima for all three enzymes were found to be about 6.5 with XynE and XynF having 50% activities at pH 9.0, while XynI had a much broader shoulder with 50% activity at pH 10.0. The temperature stabilities of XynE and XynF were very similar with at least 6 hours at 75°C, but XynF was found to be more stable at 80°C with a 75 minutes half life compared to 8 minutes for XynE. The catalytic domain of XynI was not found to be very stable at 75°C with only a 40 minutes half life and a much lower half life of 2-3 minutes at 80°C (summarized in Table 1). The *Cs. saccharolyticus* enzymes have pH optima of 5.5-6.0 but with a very broad profile in the alkaline range and they retain 50% of their activity at pH8.5.

Enzyme (activities)	Temp. optima	pH-optima	Half life	
XynE (full length) *Exo-xylanase*	75°C	6.0	@ 70°C, > 12 hours @ 75°C, > 6 hours	@ 80°C, 8 minutes @ 85°C, <1 minute
XynF (full length) *Exo-xylanase + Arabinosidase*	75°C	6.0	@ 70°C, > 12 hours @ 75°C, > 6 hours	@ 80°C, 75 minutes @ 85°C, 8 minutes
XynI (catalytic domain) *Endo-xylanase*	65°C	6.5	@ 70°C, ~170 minutes @ 75°C, ~ 40 minutes	@80°C, 2-3 minutes

Table 1. Characteristics of the full length XynE, full length XynF and catalytic domain of XynI.

The comparative biochemical characteristics of some of the *Caldocellulosiruptor* xylanases and XynA from *Thermotoga* FjSS3.B.1 have been reviewed previously *(23)*. *Thermotoga* XynA and XynB and *Dictyoglomus* XynA and B enzymes are significantly more stable than the *Caldicellulosiruptor* enzymes. *Thermotoga* XynA is significantly more stable than XynB at 95°C, having a half life of 12 hours as compared to less than 10 minutes for XynB. Of course, XynB lacks the amino- and carboxy- terminal domains associated with the native enzyme, which may confer added thermal stability (Table 2).

XynA

Temperature	Half Life
95°C	*12 hours*
90°C	*22 hours*
85°C	*No detectable loss after 16 hours*

XynB

Temperature	Half Life
95°C	*< 10 minutes*
90°C	*8 hours*
85°C	*No significant loss after 16 hours*

Table 2. Temperature stabilities of *Thermotoga* FjSS3B.1 xylanases XynA and XynB at 85, 90 and 95°C. Enzymes were incubated in 25mM Bis-Tris propane (pH 6.3) in the absence of substrate.

XynA from *D. thermophilum* Rt46B.1 showed optimal activity at 85°C,with 50% activity at 70 and 95°C. The pH optimum was determined to be 6.5, but 50% activity was retained at pH5.5 and 9.5 (Gibbs et al., 1995; and Figs. 7A and 7B).

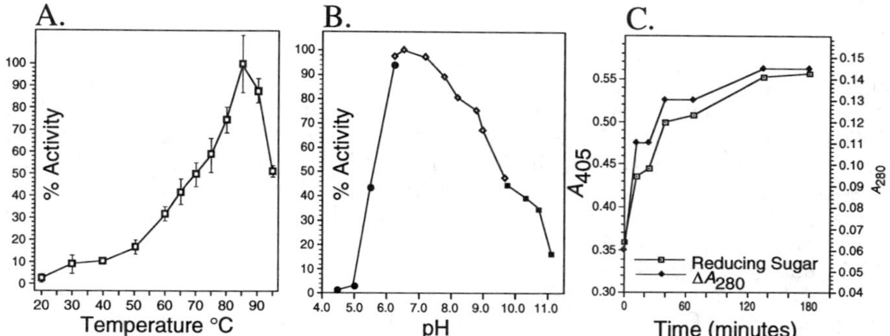

Figure 7A-C. A: Effect of temperature on activity. Xylanase activity was assayed using 0.25% Oat spelts xylan, 50 mM Bis-Tris-Propane across a range of temperatures after 10 min incubation. The level of activity at 85°C was defined as being 100%. Error bars = 2 standard deviations.
B: Effect of pH on activity of XynA. Xylanase activity was measured at 85°C for 10 min. Buffers used were; Sodium acetate pH 4.5-6.2 (•), Bis-Tris-Propane pH 6.5-9.5 (♦) and CAPS pH 9.5-11.0 (▪). All buffers were pH adjusted at 85°C, and were used at a concentration of 50 mM.
C: Comparison of the release of xylose reducing sugar (A_{405}) and lignin release (ΔA_{280}) from unbleached *Pinus radiata* kraft pulp. Assays were carried out at 85°C pH 6.8.

Figure 7C shows that there was a close correlation between the release of reducing sugars as measured by PABAH assay and lignin release by XynA from *Pinus radiata* kraft pulp as measured by $\Delta A280$. This strong correlation caused some concern but several tests confirmedthat the results in Figure 7C are a direct measure of the release of lignin which had been covalently or non-covalently associated with the xylan.

The XynB family G xylanase exhibited a pH optimum of 6.5, and a temperature optimum of 85°C, and showed 50% activity between pH 5.0 to 8.5, and between 65 and 100°C (Figure 8). XynB had a half-life of 11 minutes at 100°C.

All enzymes have been tested for the release of reducing sugars and lignin from *Pinus radiata* kraft pulp and some samples have been trialled for their ability to bleach hardwood and softwood kraft pulps. Only *Thermotoga* XynA has been tested for activity on fibre-bound substrate . Without a knowledge of the make up and degree of polymerisation of these compounds, it is impossible to calculate the percentage of total xylan that has been solubilized by XynA. However, estimates from reducing sugar release suggest that it is comparable with the release by *Trichoderma reesei* xylanases under similar conditions *(19)*. All three new enzymes from *Cs. saccharolyticus* (full length XynE, full length XynF and catalytic XynI) were also tested and found to release reducing sugars from kraft pulp but only XynF was found to release lignin from kraft pulp (Te'o, unpublished, 1996).

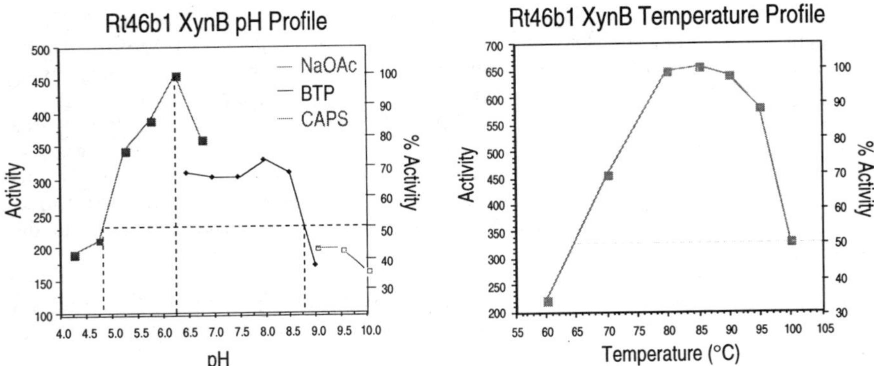

Figure. 8. pH and temperature profiles of *Dictyoglomus thermophilum* XynB. Activity shown on the left hand Y-axis is the OD405 measured from PHBAH reducing-sugar release assays. NaOAC, 250 mM sodium acetate buffer; BTP, 250 mM Bis-Tris Propane buffer; CAPS, 250 mM cyclohexylamino propanesulfonic acid. Profiles performed in 200 µl reactions comprised of 10 µl 250 mM buffer, 180 µl 0.3% Oat Spelts Xylan solution and 10 µl enzyme extract.

Two separate trials were carried out to examine the bleaching ability of Rt46B.1 on kraft pulp from several different wood species. Rt46B.1 XynA was trialed by collaborators to examine the enzyme's ability to enhance the brightness of both hardwood and softwood pulps. For comparison, a commercial enzyme was also tested. The objective of this trial was to evaluate the bleaching ability of Rt46B.1 in an ECF (elemental chlorine free) bleaching sequence to achieve a target paper brightness within the range of 84-86 ISO (an industry brightness standard). The bleaching sequence XwDED was followed. In all incubation steps the pulp was at 10% consistency (10% dry weight/v). Two separate chlorine charges were used in the initial D treatment for both hardwood and softwood pulps. Rt46B.1 XynA was used at a concentration of 2 XU per gram pulp and the commercial enzyme was used at 1 XU per gram of pulp. However, this apparent difference is units was due to the assay systems used to measure the unit concentration of each enzyme. 2 XU Rt46B.1 XynA (as defined by ourselves) showed equivalent activity to 1 XU of the commercial enzyme when assayed side by side on soluble xylan.

The results of one of these treatments are summarized in Figure 9. Several control samples were included in the trial. Pulp without enzyme was treated at the pH and temperature specified for each enzyme treated sample. Cellular protein extracted from *E. coli* was also used as a control to determine the possible effects of protein on final brightness.

Rt46B.1 XynA was observed to improve bleaching in terms of final brightness over the controls for both softwood and hardwood kraft pulps. XynA enhanced bleaching as well as, or better than the commercial enzyme. However, the control samples indicate that a large proportion of the increase in brightness observed could be attributed purely to the alkaline pH and high temperature of the treatment.

A sample of Rt46B.1 XynA was supplied to Dr Peter Nelson, CSIRO, Melbourne, Australia, for assaying the enzyme's effectiveness in removing lignin from *Eucalyptus* kraft pulp. Two separate assays were carried out, the first using 10XU/g pulp, and the second 50XU/g pulp. Enzyme was incubated with *Eucalyptus* pulp (at 6% consistency) for 2 hours at 85°C, pH6.5, then bleached following the sequence D_O-EO-D (see Table 3). Unfortunately, no detectable increase in brightness was observed for the test sheet paper produced. A second test was done, using *Pinus radiata* kraft pulp, under the same conditions but with an enzyme concentration of 69XU/g pulp. In this instance a clear increase in brightness was observed over the control sample of 2.6 points (Table 3).

The reasons for the lack of brightness increase on *Eucalyptus* kraft pulp could not be explained. It is possible that some inhibitory factor was present in the eucalypt pulp.

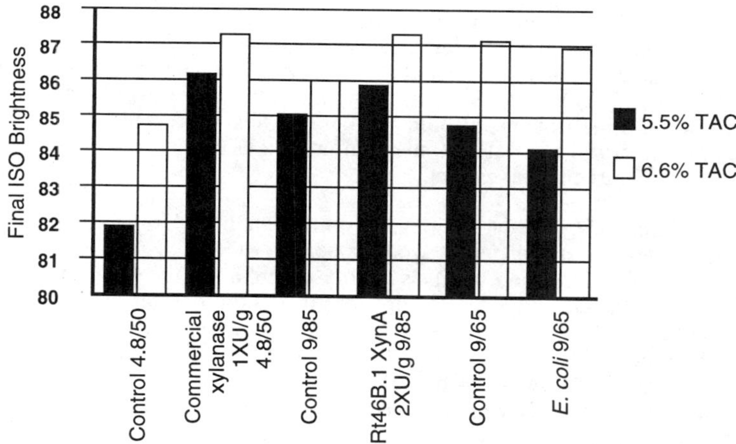

Figure 9. Bleaching of softwood kraft pulp by Rt46B.1 XynA. An enzyme treatment was followed with the chlorine charges indicated (TAC= Total active chlorine).

Pulp	Dosage (XU/gp)	Brightness (D1) Control	XynA-treated	Δ Brightness
Eucalyptus	10	86.5	85.9	-0.6
Eucalyptus	50	86.4	86.4	0
Pinus radiata	69	66.5	69.1	2.6

Table 3. Results of enzymatic treatment of kraft pulps. Rt46B.1 was incubated with the pulps shown for 3 hours at 85°C, pH6.5 in 6% pulp (dry weight). Bleach sequence kraft-X-Do(E_O)DD. Do(0.15% Cl_2 M, 70°C, 10 minutes 10% pulp) EO (1.5% NaOH, 90°C, 30 minutes, 10% pulp, 780 kPa O_2), D stages (1.32% ClO_2, 70°C, 4 hours, 10% pulp).

Discussion

We have employed genetic engineering technology to produce candidate enzymes for two reasons: first, the yields of enzymes from the native organisms are low and the organisms are often difficult to grow on a large scale, generally requiring anaerobic conditions (and in some cases, cannot be grown at all); second, it is important for the success of the programme to be able to produce pulp-bleaching enzymes free of cellulases, as these would weaken the pulp fibres and reduce pulp yields. Our research design, in addition, allows individual genes for thermostable hemicellulolytic enzymes to be cloned in a non-cellulolytic mesophilic host and allows the over-expression of their protein products.

As seen with many endoglucanases, xylanases from both families F and G may expressed as a single domain enzyme or as a modular component of a multidomain protein which will retain activity if separated from other components. The domain structures of the products of xylanase genes and the position of the xylanase domain when part of the multidomain protein have been described by Morris *et al.*, (this volume). Gilkes *et al. (16)* reported that xylanases fall into two main families, F and G. These families correspond to families 10 and 11 of the glycosyl hydrolases *(24)*. Amino acid sequence analyses strongly suggested that xylanase families F and G are structurally different *(16; 25)*, and may have evolved from unrelated progenitor enzymes. The former comprises enzymes of higher molecular weight and basic pI, the latter includes low molecular weight and acidic pI proteins. Microbial xylanases, in particular, have been found to consist of various combinations of discrete functional domains: catalytic domains (CDs), substrate binding domains (SBDs), linkers connecting such domains, and repeated sequences of amino acids *(16; 26)*.. Linkers are sequences in xylanases and cellulases *(16)* which are thought to function as flexible hinges between the catalytic and substrate binding domains and they may also have a role analogous to that of introns, by enabling sequences encoding discrete domains to be excised and fused to other genes, thus generating novel hybrid enzymes *(26)*. Many multidomain enzymes do not contain obvious linker sequences between domains, and enzymes may not necessarily require linkers between domains to maintain their individual structure and function *(16; 27)*.

Substrate-binding domains are required by many enzymes to bind to xylan and cellulose substrates. The occurrence of xylan-binding domains (XBDs) found in xylanases is very low compared to the number of cellulose binding domains (CBDs) *(16; 26)*. A reason generally accepted for the very low numbers of XBDs found (if any) is the heterogeneity of the polysaccharide substrate, which might preclude the evolution of a protein domain which binds to all xylans, irrespective of their source *(26)*. Interestingly, the XBD identified in xylanase A from *Thermomonospora fusca (28)* exhibited significant affinity for xylan as well as cellulose. The only xylanase found so far to contain a xylan-specific binding domain is XylD from *Cellulomonas fimi (29)*. The XBD of XylD was found to bind to xylan with high specificity and only negligible affinity for cellulose and could function independently of the rest of the xylanase. In contrast, many of the xylanases identified and characterized contain CBDs instead. For example, the multidomain XynA of the hyperthermophilic bacterium *Thermotoga maritima* was found to contain a ~170 amino acid repeated C-terminal domain which facilitated binding of the 120kDa full length xylanase to cellulose *(20; 30)*. Likewise, the xylanase XylB and an arabinofuranosidase XylC from *Pseudomonas fluorescens* subsp. *cellulosa* contained identical CBDs which were both found to bind to Avicel *(31)*.. Considerable homology exists between the cellulose binding domains of several xylanases and cellulases *(16)*, of which the following features are highly conserved: (1) low contents of charged amino acids; (2)

two cysteines are present close to the N- and C-termini respectively; (3) there are four very highly conserved tryptophan residues in addition to glycine and asparagine amino acids.

The catalytic domains (CDs) of many xylanases and cellulases have been delineated by proteolytic studies *(16; 26)*. The xylanases in particular vary widely in the numbers of amino acids they contain, but their catalytic domains tend to be more uniform in size *(16)*. However, relatively few catalytic domains have been identified other than by sequence relatedness to known domains. In two cases, multicatalytic enzymes are seen. CelB from *Cs. saccharolyticus* possesses an N-terminal xylanase and a C-terminal endoglucanase, while XynA from *Ruminococcus flavifaciens* has an N-terminal family G xylanase and a C-terminal family F xylanase domain. Domains are often separated by amino acid repeats thought to act as linkers or hinges between functional domains. Multidomain/multi-functional cellulose/hemicellulose-degrading enzymes are a common arrangement in *Cs. saccharolyticus*. All the cellulase/hemicellulase enzymes from *Cs. saccharolyticus* (CelA, ManA, CelB and CelC/ManB) contained at least one cellulose binding domain with two to three proline-threonine rich linkers, and two catalytic domains (see Figure 4). However, only XynF (neither family F nor G xylanase) from the xylanase cluster located on the 18.5kb *Bam*HI fragment of pNZ1084 appears to have a multidomain/multi-functional arrangement (see Figure 3).

In the case of orf3/4 of *Cs. saccharolyticus* and xynC of *Thermatoga* FjSS3.B.1, and in other examples we have not described here, we have found evidence for non-functional genes (pseudogenes) on the genomes of thermophilic bacteria. Pseudogenes have been reported from higher eukaryotes but their occurrence in prokaryotes is unusual. Presumably, a pseudogene could be maintained only because of the presence of multiple copies of the gene. It is widely recognized that β-glycanases are modular enzymes, generally comprised of discrete catalytic and non-catalytic domains which are arranged in different combinations in different enzymes (18). It has been found that the catalytic domains of β-glycanases fall into distinct families, and that a given organism may contain representatives from several of these families. These observations have led to the proposal that domains involved with carbohydrate metabolism have evolved through the duplication, and subsequent modification, of progenitor sequences - the acquisition of new catalytic specificities and the optimisation of existing specificities has presumably come about through the process of di- and convergent evolution *(24)*. An inevitable consequence of such evolutionary mechanisms would be the accumulation of pseudogenes by virtue of non-productive gene rearrangements. Therefore, it is reasonable to expect that some of these pseudogenes would persist in the genomes of saccharolytic organisms if they were closely linked to metabolically important genes, as would be the case in multigene clusters. In this context, it is important to note that both of the *Cs. saccharolyticus* pseudogenes are members of multigene clusters. Ultimately, it may be found that β-glycanase pseudogenes are quite widely distributed, especially in those organisms containing gene clusters. However, their identification may require a systematic examination of the entire β-glycanase system of an organism, as any pseudogenes would be overlooked by standard techniques used for plate assays of genomic expression libraries.

Recombinant DNA technology has allowed us to produce cellulase-free hemicellulases for trialing as pulp bleaching aids. Preliminary trials of *Cs. saccharolyticus* ManB (not reported here) and *Dictyoglomus thermophilum* sp. Rt46B.1 XynA for their ability to hydrolyse pulp hemicellulose and ability to enhance paper brightness have shown that they have a potential use as pulp bleaching aids.

For both enzymes, results were comparable to the mesophilic enzymes trialed at the same time, but in addition, Rt46B.1 XynA and *Cs. saccharolyticus* ManB worked at higher temperatures, and in the case of Rt46B.1 XynA, at a higher pH than the mesophilic enzymes.

One of the major considerations that requires investigation is the relative abilities of different enzymes to retain their activity during prolonged treatment of kraft pulps. Morris *et al. (11)* have shown that *Cs. saccharolyticus* ManB is rapidly inactivated following its addition to unbleached kraft pulp, with the majority of mannan hydrolysis occurring within the first 10-20 minutes after addition to pulp. The stepwise addition of fresh enzyme to kraft pulp was shown to release equivalent amounts of reducing sugar at each stage, indicating that the initial enzyme dose is inactivated well before the available substrate is completely removed. Similar results have been observed in recent trials on kraft pulp using Rt46B.1 XynA and XynB (D. Morris, personal communication). The kraft pulp bleaching assays and reducing sugar release assays reported are more likely to be indications of the relative abilities of the various enzymes to withstand inactivation during treatment, rather than their respective abilities to facilitate the removal of lignin from pulp. The relative abilities of different enzymes to withstand inactivation during the hydrolysis of pulp-bound hemicellulose is likely to be the most important factor in determining their usefulness as pulp bleaching aids. The very high enzyme loadings used for the commercial pulp bleaching (of up to 10,000 nkatals of enzyme per gram of pulp) may negate any loss of enzyme activity due to inhibition. However, a better understanding of the causes of enzyme inactivation, and the relative abilities of different hemicellulases to withstand inactivation should allow lower enzyme doses to be used in pulp treatments, and subsequently, lower the costs for enzyme treatment.

Acknowledgments. This work was supported in part by the University of Auckland Research Grants Committee, Pacific Enzymes Ltd., the Foundation for Research, Science and Technology.and a Macquarie University Research Grant.

Literature cited

1. Kantelinen, A., Hortling, B., Sundquist, J., Linko, M. & Viikari, L. *Holzforsh.* **1993,** *47,* 318.
2. Tremblay, L. & Archibald, F. *Can. J. Microbiol.* **1993,** *43,* 853.
3. Rättö M., Mathrani, I. M., Ahring, B. & Viikari, L. *Appl. Microbiol. Biotechnol.* **1994,** *41,* 130.
4. Viikari, L., Kantelinen, A., Buchert, J. & Puls, J. *Appl. Microbiol. Biotechnol.* **1994,** *41,* 124.
5. Wolfe, R.S. In *The Prokaryotes,* ed. Starr, M.P., Stolp, H., Truper, H.G., Balows, A. & Schlegel, H.G. Springer-Verlag, Berlin. **1981**, Vol. 1; p.v-vi.
6. Brock, T.D. *Symp. Soc. Gen. Microbiol.* **1987,** *41,* 1-17.
7. Wayne, L.G., Brenner, D.J., Colwer, R.K., Grimont, P.A.D., Kandler, O., Krichivsky, M.I., Moore, L.H., Moore, W.E.C., Murray, R.G.E., Stackebrandt, E., Starr, M.P. & Truper, H.G. *Int. Jour. Syst. Bacteriol.* **1987**, *37* , 463-464.
8. Sambrook, J., Frisch, E.F. & Maniatis, T., Eds.; *Molecular Cloning: A Laboratory Manual,* Cold Spring Harbor Press, Cold Spring Harbor, New York (1989).
9. Short, J.M., Fernandez J.M., Sarge J.A. & Huse, W.D. *Nucl. Acids Res.* **1988,** *16,* 7583.
10. Teather, R.M. & Wood, P.K. *Appl. Environ. Microbiol.* **1982,** *43* 777.
11. MorrisD. D., Reeves, R. A., Gibbs, M. D., Saul, D. J. & Bergquist, P. L. *Appl. Environ. Microbiol.* **1995,** *61,* 2262.

12. Schauder, B., Blöcker, H., Frank, R. & McCarthy, J. E. G. *Gene* **1987,** *52*, 279.
13. Lüthi, E., Love, D. R., McAnulty, J., Wallace, C., Caughey, P. A., Saul, D. J. & Bergquist, P. L. *Appl. Environ. Microbiol.* **1990,** *56,* 1017.
14. Lüthi, E., Jasmat, N. B., Grayling, R. A., Love, D. R. & Bergquist, P. L. *Appl. Environ. Microbiol.* **1991**, *57*, 694.
15. Devereux, J., Haeberli, P. & Smithies, O. *Nucl. Acids Res.* **1984,** *12,* 387.
16. Gilkes, N. R., Henrissat, B., Kilburn, D. G., Miller, M. C. & Warren, R. A. J. *Microbiol. Reviews.* **1991**, *55*, 2303.
17. Rainey, F., Ward, N., Morgan, H., Toalster, R. & Stackebrandt, E. *J. Bacteriol.* **1993,** *175,* 4772.
18. Dwivedi, P. P., Gibbs, M. D., Saul, D. J. & Bergquist, P. L. *Appl. Microbiol. Biotechnol.* **1996,** *45,* 86.
19. Saul, D. J., Williams, L. C., Reeves, R. A., Gibbs, M. D. & Bergquist, P. L. *Appl. Environ. Microbiol.* **1995**, 61, 4110.
20. Winterhalter, C. & Liebl, W. *Appl. Environ. Microbiol.* **1995,** *61,* 1810.
21. Gibbs, M. D., Reeves, R. A. & Bergquist, P. L. *Appl. Environ. Microbiol.* **1995**, *61,* 4403.
22. Lüthi, E., Jasmat, N. B. & Bergquist, P. L. *Appl. Environ. Microbiol.* **1990**, *56,* 2677.
23. Bergquist, P. L., Gibbs, M. D., Saul, D. J., Te'o, V. S. J., Reeves, R. & Morris, D. *Biotechnology in the Pulp and Paper Industry: Recent Advances in Applied and Fundemental Research.* Ed. Srebotnik, E. Facultas-Univ.-Verl., Vienna, Austria. **1996,** p497 - 502.
24. Henrissat, B. *Biochem. J.* **1991,** *280,* 309.
25. Henrissat, B., Claeyssens, M., Tomme, P., Lemesle L. & Mornan, J. *Gene* **1989,** *81,* 83.
26. Gilbert, H. J. & Hazlewood, G. P. *J. Gen. Microbiol.* **1993,** *139,* 187.
27. Ferreira, L. M. A., Durrant, A. J., Hall, J., Hazlewood, G. P. & Gilbert, H. J. Biochem. J. **1990,** *269,* 261.
28. Irwin, D., Jung E. D. & Wilson, D. B. *Appl. Environ. Microbiol.* **1994,** *60,* 763.
29. Black, G. W., Hazlewood, G. P., Millward-Sadler, S. J., Laurie, J. I. & Gilbert, H. J. *Biochem. J.* **1995.** *307,* 191.
30. Winterhalter, C., Heinrich, P., Candussio, A., Wich G. & Liebl, W. *Mol. Microbiol.* **1995,** *15,* 431.
31. Kellett, L. E., Poole, D. M., Ferreira, L. M. A., Durrant, A. J., Hazlewood, G. P & Gilbert, H. J. *Biochem. J.* **1990,** *272,* 369.

Chapter 8

Cloning of a Family G Xylanase Gene *(xynB)* from the Extremely Thermophilic Bacterium *Dictyoglomus thermophilum* and Activity of the Gene Product on Kraft Pulp

D. D. Morris[1], M. D. Gibbs[2], and P. L. Bergquist[1,2]

[1]School of Biological Sciences, University of Auckland, Level 4 Thomas Building, 3A Symonds Street, Auckland, New Zealand
[2]School of Biological Sciences, Macquarie University, North Ryde, Sydney, New South Wales 2109, Australia

Xylanases can be grouped into two unrelated families, namely family F and family G. We report here the cloning of a family G xylanase gene (*xynB*) from the *Dictyoglomus thermophilum* strain Rt46B.1 and the characterisation of the expressed gene product (229B). Novel consensus-PCR and genomic-walking PCR techniques were used to isolate the *xynB* gene from *Dictyoglomus thermophilum* genomic DNA. Various 229B xylanases produced from *xynB* expression constructions had pH optima of 6.5, and temperature optimum ranges of between 70 and 85°C. The 229B xylanase was active on kraft pulp as shown by the release of reducing sugars.

The enzymes involved in the metabolism of plant carbohydrate polymers have been grouped into 35 different families on the basis of primary and tertiary sequence homologies (*1*). The endo-1,4-ß-D-xylanases comprise family 10 (F) and family 11 (G). The only similarity between the members of the two xylanase families are their ability to hydrolyse the acetyl-methylglucuronoxylans of hardwoods and arabino-methylglucuronoxylans of softwoods: biochemically and structurally the two families are unrelated. Xylanases, like most other cellulolytic and hemicellulolytic enzymes, are highly modular in structure and can be composed of either a single domain or a number of distinct domains, broadly classified as catalytic and non-catalytic domains. Linker peptides typically delineate the individual domains of multidomain enzymes into discrete and functionally-independent entities, as demonstrated by both proteolysis of intact enzymes and through genetic manipulation of the genes encoding multidomain enzymes (*3*). It is the catalytic domain of a xylanase which encompasses the hydrolytic activity, and therefore governs the classification of the enzyme as either a family F or family G xylanase.

0097–6156/96/0655–0101$15.00/0
© 1996 American Chemical Society

Biochemical Features of Family F and G Xylanase Domains. The catalytic domains of family F and G xylanases differ in their molecular weight, net charge, ratio of uncharged basic to acidic amino-acids, and consequently, their isoelectric points (pI value) as listed in Table I. Both F and G xylanases have an excess of uncharged acidic residues (Glutamine and Asparagine) to uncharged basic residues (Histidine), however the ratio is higher for the family F enzymes.

Table I. Characteristics of F and G Xylanase Domains.

Family	*Average MW*	*Average pI*	*Average Net Charge*	*Average H:Q+N* [a]
F	**37Kd** (32Kd to 40Kd)	**5.7** (4.7 to 6.8)	**-9.0** (-3.0 to -19.0)	**0.3** (0.1 to 0.5)
Gi •Acidic pI Subfamily	**21Kd** (17Kd to 24Kd)	**5.0** (3.8 to 6.0)	**-5.4** (-2 to -15)	**0.11** (0.03 to 0.24)
Gii •Basic pI Subfamily	**21Kd** (20Kd to 24Kd)	**9.5** (8.8 to 10.0)	**+2.5** (+1.0 to +3.0)	**0.09** (0.04 to 0.16)

[a]Ratio of His:Gln+Asn (uncharged acidic to basic amino-acid ratio)

The general characteristics listed in Table I apply to most F and G xylanases, however there are exceptions. For example the family F xylanases from the rumenal bacterium *Bacteroides ovatus* (XynA) and the rumenal fungus *Neocallimastix patriciarum* (XynA) have net positive charges (+3 and +5, respectively) and pI values in the alkaline pH range (8.7 and 8.5, respectively).

Family G xylanases can be divided into two subfamilies: the Acidic pI subfamily and the Basic pI subfamily (*23*). The former have pI values in the acidic range due to an increased number of Glu and Asp residues which give the enzyme a net negative charge. The ratio of uncharged acidic to basic residues remains more or less unchanged between the two subfamilies. The G xylanases of the Acidic pI subfamily are represented predominantly by the fungal enzymes from the *Trichoderma, Aspergillus, Emericella, Neocallimastix, Aureobasidium* , *Penicillium* and *Schizophyllum* genera, although the three *Ruminococcus flavefaciens* G xylanases (XynA, XynB and XynD), the *Fibrobacter succinogenes* G xylanase (XynC) and the *Clostridium stercorarium* G xylanase (XynA) are also members.

Structural Architecture of F and G Xylanases. The sequence data of nearly 80 xylanases reside in the GenBank, EMBL and SwissProt databases. Half of these xylanase sequences are from family G enzymes, which can be further divided more or less equally into the Acid pI and Basic pI subfamilies. The structural architecture of the 75 F and G xylanases presently in the GenEMBL/SwissProt database are shown in Table II.

The xylanases featuring multidomain architecture listed in Table II are produced by organisms which can be segregated into either of the following general groups: (i) rumenal bacteria (*Butyrivibrio, Ruminococcus, Fibrobacter, Bacteroides*) and rumenal fungi (*Neocallimastix*); (ii) thermophilic bacteria (*Clostridial* thermophiles, *Thermotoga*), or; (iii) soil bacteria from the *Actinomycetes* (low %GC gram-positive; *Cellulomonas, Streptomyces, Actinomadura, Thermononospora*) and *Pseudomonadaceae* (gram-negative; *Pseudomonas, Cellvibrio*) groups. However, it should be noted that these organisms also produce single domain xylanases, with the exception of *Pseudomonas fluorescens, Ruminococcus flavefaciens* and *Neocallimastix patriciarum*, which to date produce exclusively multidomain xylanases.

The xylanases produced by the strains of Bacilli and *Aspergillus* are almost exclusively single domain enzymes (11/12 and 8/8 of the enzymes listed in Table II, respectively). In contrast, 10 of the 13 xylanases from rumenal organisms and 14 of the 18 xylanases from thermophilic bacteria are multidomain enzymes. In addition, 80% of the xylanases produced by the soil organisms listed previously are multidomain enzymes. For many of the multidomain xylanases described in Table II, no known function can be attributed to the non-catalytic domains. Those non-catalytic domains from the examples which have been characterised fall into three groups: (i) substrate-binding domains; (ii) thermostabilising domains, and; (iii) cellulosome docking domains.

Cellulose-Binding Domains. Cellulose-binding domains (CBDs) are relatively widespread throughout the cellulases, and in most cases their presence facilitates the activity of the respective enzymes on insoluble cellulosic substrates (*2-4*). CBDs have been divided into five principal families (I-V), although it is clear that several further families now exist (*5,24*). Type-II and III CBDs (100 - 140 amino-acids in length) are predominantly of bacterial origin, while those of Type-I are exclusively fungal (36 amino-acids in length). Given the close proximity of xylan to cellulose in native cellulosic materials, it is of little surprise to find that some of the endoxylanases are also endowed with discrete cellulose-binding domains. Presumably, by adhering xylanases to the plant wall cellulosic components, the CBDs are able to position the xylanase domains in the immediate vicinity of the fibre-bound xylan substrates, and thereby enhance the activity of the enzyme.

Three family F xylanases have been found to contain domains with confirmed cellulose-binding activity. The C-terminal domain of Cex from *Cellulomonas fimi* (*6*) and the N-terminal domain of XynB from *Pseudomonas fluorescens* (*7*) constitute type-II cellulose-binding domains. By homology (67%), the respective N-terminal domain of *Pseudomonas fluorescens* XynA also appears to be a type-II CBD. The central domain partitioning the F xylanase domain and endoglucanase domain of the novel multi-functional CelB enzyme from *Caldicellulosirptor saccharolyticus* (formerly "*Caldocellum saccharolyticum*") is a type-III bacterial CBD (*8*). Identical type-III CBDs are also included (in a central position) in the three other multidomain/multi-functional enzymes of the *Caldicellulosirptor saccharolyticus* gene-cluster. Lastly, the vaguely duplicated domains at the C-terminus of XynA from *Thermotoga maritima* represent a novel family of CBD (*9*). The two other *Thermotoga* strains from which xylanases have been cloned and sequenced (*T. neopolitana* and *T.* spp. Fjss3.B1) also produce XynA-type enzymes (XynA, and XynB/C, respectively), hence these enzymes will also incorporate the novel *T. maritima* XynA-type CBDs at their C-termini (*10*). By homology to the *T. maritima* C-terminal CBDs, the duplicated C-terminal regions of XynA from *Thermoanaerobacter saccharolyticum* (63% homology) and XynX form *Clostridium thermocellum* (60% homology) also appears to belong to this novel family of CBD (*9*).

Three family G xylanases contain CBDs with confirmed cellulose-binding capabilities. The two repeat domains at the C-terminal region of XynA from *Clostridium stercorarium* constitute a novel family of CBD (*11*). Equivalent domains are also present in the XynZ F xylanase from *C. thermocellum* (67% homology) and the novel arabinofuranosidase/endoxylanase XynD from *Bacillus polymyxa* (63% homology). Lastly, the C-terminal domains from *Streptomyces lividans* XynB (*12*) and *Thermomonospora fusca* XynA (*13*) constitute yet another unique family of CBD which may also have xylan-binding activity (65% homology).

Thermostabilising Domains. There are two examples of specific domains in multidomain F xylanases which appear to provide a thermostabilising function to the enzyme. The duplicated N-terminal domains ("D1a" and "D1b", which share 50%

Table II. F and G Xylanases in the GenEMBL /SwissProt Databases.

1. BACTERIAL XYLANASES:

Organism	Gene	Domain Architecture
Family F:		
Actinomadura sp. FC7	xyl II	
Bac. stearothermophilus 21	XYN	
Alk. Bacillus C-125	xynA	
Butyrivibrio. fibriosolvens (R)	xynB	
Butyrivibrio . fibriosolvens (R)	xynA	
Bacteroides ovatus (R)	xyn	
Bacillus sp. 137	xynA	
Bac. stearothermophilus	xynT6	
Caldicellulosiptor sacc.	xynA	
Caldicellulosiptor sacc.	celB	CBD A1
Cellulomonas fimi (M15824)	Cex	
Cellulomonas fimi	xynC	TSD
C. thermocellum	xynX	TSD CBD
C.stercorarium	XYN	
C.thermocellum	xynZ	CBD X
C. thermocellum	xynY	CBD X
Cellvibrio mixtus •	xynB	
Rt46B.1 (Dictyoglomus therm)	xynA	
Eubacterium ruminantium (R)	xynA	
Pseudomonas fluorescens •	xynA	CBD
P.fluorescens •	xynB	CBD
P.fluor. ssp. cell. •	xynF	
Prevotela ruminicola R	xynA	
R. flavefaciens (R)•	xynA	G
Rt8B.4 •	xynA	TSD
Streptomyces lividans	xlnA	
Thermotoga maritima	xynA	TSD TSD CBD CBD
Thermotoga neapolitana	xynA	TSD TSD CBD CBD
Thermotoga spp. fjss	xynA	
Thermoanaerobacter sacch.	xynA	TSD TSD CBD CBD
Family G:		
Aeromonas caviae	xynA	
Bac.subtilis	xynA	
Bac. circulans	xlnA	
Bacillus pumilus	xynA	
Bac. stearothermophilus	xynA	
Bac. subtilis 168	xynA	
Bacillus sp.	xynS	
Bacillus sp.	xynY	
C.stercorarium	xynA	CBD CBD
C.acetobutylicum	xynb	
Cellvibrio mixtus •	xynA	
Fibrobacter succinogenes (R)	xynC	
P.fluor. ssp. cell. •	xynG	
R. flavefaciens (R)•	xynD	ß-1,3-1,4
R. flavefaciens (R)•	xynB	
Streptomyces sp	XYN	
Streptomyces lividans	xlnB	CBD

Table II. Continued

Organism	Gene	Domain Architecture
Family G continued:		
Streptomyces lividans	xlnC	
Thermomonospora fusca YX	xynYX	CBD
2. FUNGAL XYLANASES		
Family F:		
Emericella nidulans	xlnC	
Aspergillus kawachii	xynA	
Cryptococcus albidus	xynA	
Magnaporthe grisea	xyn33	
N. patriciarum (R)•	xynB	
Penicillium chrysogenum	xynA	
Thermoascus aurantiacus	xynA	
Fusarium oxysporum	xyn	
Family G:		
Emericella nidulans	xlnA	
Emericella nidulans	xlnB	
Aureobasidium pullulans	xynA	
Aspergillus tubigensis	xlnA	
Aspergillus kawachii	xynB	
Aspergillus niger	xynB	
Aspergillus kawachii	xynC	
Cochliobous carbonum	XYN	
Magnaporthe grisea	xyn22	
N.frontalis (R)	xynI	
N.frontalis (R)	xynII	
N.patriciarum (R)•	xynA	
Penicillium purpurogenum	xynB	
Aspergillus kawachii	xynC	
Trichoderma reesei (C30)	xynI	
Trichoderma reesei (C30)	xynII	
Trichoderma reesei (VTT)	xynII	
Trichoderma reesei (mRNA)	xynII	
Schizophyllum commune	xynA	

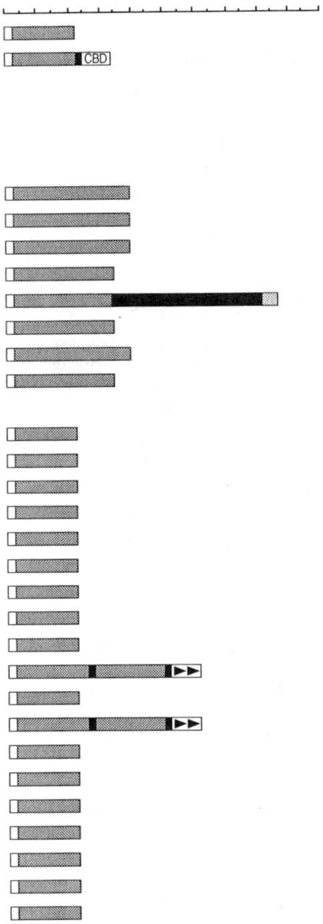

Key:

- "A1 or G or ß-1,3-1,4" denote the classification of the additional ß-glycanase domains in multidomain/multifunctional enzymes
- "CBD" = Cellulose Binding Domain
- "TSB" = ThermoStabilising Domain
- "X" = cellulosomal docking motif
- ► = repeated domain of unknown function
- "(R)" denotes rumenal organism
- "Bac" = Bacillus"
- "C." = Clostridium
- Bullet (•) denotes organsims with exclusive multidomain xylanases
- Acidic pI family G xylanases classified as Gi
- Thermophilic organisms are highlighted by light-grey shading

homology) of XynA from *Thermoanaerobacter saccharolyticum* (*14*) and Domain 3 of XynY ("XynY D3") from *Clostridium thermocellum* (*15*) are both reported to increase the temperature optimum of their respective enzymes. Moreover, it has been demonstrated that the latter thermostabilising domain is able to enhance the thermal tolerance of the catalytic domain from a mesophilic ß-glycanase (*15*). The XynY D3 and XynA D1b share 43% sequence homology. It is intriguing to note that almost every thermophilic family F multidomain xylanase reported to date (whose nucleotide/peptide sequences have been deposited in the GenEMBL/SwissProt databases) contains a region homologous to the *Thermoanaerobacter saccharolyticum* thermostabilising domain (XynA D1a or D1b) immediately upstream of the family F catalytic domain. The *Clostridium thermocellum* XynY thermostabilising domain is the only example which is located at the carboxy-terminal side of the F xylanase domain (this enzyme is also exceptional on the basis that it is located within a multi-subunit complex termed a "cellulosome" as discussed below). Oddly, the XynC F xylanase from the mesophilic bacterium *Cellulomonas fimi* also contains an apparent thermostabilising domain immediately upstream of the catalytic domain.

The sequences of only two family G xylanases derived from thermophilic organisms have been reported to date (*Clostridium stercorarium* XynA and *Thermomonospora fusca* XynA), and both of these enzymes exhibit a multidomain architecture. We have determined the sequence of an additional thermophilic family G xylanase (456D, unpublished) which is also a multidomain enzyme of similar overall structure to the xylanase from *Thermomonospora fusca*. These thermophilic G xylanases do not contain any apparent thermostabilising domains. As discussed above, the non-catalytic domains from the *C. stercorarium* and *T. fusca* enzymes have been characterised as substrate-binding domains. Presumably, the compact tertiary structure of the G xylanase domain can be rendered sufficiently thermostable through mutations which strengthen the interactions at key structural locations within the domain, and thereby preclude the necessity for domains which harbour additional thermostabilising functions.

Cellulosome-Docking Domains. The cellulases and xylanases of numerous anaerobic cellulolytic bacteria aggregate to form a highly active cell wall-associated quaternary complex called a "cellulosome" (*17*). These cellulosomal structures harbour a wide repertoire of cellulolytic and xylanolytic activities to enable the bacterial cells to break down complex cellulosic substrates. Additionally, the cellulosomes mediate the adsorption of the bacterial cells to native cellulose substrates and thereby allow the rapid and efficient uptake of the hydrolysis products. Analogous to keys on a key-ring, the different hydrolytic components of the cellulosome are attached via "docking" domains to the central anchor peptide which harbours the cellulose-binding function. Hence the cellulosomal cellulases and xylanases from the anaerobic bacterial strains like *C. thermocellum* and *C. cellulovorans* contain a specific docking peptide necessary for cellulosome-association. The F xylanases XynY and XynZ from *C. thermocellum* (listed in Table II) are both cellulosome-associated enzymes, and contain a duplicated segment of 22 amino-acids which serve as a cellulosome-docking motif (*17*).

Non-Catalytic Domains of Unknown Function. Given the abundance of multidomain xylanases from unrelated organisms in similar environments, it is unlikely that the large number of non-catalytic domains from multidomain xylanases which have no apparent physiological function are indeed non-functional, *in their natural environments*. Presumably, multidomain ß-glycanases have evolved to allow organisms to make efficient use of the carbohydrate resources which are present in the unique environments in which they inhabit. For example, rumenal organisms may have evolved unique non-catalytic domains which work in synergy with the catalytic

domains to make best use of the unique cellulosic substrates *within* their rumenal environment. It is therefore possible that the actual physiological function of such non-catalytic domains would not be apparent in the artificial environment of a test-tube, especially if laboratory-grade model substrates are being used to characterise the enzymes. Furthermore, some organisms may have evolved multidomain ß-glycanases equipped with non-catalytic domains whose function is to assist/stabilise the catalytic domains only under sub-optimal physiological conditions. For example many thermophilic bacteria live in environments which can fluctuate in temperature, pH, mineral composition and substrate availability. The bacterial cells can be protected to a large extent from such environmental changes by living within the protective milieu of a biofilm. The extracellular ß-glycanases however would need to be intrinsically stable, and preferentially active, under these fluctuating conditions. The incorporation of stabilising non-catalytic domains into such enzymes could be one possible mechanism adopted by thermophilic bacteria to adapt to short-term changes in their physical environment. Elucidation of the function of these non-catalytic domains would require the characterisation studies to be performed under the physical conditions, and using the substrates present, in the actual thermal environments.

It is noteworthy that the non-catalytic domains present in the multidomain xylanases exhibit a very high degree of sequence variability. However it is likely that more sensitive sequence comparison techniques which detect tertiary structural homologies (for example hydrophobic cluster analysis), rather than basic pair-wise sequence homologies, are likely to narrow the margin of apparent variability between the non-catalytic domains. Nonetheless, it is clear that the non-catalytic domains of ß-glycanases have enjoyed a far greater degree of evolutionary freedom than the catalytic domains, whose structures are bound by the strict requirements for hydrolysis of glycosyl-linkages. Evidently, the physiological 'functions' of the non-catalytic domains may only require the interaction of a few key structural motifs - the amino-acid residues comprising the intervening peptide sequences may only need to conform to a rather loose set of physical parameters, which may explain the high degree of apparent variability within the non-catalytic domains.

Novel Consensus-PCR and Genomic-Walking PCR Approach to Cloning of the Rt46B.1 *xynB* Gene. Traditionally, the identification, cloning and sequencing of family F and family G xylanase genes from bacteria involves the screening of bacterial gene libraries for either expression of endoxylanase activity, or hybridisation to oligonucleotide or DNA-based probes. Following the identification of xylanase-positive library colonies (plasmid-based libraries) or plaques (bacteriophage lambda-based libraries), a relatively extensive sub-cloning regimen is then required to pinpoint the exact location of the xylanase gene along the DNA fragment of the library recombinant. Further sub-cloning is then required to generate DNA fragments of appropriate size for nucleotide sequencing. Typically, these traditional cloning approaches are time-consuming, and may involve several months for completion (depending upon the size of the gene).

We have developed a novel approach to the cloning of family F and G xylanase genes which involves (i) the amplification of xylanase consensus fragments (XCFs) from the internal region of family F and G xylanase genes using consensus PCR primers targeted at either F or G xylanase genes, and then (ii) genomic-walking PCR to generate DNA fragments up- and downstream of the XCFs. Using this approach, we were able to identify and fully sequence the *xynB* gene from Rt46B.1 within at least half the time as would have been required using the traditional cloning approaches.

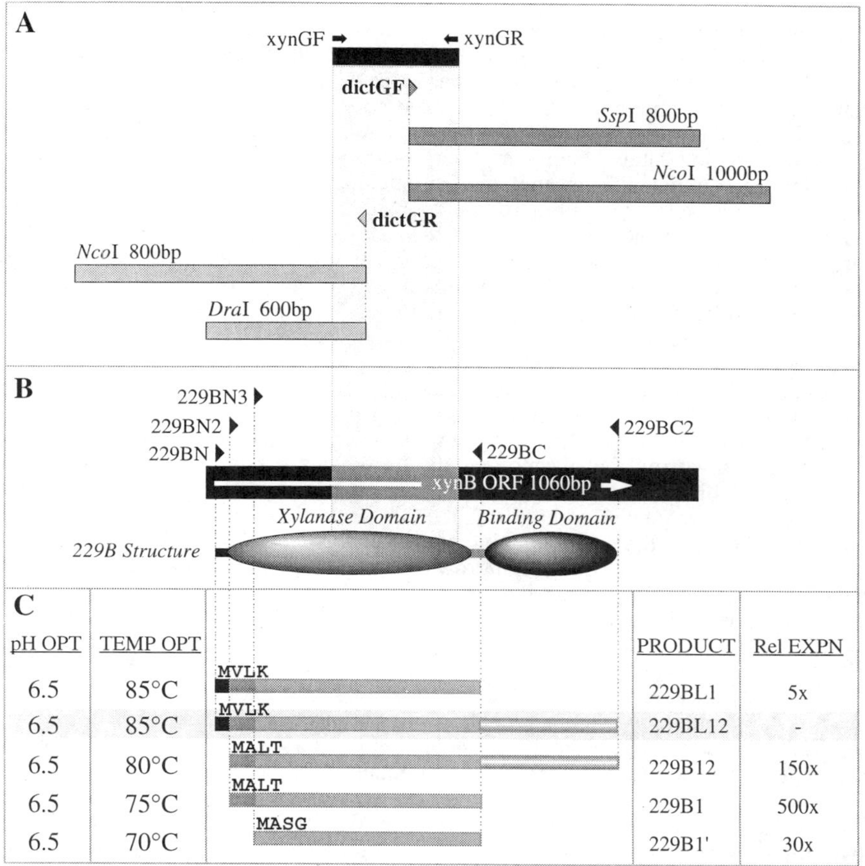

Figure 1. Overview of the cloning and expression of the Rt46B.1 *xynB* gene. (A) DNA fragments used for the sequencing of the xynB gene: the genomic-walking PCR-products from the dictGF and dictGR walking primers are shown in grey, whilst the *xynB* xylanase consensus fragment amplified by the xynGF/xynGR consensus primers is shown in black. (B) Domain architecture of the 229B xylanase encoded by the Rt46B.1 *xynB* gene. The locations of the three forward and two reverse PCR primers used for the expression studies are indicated. The light grey box shows the region covered by the *xynB* xylanase consensus fragment. (C) Summary of the pH and temperature optima of the five 229B xylanases. The domain structure, name and expression level of each construct relative to the 229BL12 xylanase is shown. The first four amino-acids of each 229B xylanase are indicated.

Cloning of the Rt46B.1 *xynB* Gene. Rt46B.1 (isolate TG229) is a strain of *Dictyoglomus thermophilum* (*19*) which was isolated from a thermal spring in Rotorua, New Zealand, by Patel and workers in 1987 (*22*). The existence of a family G xylanase gene (*xynB*) in the *Dictyoglomus thermophilum* strain Rt46B.1 genome was detected through the use of a pair of consensus PCR primers (xynGF and xynGR) which were specific to the conserved regions of family G xylanase genes.

The 300bp consensus PCR products that were amplified from Rt46B.1 genomic

DNA by the xynGF and xynGR consensus primers were cloned into M13mp10 and sequenced in order to confirm the authenticity of the fragments, and to determine the number of family G xylanase genes present. Only one species of xylanase consensus fragment was observed during the sequencing of a representative sample of the Family G consensus fragments amplified from Rt46B.1, which suggested that Rt46B.1 contained only one Family G xylanase gene. This new gene was named *xynB*, in keeping with the naming of the Family F xylanase gene previously isolated from Rt46B.1 as *xynA*.

The next step in the isolation of the Rt46B.1 *xynB* gene involved amplifying DNA fragments upstream and downstream of the *xynB* xylanase consensus fragment using genomic-walking PCR from which we could generate the nucleotide sequence of the complete *xynB* gene. Once armed with the *xynB* sequence data, we could design PCR primers to transfer the *xynB* gene into an appropriate plasmid vector for expression of the xylanase (229B) encoded by *xynB*. The genomic-walking PCRs (GWPCRs) were performed as described previously (*18*) using forward and reverse genomic-walking primers (dictGF and dictGR, respectively) which were designed from the sequence of the Rt46B.1 *xynB* xylanase consensus fragment. Of the genomic-walking fragments that were amplified from the Rt46B.1 genome, four were sufficient in size to encompass the *xynB* gene (Figure 1.). The GWPCR products shown in Figure 1 are named according to their size in base-pairs, and the restriction endonuclease which was used to generate the restriction-fragment/linker library from which the GWPCR product was amplified (seven different linker-libraries were prepared for the *xynB* GWPCRs using the *Nco*I, *Ssp*I, *Dra*I, *Eco*RV, *Hpa*I, *Hinc*II and *Pvu*II restriction endonucleases). In total, 1190bp of sequence data was derived from the upstream and downstream *xynB* GWPCR products, of which 1080bp comprised the *xynB* open-reading frame. Given that the catalytic domain of a family G xylanase is encoded on average within 700bp of DNA, it was apparent that the Rt46B.1 *xynB* gene encoded a multidomain family G xylanase (229B).

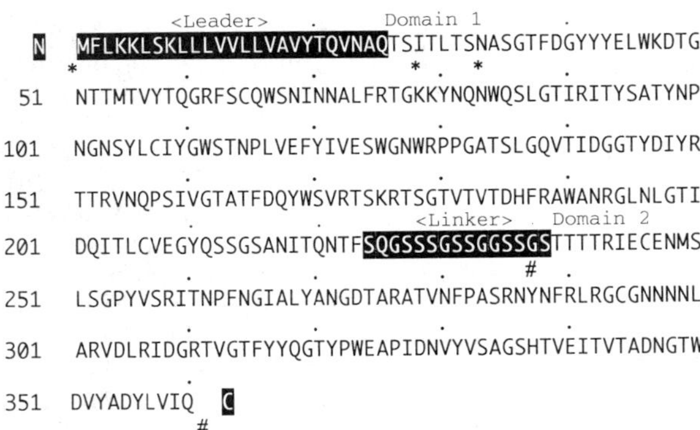

Figure 2. Annotated 229B peptide sequence. The putative leader-sequence and the interdomain linker peptide are shown in reverse-font. Asterisks indicate start-positions of the various 229B expression constructs, while the hashes indicate the end-points.

Analysis of the Rt46B.1 *xynB* Gene. The deduced peptide (229B) encoded within the *xynB* gene was 360 amino-acids (aa) in length and consisted a 223aa catalytic domain (incorporating the leader peptide of ca. 25aa's) followed by a short

linker peptide of 15 amino acids rich in glycine and serine residues, and a 122aa C-terminal domain as depicted in Figure 2.

The 229B xylanase had an almost identical structural architecture (with 85% amino-acid homology) to a family G xylanase (456D, unpublished) isolated by us previously from a Clostridial thermophilic bacterium using GWPCR techniques. However, whereas the isolated N-terminal catalytic domain of 456D showed optimal activity at 70°C, pH6.5, the respective domain from 229B was expected to show optimal activity at 80 - 85°C (the optimal temperature for growth of Rt46B.1). The C-terminal domain of 229B shared 60% homology to the uncharacterised C-terminal region of the novel multi functional xylanase/arabinofuranosidase (XynD) from *Bacillus polymyxa* .

Expression of the Rt46B.1 *xynB* Gene. Various PCR primers were designed to allow for expression of *xynB* fragments in the heat-inducible pJLA602 expression vector. Three different N-terminal primers were designed to amplify *xynB* fragments which encoded 229B xylanases starting from either : (i) the beginning of the leader peptide; (ii) 3 residues downstream of the putative leader peptide cleavage point, or; (iii) 8 residues downstream of the putative leader peptide cleavage point. These start points are indicated by asterisks Figure 2. The latter two N-terminal primers represented two approximate positions where leader peptide cleavage may have occurred. Two C-terminal primers were designed to be used in conjunction with the N-terminal primers to amplify *xynB* fragments with or without the C-terminal domain. Five *xynB* constructs were prepared (229BL, 229BL12, 229B12, 229B1 and 229B1') as described in Figure 1. The "1" and "2" in the enzyme nomenclature represent domains 1 and 2 from 229B; the "L" signifies the inclusion of the leader peptide in the expression construct, whilst the prime in "229B' " shows that the encoded 229B xylanase contained an 8 residue N-terminal deletion as a result of being prepared using the 229BN3 PCR primer.

The level of 229B expression from the five different *xynB*:pJLA602 recombinant plasmids varied considerably (Figure 1C) and was found to be effected by both the length of the *xynB* fragments and by the sequence at the immediate 5' end of the *xynB* fragments. However, the latter variable appeared to have the more significant effect on the levels of xylanase expression from the *xynB*:pJLA602 recombinant plasmids. The full-length xylanases, 229BL12 and 229B12, were expressed at a lower level as compared to the shorter 229BL1 and 229B1 xylanases. This observation could readily be explained in terms of the time and energy demands required to process the longer 229BL12 and 229B12 open-reading frames. The effect of the 5' *xynB* sequence on the 229B expression level however was not obvious. The 229B xylanases which were expressed at the lowest apparent level were the 229BL1 and 229BL12 xylanases. It is possible that the presence of the 229B leader peptide in 229BL1 and 229BL12 effected the apparent expression levels of these xylanases. The 229B leader peptide was not targeting the 229BL1 and 229BL12 xylanases for secretion in *Escherichia coli*, hence the low apparent expression levels of these xylanases was not simply due to the fact that these enzymes were present at an extracellular location (the 229B xylanases were isolated from *E. coli* cell pellets by mechanical lysis in a French Pressure device). The 229B leader peptide present on the 229BL1 and 229BL12 xylanases may have increased the toxicity of these enzymes in *E. coli*, which would have resulted in lower overall expression levels.

With the exception of the leader peptide, the only significant differences between the 229BL1/229BL12 xylanases and the more highly expressed 229B1 and 229B1' xylanases was the nature of the penultimate amino-acid. This amino-acid was a valine residue in the 229BL1/229BL12 xylanases, and an alanine residue in both 229B1 and 229B1'. However, given that an N-terminal valine residue does not appear to be one

of the amino-acids which target a protein for rapid degradation in E. coli (27), and that the 229B1 and 229B1' xylanases were expressed at quite different levels, the nature of the penultimate amino-acid did not appear to effect the apparent levels of expression of the 229B xylanases.

Characterisation of the 229B Xylanases. As listed in Figure 1C, the 229BL1, 229BL12, 229B12, 229B1 and 229B1' xylanases all displayed optimal activity at pH6.5, consistent with most other bacterial xylanases. However, a significant variation in the temperature optimum existed between the four enzymes: only the 229BL1 and 229BL12 xylanases showed optimal activity at the expected temperature of 85°C; the temperature optima of the three remaining 229B xylanases were significantly reduced, and dropped to a remarkably low 70°C for the 229B1' construct. Presumably, the 229BL1/229BL12 xylanases recorded the highest temperature optimum because the leader peptides were processed by the *E. coli* bacterial host to yield optimally cleaved, and hence stable, enzymes. Therefore, it followed hat the lower temperature optima of the three other 229B xylanases was a result of sub optimal placement of the N-terminal PCR primers used in the preparation of the 229B1, 229B12 and 229B1' expression constructions. It is interesting to note that the inclusion of the second domain on the 229B12 xylanase resulted in an increase in the temperature optimum of the enzyme from 75°C to 80°C, presumably because the C-terminal domain was able to compensate to some extent the reduced stability at the N-terminus of the enzyme. However, it is clear from the temperature optima of the 229BL1 and 229BL12 enzymes that the C-terminal domain does not provide a thermostabilising function, *per se*.

Kraft Pulp Hydrolysis by 229B. The release of reducing sugars was measured during the hydrolysis of 0.4g washed kraft pulp by 20XUs of 229B to confirm that 229B could hydrolyse the fibre bound xylan substrates present in kraft pulp, . The 229B1 xylanase was selected for the kraft pulp hydrolysis assays because of the very high levels of expression obtained from the 229B1 pJLA602 construction. To allow a comparison between the family F and G xylanases from Rt46B.1, a duplicate assay was performed using 20XUs of the 229A family F xylanase from Rt46B.1. Both enzymes were able to generate measurable amounts of reducing sugar from the kraft pulp. After six hours of hydrolysis, 229A released the equivalent of 0.035µmoles of reducing sugar (at 85°C incubation) from 0.4g of kraft pulp, and 229B released 0.025µmoles of reducing sugar (at 75°C incubation). However, whilst 229A hydrolyses xylan substrates to produce mainly xylose and xylobiose, the 229B xylanase releases xylotriose as the lowest molecular weight oligomer from xylan substrates, as indicated by thin layer chromatography of the hydrolyses products from oat spelts xylan. Based on this observation, the 229B xylanase released more total sugar from kraft pulp as compared to the 229A xylanase, assuming that 229A and 229B were hydrolysing the fibre-bound xylan substrates in a similar fashion to the soluble model xylan substrates.

Additional assays were performed to compare the action of the 229B1 and 229B12 xylanases on kraft pulp. Both enzymes released exactly the same amount of reducing sugar from kraft pulp. Studies presently are being done to ascertain whether or not the observed release of reducing sugars from kraft pulp by 229B1 results in either a reduced chlorine demand during the conventional bleaching, or a higher final pulp brightness in TCF bleaching following an enzymatic pre-treatment.

Preliminary Studies of the 229B C-terminal Domain. The 229B xylanase has a very similar overall domain architecture to the family G xylanases from *Streptomyces lividans* (XynB) and *Thermomonospora fusca* (XynYX). In both these enzymes the C-terminal domain is a cellulose-binding domain (CBD) of 80-90 amino acids in length which is separated from the catalytic domain by a linker peptide of approximately 20 amino-acids. Comparative hydrolysis, binding and kinetic assays between the 229BL1 and 229BL12 xylanases have indicated that the 229B C-terminal domain functions as a xylan-binding domain.

The 229B C-terminal domain was found to have a very dramatic effect on the degree of hydrolysis of agarose-immobilised oat spelts xylan: whereas the 229BL1 xylanase hydrolysed agarose-immobilised xylan to produce a very diffuse and faint clearing of substrate, the respective clearing produced by an equal amount of 229BL12 was considerably smaller, sharper and more intense. These observations suggested that the C-terminal domain present in the 229BL12 xylanase was both slowing down the diffusion of the enzyme through the xylan, and also facilitating the hydrolysis of the xylan. A xylan-binding role of the 229B C-terminal domain would explain the lowered diffusion and enhanced hydrolysis of the 229BL12 xylanase. It was later demonstrated that the 229BL1 xylanase could be freely eluted from agarose-immobilised xylan, whereas the 229BL12 xylanase was mostly retained in the xylan. In addition, kinetic assays indicated that the 229BL1 xylanase had a three-fold higher Km value than the 229BL12 xylanase when hydrolysing oat spelts xylan. These observations all suggest a xylan-binding role for the 229B C-terminal domain. No apparent binding to crystalline cellulose (Avicel) could be observed for the 229BL12 xylanase, therefore it appeared that the 229B C-terminal domain was a specific xylan-binding domain as apposed to a broad specificity cellulose/xylan binding domain.

The 229B C-terminal domain was 60% similar to the C-terminal region of the XynD enzyme from *Bacillus polymyxa*. It is interesting to note that the 229B-like domain in the XynD enzyme is immediately downstream of a type VI cellulose-binding domain. This observation lends support to the role of the 229B C-terminal domain as a specific xylan-binding domain, rather than a general CBD.

Concluding Remarks. We described here a two-step PCR regime to identify and isolate family G xylanase genes from the genomic DNA of a micro-organism. This technique has proven to be a very expedient means for the isolation of family G xylanase genes, and can be extended to family F xylanase genes when used in conjunction with the family F xylanase consensus primers. The procedure also has potential applications for the identification and isolation of F and G xylanase genes from complex DNAs obtained from mixed or unculturable bacterial sources. As the technique is purely PCR-based, only very small amounts of substrate DNAs are required, however, it would be desirable to have at least 1μg of DNA for the preparation of the substrate DNAs for the genomic-walking PCRs.

The observed differences in the temperature optima and apparent expression levels of the genetically-altered derivatives of the 229B xylanases demonstrates how subtle differences in the design of an expression system can have a significant effects on the nature of the final product. The results from the 229B expression work suggest that it is desirable, in the absence of any N-terminal sequence information, to incorporate the leader peptide into a G xylanase expression construction. It appears that the *Escherichia coli* recombinant host can then cleave the leader peptide to generate an enzyme with an optimally positioned N-terminus. As seen with the 229B1 and 229B1' xylanases, sub optimal placement of the N-terminus can effect the stability of the enzyme. However, the incorporation of the 229B leader peptide did appear to significantly reduce the level of expression of the 229B xylanase. Therefore, whilst the inclusion of the leader peptide may be essential for accurate characterisation of the

enzyme, it's absence may be more appropriate subsequently if large amounts of enzyme are required.

The C-terminal domain of 229B was shown to have a profound effect on the ability of the N-terminal G xylanase domain to hydrolyse oat spelts xylan. Subsequent analysis revealed that the 229B C-terminal domain was able to bind the enzyme to the xylan. As no binding to crystalline cellulose could be observed, the 229B C-terminal domain appeared to be a specific xylan-binding domain. Such a binding domain has not been reported to date. The lack of any sequence homology of the 229B C-terminal domain to any of the present cellulose-binding domain families, and the presence of a homologous domain immediately downstream of a cellulose-binding domain in the *Bacillus polymyxa* xylanase lends some support to the designation of the 229B C-terminal domain as a specific xylan-binding domain. However, further work must be done to rule out the possibility that the 229B C-terminal domain has additional cellulose-binding activity.

Methods and Materials. The molecular biological techniques described in this chapter involving genomic-walking PCR, DNA cloning and sequencing are identical to those described previously by Morris *et al.* (18). Consensus PCRs using the xynGF (5' TAT NTG RST NTM TAT GGW TGG 3') and xynGR (5' CCG CTN CTT TGG TAN CCT TC 3') primers were performed in standard 50µl reactions for 35 cycles using the following profile: 94°C 30 seconds, 45°C 30 seconds, 72°C 30 seconds. The xynGF and xynGR primers were on average 83% and 96% similar to the family G xylanase sequences in the GenBank/EMBL databases.

Crude enzyme samples were prepared for the characterisation studies as follows: 1200ml LB-broth supplemented with ampicillin to 100µg/ml was inoculated with 12ml of an overnight culture of Escherichia coli JM101 harbouring the desired xynB:pJLA602 recombinant plasmid grown at 30°C. The culture was incubated with shaking at 30°C until the culture attained an optical density of 1.0 at 600nm. The culture was then transferred to a 42°C shaking water bath and incubated for a further 3 hours. The cells were harvested by centrifugation, resuspended in 30ml of TE buffer (10mM Tris-HCl, 1mM EDTA, pH8.0), and then lysed by passage through a French Pressure cell. The cell extracts were partially purified by heating the extracts at 75°C for 30 minutes and removing the denatured host proteins by centrifugation. Enzyme samples were diluted in TE buffer to an enzyme concentration of approximately 0.005nkat/µl for the pH and temperature optima assays.

Quantitative assays for endoxylanase activity were performed in triplicate 200µl reactions comprising 180µl of 0.25% oat-spelts xylan, 10µl of 250mM pH buffer and 10µl of a 229B solution at a concentration of 0.005nkat/µl. Reactions were incubated for 15 minutes and the reducing sugars released were detected using the *p*-hydroxybenzoic acid hydrazide (PHBAH) colorimetric assay described by Lever (25). Sodium acetate buffer was used for pH values between 4.0 and 6.0, BTP Buffer (Sigma) between pH 6.0 and 9.0, and CAPS Buffer (Sigma) for pH 9.0 and above. The pH buffers were pH equilibrated for the appropriate reaction temperature.

Qualitative assays for endoxylanase activity were performed using agarose-immobilised xylan. Approximately 1ml of molten 1% agarose/0.5% oat spelts xylan solution was poured into a standard 120mm petri-dish and allowed to set. An aliquot of xylanase solution (5-20µl enzyme) was applied to the surface of the agar/xylan. Once the xylanase solution had dried onto the surface of the agar the petri-dishes were covered and tightly sealed with tape, and the plates were transferred to a 70°C oven. The action of the xylanase on the agarose-immobilised xylan could be monitored by visualising the degree of substrate clearing around the enzyme samples. After a sufficient amount of hydrolysis had occurred, the petri-dishes were allowed to cool

and then stained for 10 minutes with a solution of 2% Congo Red (Sigma) and destained as required in 1M NaCl.

Binding assays of xylanase samples to agarose-immobilised xylan was carried out as follows: 100μl xylanase solution (approximately 0.01nkat/μl) was mixed with 100μl of a molten solution of 1% agarose/1% oat spelts xylan/12.5mM BTP buffer, pH $6.5^{70°C}$ and incubated for 10 minutes at 70°C. The mixture was then applied to a 0.5ml microcentrifuge tube containing 100μl of glass beads overlaying a small aperture in the base of the tube. The agarose was allowed to set firm, and the 0.5ml tube was inserted into a standard 1.5ml microcentrifuge tube. The setup was centrifuged at 13 000rpm for approximately 10 seconds to allow the solute to pass from the agarose through the aperture in the base of the 0.5ml microcentrifuge tube and into the 1.5ml microcentrifuge tube. The residual xylanase activity in the solute was measured using quantitative PHABAH reducing sugar release assays described above.

Acknowledgements. This work was supported by a grant from the Foundation for Research, Science and Technology, New Zealand. Bacterial strains were provided by Professor Hugh Morgan, Thermophile Research Unit, University of Waikato, New Zealand.

Literature Cited.

1. Henrissat, B., Biochem. J., **1991**, *280*, 309-316.
2. Beguin, P. Annu. Rev. Microbiol., **1990**, *44*, 219-248.
3. Gilkes, N. R.; Henrissat, B.; Kilburn, D. G.; Miller, R. C. Jr.; Warren, R. A. J., Microbiol. Rev., **1991**, *55(2)*, 303-315.
4. Millward-Sadlert, S. J.; Poole, D. M.; Henrissat, B.; Hazlewood, G. P.; Clarke, J. H.; Gilbert, H. J., Mol. Microbiol., **1994**, *11(2)*, 375-382.
5. Coutinho, J. B.; Gilkes, N. R.; Warren, D. G.; Miller, R. C. Jr., Mol. Microbiol., **1992**, *6(9)*, 1243-1252.
6. White, A.; Withers, S. G.; Gilkes, N. R.; Rose, D. R., Biochemistry, **1994**, *33*, 12546-12552.
7. Kellett, L. E.; Poole, D. M.; Ferreira, L. M.; Durrant, A. J.; Hazlewood, G. P.; Gilbert, H. J., Biochem. J., **1990**, *272*, 369-376.
8. Saul, D. J.; Williams, L. C.; Grayling, R. A.; Chamley, L. W.; Love, D. R.; Bergquist, P. L., Appl. Environ. Microbiol., **1990**, *56*, 3117-3124.
9. Winterhalter, C.; Heinrich, P.; Candussio, A.; Wich, G.; Leibl, W., Mol. Microbiol., **1995**, *15(3)*, 431-444.
10. Dakhova, O. N.; Kurepina, N. E.; Zverlov,V. V.; Svetlichnyi, V. A.; Velikodvorskaya, G.A., Biochem. Biophys. Res. Commun., **1993**, *1949*, 1359-1364.
11. Sakka, K.; Kojima, Y.; Kondo, T.; Karita, S.; Ohmiya, K.; Shimada, K., Biosci. Biotech. Biochem, **1993**, *57(2)*, 273-277.
12. Sharak, F.; Roy, C.; Yaguchi, M.; Morosoli, R.; Kluepfel, D., Gene, **1991**, *107*, 75-82.
13. Irwin, D.; Jung, E. D.; Wilson, D. B., Appl. Environ. Microbiol., **1994**, *60(3)*, 763-770.
14. Lee, Y.; Lowe, S. E.; Henrissat, B.; Zeikus, G. J., J. Bact., **1993**, *175(18)*, 5890-5898.
15. Fontes, C. M. G. A.; Hazlewood, G. P.; Morag, E.; Hall, J.; Hirst, B. H.; Gilbert, H. J., Biochem. J., **1995**, *307*, 151-158.

16. Dakhova, O. N.; Kurepina, N. E.; Zverlov,V. V.; Svetlichnyi, V. A.; Velikodvorskaya, G.A., Biochem. Biophys. Res. Commun., **1993**, *1949*, 1359-1364.
17. Beguin, P.; Millet, J.; Aubert, J. P., FEMS Microbiol. Lett., **1992**, *100*, 523-528.
18. Morris, D. D.; Reeves, R. A.; Gibbs, M. G.; Bergquist, P. L., Appl. Environ. Microbiol., **1995**, *61*, 2262-2269.
19. Gibbs, M. G; Reeves, R. A; Bergquist, P. L., Appl. Environ. Microbiol., **1995**, *61*, 4403-4408.
20. Senior, D. J.; Mayers, P. R.; Saddler, J. N., Biotech. Bioeng., **1991**, *37*, 274-279.
21. Tenkanen, M.; Buchert, J.; Viikari, L., Enzyme and Microbial Technology, **1995**, *17*, 499-505.
22. Patel, B.; Morgan, H.; Weigel, J.; Daniel, R., **1987**, Arch. Microbiol., *147*, 21-24.
23. Torronen, A.; Rouvinen, J., **1995**, Biochemistry., *34*, 847-856.
24. Tomme, P.; Warren, R. A. J.; Gilkes, N. R., **1995**, Adv. Microb. Physiol., *37*, 1-81.
25 Lever, M., **1973**, Biochem. Med., *7*, 274-281.
26. Tobias, J. W.; Shrader, T. E.; Rocap, G.; Varchavsky, A., **1991**, Science, *254*, 1374-1377

Chapter 9

Structural and Functional Modifications of a Xylanase from *Streptomyces lividans* Belonging to Glycanase Family 10

C. Dupont, M. Roberge, R. Morosoli, F. Shareck, A. Moreau, and D. Kluepfel[1]

Centre de Recherche en Microbiologie Appliquée, Institut Armand-Frappier, Université du Québec, 531 boulevard des Prairies, Laval, Québec H7N 4Z3, Canada

Site-directed mutagenesis in concert with kinetic characterization was used to investigate the role and function of highly conserved amino acid residues among xylanases of family 10 using xylanase A from *Streptomyces lividans* (XlnA). As an example, mutants with modified catalytic properties, thermostability and pH profile were obtained. The three-dimensional structure of XlnA was used to rationalize the effects of the mutations.

Xylanases (1,4-β-D-xylan xylanohydrolase, EC 3.2.1.8) have raised considerable interest in the past decade, especially for their application in an environmentally sound bio- or pre-bleaching process in the pulp and paper industry. Xylanase could also be used for the conversion of hemicelluloses by the agricultural and food industries. These industrial possibilities generated numerous studies in search of better enzymes, and more recently genetic and protein engineering were used to tailor enzymes with improved properties.

Xylanase A of *Streptomyces lividans.*

Streptomyces lividans produces three different xylanases and the genes have been cloned and the corresponding enzymes characterized (*1*,*2*). Based on sequence homologies and hydrophobic cluster analysis, xylanases were originally classified into two families, 10 and 11 (*3*,*4*) [formerly families F and G (*5*)]. Xylanase A of *S. lividans* (XlnA) belongs to the

[1]Corresponding author

0097–6156/96/0655–0116$15.00/0
© 1996 American Chemical Society

family 10 while xylanases B and C were classified as part of family 11. An *in vitro* site-directed mutagenesis program was initiated after determination of DNA sequence homologies of xylanases from *S. lividans* and other xylanases (*1*). Mutagenesis was performed according to Kunkel (*6*) using the phagemid pIAF217 (*7*). With this phagemid, the *xlnA* gene is under the control of the p*lac* promotor which allowed sufficient amount of enzyme to be expressed in *Escherichia coli* for preliminary screening using crude periplasmic extracts (*7*).

Mutations were carried out on the structural gene of XlnA, targeting highly conserved amino acid residues identified from the known sequences of family 10 xylanases. These homologies are clustered in five areas of the xylanase molecule of this family, and with the exception of one, are all located on β strands of the enzyme (Figure 1). Two glutamic acid residues in xylanases of both family 10 and 11 were assigned as putative active site residues based on crystallographic data obtained from *Bacillus pumilus* xylanase (*8*). Replacement of these glutamic acids of xylanase A of *S. lividans* in positions 128 and 236 by glutamine residues abolished the enzymatic activity towards xylan and *p*-nitrophenyl-ß-1,4-cellobioside (*p*NPC) (*7*). Substitutions were made also for 23 conserved residues and for 22 residues frequently found in xylanases of family 10. The mutant proteins were screened for their biochemical properties.

Three-dimensional Structure of XlnA. In the past two years, three-dimensional structures from xylanases of both families have been solved (Table I). The structures of all four xylanases belonging to family 10 showed a similar fold consisting of an $(\alpha/\beta)_8$ barrel motif with their catalytic site invariably situated at the carboxy-terminal end of the β-barrel (Figure 2). In the case of XlnA produced by *S. lividans*, the barrel has a long and a short axis (17.5 and 9.5 Å) making it one of the most elliptical barrels known (*9*). The structure of the XlnA can explain some of the results obtained by site-directed mutagenesis of the enzyme.

Mutations Affecting Catalytic Activity. As shown in Table II, a series of mutants of the enzyme, H86K, S87A, Q88E, and N173D, screened in *E. coli*, showed an increased activity on xylan and *p*NPC. The mutants D50N and L83 were more active towards *p*NPC. H86W, F155Y, and R156K were more active toward birchwood xylan than the wild-type enzyme. Other mutants showed a marked decrease in activity on the two substrates (e.g. L167K, N170D, D171N, K181R, D200N, and H207E; results not shown). As expected, E128Q and E236Q representing replacement of the catalytic glutamic acid residues, were completely devoid of activity (Table II).

While xylanases from families 10 and 11 have very different overall structures, they use the same double displacement mechanism for the hydrolysis of the xylosidic bond. Xylanases hydrolyse β-1,4 bonds in an endo fashion with net retention of the configuration of the anomeric carbon. This mechanism involves two catalytic residues. One plays the role of an acid/base catalyst while the other is the nucleophile of the reaction which stabilizes the oxocarbonium intermediate (*17*). Before the reaction begins, the acid/base Glu has to be protonated while the nucleophile Glu is in an ionized form required for the double displacement mechanism to occur. Those two catalytic residues were proven by site-directed mutagenesis of E128 and E236 of XlnA (*7*). The three-dimensional structure of XlnA fully supports these assignments and shows the implication of other amino acids in the catalytic reaction (*9*, Figure 3).

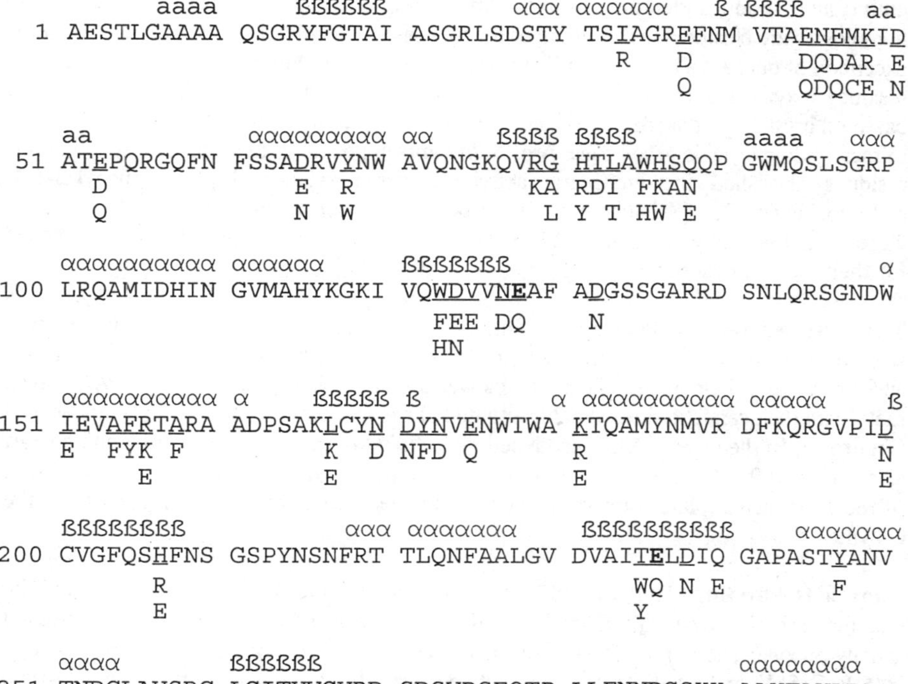

Figure 1. *Sequence of catalytic domain of XlnA. Amino acid residues mutated are underlined and the corresponding substitutions made are indicated below. Catalytic amino acids (E128 and E236) are in bold characters. α: indicates amino acid residues found in α-helix; β: indicates amino acid residues found in β strand; a: indicates amino acid found in small left-handed helix.*

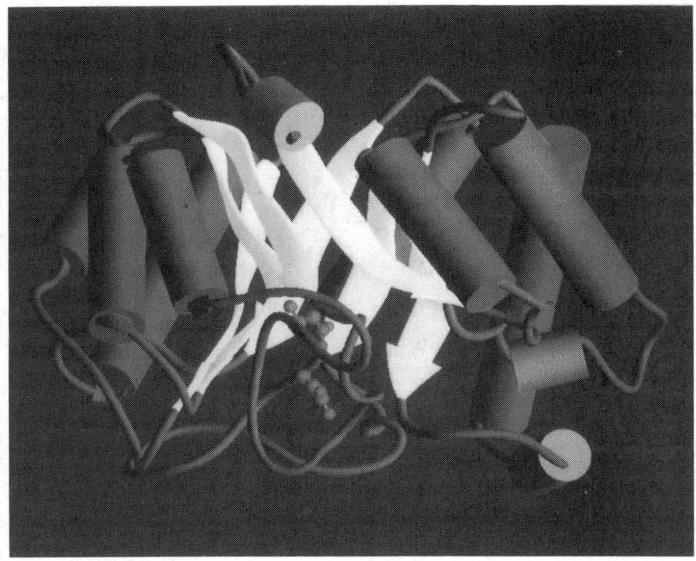

Figure 2. *A schematic representation of the three-dimensional structure of XlnA. The two catalytic residues (E128 and E236) are shown in ball and stick representation.*

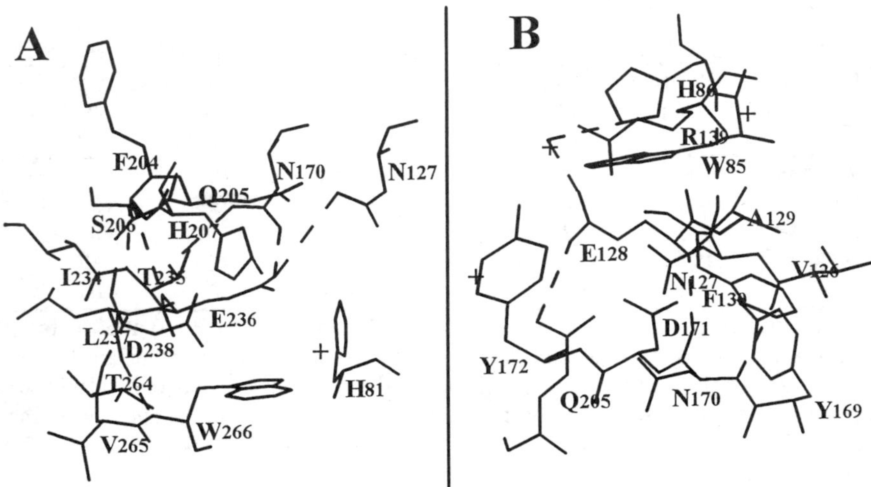

Figure 3. *Surrounding at 5Å radius of catalytic residues in Streptomyces lividans XlnA. Hydrogen bonds are shown as dashed lines (A) proposed nucleophile Glu236. (B) Proposed acid/base catalyst Glu128. (Reproduced with permission from ref. 10. Copyright 1996 Birkhäusser Verlag)*

Table I. Three-dimensional structure of xylanases

Xylanase		Mutant	Ligand	Resolution (Å)	Reference
FAMILY 10[a]					
Streptomyces lividans	XlnA	Wild type[b]		2.6	(*9*)
Cellulomonas fimi	Cex	Wild type[b]		1.8	(*11*)
Pseudomonas fluorescens	XynA	Wild type[b]		3.0	(*12*)
		E246C[b]	xylopentaose	2.5	(*12*)
Clostridium thermocellum	XynZ	Wild type[b]		1.4	(*13*)
FAMILY 11[a]					
Bacillus pumilus	Xyn	Wild type		2.2	(*8*)
Bacillus circulans	XynA	Wild type		1.6	(*14*)
		S100C		2.3	(*14*)
		E172C	xylotetraose	1.8	(*15*)
		S100C/N148C		1.6	1XNC[c]
Trichoderma harzianum	Xyn	Wild type		1.8	(*16*)
Trichoderma reesei	XynII	Wild type		1.8	(*17*)
	XynI	Wild type		2.0	(*17*)

[a]according to references (*3*) and (*4*);
[b]Catalytic domain
[c]Protein Data Bank (PDB) accession number

Table II. Relative activity of wild-type and some mutant xylanases

Enzyme	Relative activity[a] xylan (%)	Relative activity[a] *p*NPC (%)	Enzyme	Relative activity[a] xylan (%)	Relative activity[a] *p*NPC (%)
Wild type	100	100	Wild type	100	100
D50N	100	182	Q88E	149	161
L83I	103	132	E128Q	0	0
H86K	173	183	N173D	125	117
S87A	124	168	E236Q	0	0

[a]: Activity was determined with 1% xylan in 50 mM citrate buffer pH 6.0 at 60°C for 10 minutes using 400 μg protein from the periplasmic extract of the recombinant bacteria (mean of three assays). The amount of xylanase in the extract was evaluated by laser scan densitometry of Western blot autoradiograms, and the specific activity was then determined.

Examination of the local environment of each residue showed that the proposed acid/base catalyst is surrounded by hydrophobic residues (Y172, W85, Y169, F130), which could provide an hydrophobic environment responsible for the elevated pK_a of this residue (Figure 3B). On the other hand, the proposed nucleophile, E236 is located within a cluster

of charged and polar residues (N170, N127, Q205, E238) which contribute to a more hydrophilic environment. Furthermore, this residue accepts two hydrogen bonds to its carboxy side chain from H207 and N170 (Figure 3A). Therefore, there is little doubt that this residue is in an ionized state. Finally, the distance between the two catalytic side chain residues is 7.0 Å for XlnA, which is a distance consistent between two residues involved in a double displacement mechanisms with retention of the anomeric configuration of the leaving group (*19*).

Mutations Affecting the Mode of Cleavage. The mutants were also evaluated in regard to their bond-cleavage frequencies using tritiated xylopentaose as substrate. While the wild-type hydrolyses the second and third xylosyl linkages with the same frequency, the mutant N173D cleaved preferentially >90% at the second position (Figure 4). Furthermore, HPLC analysis of the degradation products showed that the transglycosylation activity of mutant N173D differed from that of the wild type enzyme (*20*). Kinetic studies indicated that residue N173 should be located at the substrate-reducing end in the catalytic site at a distance of three xylosyl residues of the cleavage point. When examining the three-dimensional structure of the uncomplexed XlnA, a distance of 11.7 and 10.0 Å can be observed between residue N173 and the two catalytic residues, E128 and E236. This distance is sufficient to accommodate more than two xylosyl residues, a fact which corroborates the location of the residue N173 in relation to the active site and supports a direct interaction of N173 with the aglycone part of the substrate during hydrolysis of xylopentaose yielding xylotriose and xylobiose. N173D and the wild-type enzyme differ also in their transglycosylation mechanism. During xylopentaose hydrolysis, the wild-type enzyme liberates appreciable quantities of xylooligosaccharides of a higher degree of polymerization than xylopentaose, whereas the mutant generates only very small quantities of these compounds. Since N173 could interact directly with the xylooligosacharide, in the presence of large amounts of substrate, xylopentaose will bind to the active site by interaction with N173 while the xylobiose carbonium intermediate is still in place in the active site which will lead to longer xylooligosaccharides. The mutant enzyme N173D will position the xylopentaose in such way that an aglycone of only two xylosyl residues can be accommodated. This arrangement prevents transglycosylation. Given the relative position of E236, E128 and N173, the three-dimensional structure of XlnA fully supports this hypothesis.

Mutations Affecting the Temperature Optimum and Thermal Stability of XlnA. Positions R156 and N173 of XlnA were investigated for their effect on the enzyme's thermostability (Table III). The mutation R156E however increased the optimal temperature by 5°C and the enzyme had a slightly higher half-life at 60°C. However mutant N173D had the same temperature optimum as the wild-type but showed a significant increase in half-life at 60°C. Finally, the double mutant R156E/N173D, while doubling the enzyme's half-life, showed a decreased catalytic activity by 53% (*21*).

These results illustrate the difficulties and ambiguities encountered in protein engineering when more than one property is targeted for improvement. However, from the three-dimensional structure of XlnA, it was possible to locate R156 at the C-terminal portion of the fourth α-helix of the enzyme with its side chain pointing toward the solvent. By its location, this residue would certainly contribute to the global surface charge of the protein and therefore, replacement by glutamic acid would modify this parameter. This could be

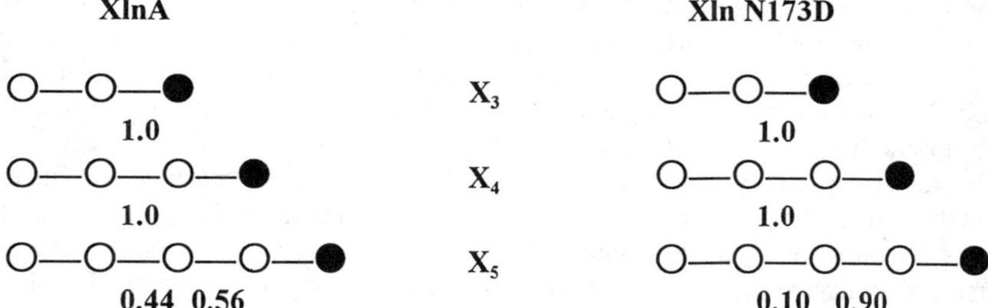

Figure 4. *Bond-cleavage frequency of XlnA and XlnN173D. The values indicate the bond-cleavage frequencies between D-xylose units linked by β-1,4 linkage. The black circle represents the reducing end of xylooligosaccharides labelled with ^3H (X_3: xylotriose; X_4: xylotetraose; X_5: xylopentaose). (Adapted from ref. 20.)*

one of the factors influencing the increased optimum temperature of R156E (Table III). However, no direct structural evidence can explain the increase in relative activity observed for the mutants R156K and R156E, since the residue R156 is very distant from the two catalytic residues. The possibility of a long range interaction would have to be verified by further kinetic and mutagenesis experiments. Finally, no obvious explanation can yet be given for the increased half-life at 60°C and relative activity observed for the mutation N173D.

Table III. Thermal activity and stability of wild-type and mutant xylanases

Enzyme	Relative activity[a] (%)	Optimum temperature[b] (°C)	Half-life[c] (min)
Wild type	100	65	110
R156E	110	70	116
R156K	150	65	110
N173D	125	60	150
R156E/N173D	78	65	220
R156K/N173D	53	65	110

[a] Activity was determined with 1% xylan in 50 mM citrate buffer pH 6.0 at 60°C for 10 minutes using 400 μg protein from the periplasmic extract of the recombinant bacteria (mean of three assays). The amount of xylanase in the extract was evaluated by laser scan densitometry of Western blot autoradiograms, and the specific activity was then determined.

[b] Optimum temperatures were determined using 1% xylan in 50 mM citrate buffer, pH 6.0, for 10 minutes at various temperatures.

[c] Thermostabilities were measured by incubating the enzyme preparations without substrate at 60°C. At given time, aliquots were withdrawn and the remaining activity was measured at 60°C with 1% xylan in 50 mM citrate buffer, pH 6.0 for 10 minutes.

Role of the Conserved Histidines. From the three-dimensional structure of XlnA it was possible to identify three histidines (H81, H86 and H207) which interact with the two catalytic residues (Figure 3). From the amino acid sequences, these histidines are highly conserved among the xylanases from family 10 (Figure 5). The same arrangement where the nucleophile is directly hydrogen-bonded to an histidine (Figure 3), is found in other three-dimensional structures of xylanases such as Cex from *C. fimi* or XynA from *P. fluorescens*. Results of site-directed mutagenesis of two of the three histidine are shown in Table IV.

Two substitutions were made, H86K and H86W. For these mutations, the gene was cloned from pIAF217 into pIAF18 and the mutant proteins expressed in *S. lividans*. This allowed overexpression and purification to homogeneity of the corresponding mutated enzymes (*2*). In both cases, the kinetic characterization showed a slight increase in specific activity. The only other noticeable difference was a three-fold increase in the K_M for the H86W mutant. However, compared to the wild-type, this mutant showed a major difference in enzyme stability as measured by its half-life values at 60°C. A drop of 90% was observed when measured in absence of substrate. On the other hand, when the experiment was conducted in presence of xylan, 60% of the wild-type value was retained. Thus comparing

```
                ▼    ▼                                     ▼
SlivA 121-GHTLAWHS..QQPGW-133 | 239-PIDCVGFQSHFNSG..SPYN-256
Acti2 118-GHTLAWHS..QQPGW-130 | 236-PIDCVGFQSHFNSG..SPYN-253
AawaA 108-GHTLVWHS..QLPSW-120 | 226-PIDGIGSQTHLSAGGGAGIS-245
BaccA 130-FHTLVWHS..QVPEW-142 | 263-PIDGVGHQSHIQI.GWPSI.-280
BsteA 121-FHTLVWHS..QVPQW-134 | 255-PIDGIGHQSHIQI.GWPSE.-272
BsteX  71-GHTLVWHN..QTDSW-83  | 202-PIHGIGMQAHWSL.NRPTLD-230
BovaI 106-GHCLIWHS..QLAPW-118 | 227-RIDAIGMQGHIGM.DYPKI.-244
BfibA 139-GHTLVWHS..QTPTW-151 | 273-VCAGVGMQSHLGTGFP....-288
BfibB  89-GHTLVWHN..QTPKW-101 | 217-LIDGMGMQSHLLMDHP....-232
CsacA  90-GHTFVWHN..QTPGW-102 | 213-PIDGIGIQAHWNIWDKNLV.-231
Csac4  42-GHVLVWHN..QTPEW-55  | 173-LIDGLGLQPTVGLNYPELDS-192
Cfimi 120-GHTLVWHS..QLPDW-132 | 237-PLDCVGFQSHLIVG...QVP-253
CsteB 120-FHTLVWHN..QTPTG-132 | 255-PIDGVGHQTHIDI.YNPPV.-272
CtheZ 595-GHTLIWHN..QNPSW-607 | 714-PIDGVGFQCHFINGMSPEYL-733
CtheX 285-GHTLLWHN..QVPDW-298 | 415-PISGIGMQMHINI..NSNI.-431
DtheA 107-GHTLVWHN..QTPGW-119 | 224-PIHGIGIQGHWTL.AWPT..-240
EnidC 105-GHTLVWHS..QLPSW-117 | 222-PIDGIGSQAHYSS.SHWSS.-239
ErumX 344-GHVLVWHS..QAPEW-356 | 496-RIDGFGMQGHYSV.NAPTVD-514
PchrP 112-GHTLVWHS..QLPSW-124 | 230-PIDGIGSQTHLGAGAGAAAS-249
PfluA 342-GHALVWHPSYQLPNW-356 | 470-PIDGVGFQMHVMNDYPS..I-487
PfluB 384-AHTFVWGA..QSPSW-395 | 492-YIDAVGLQAH...ELKGMTA-508
RflaA 712-GHTFVWYS..QTPDW-724 | 846-YIDGIGMQSHLATNYP....-861
TsacA 433-GHTLLWHN..QVPDW-445 | 563-PIDGIGMQMHINI..NSNI.-579
TaurX  78-GHTLVWWS..QLPPW-90  | 166-PIIGIGNQTARAAITVWGVA-185
TmarA 448-GHTLVWHN..QTPDW-460 | 570-LIDGIGMQCHISLA..TDI.-586
TneaA 444-GHTLVWHN..QTPEW-456 | 566-LIDGIGMQCHISLA..TDI.-582
TbacA 430-GHTLVWHQ..QTPSW-432 | 559-PVHGVGLQCHISL.DWPDV.-576
```

Figure 5. *Amino acid alignment of xylanases from family 10. The arrow indicates the highly conserved histidines shown in bold characters (corresponding to H81, H86 and H207 of S. lividans mature XlnA). Numbering are from the translational methionine of each enzyme. Sequence accession numbers from GenBank, SwissProt or NCBI are indicated in bracket with the following identification key: SlivA: Streptomyces lividans XlnA [P26514]; Acti2: Actimomadura sp. FC7 XilII [U08894]; AawaA: Aspergillus awamori IFO4308 XynA[P33559];BaccA:Bacillus sp. C-125 XynA [P07528]; BsteA: Bacillus stearothermophilus T-6 XynA [Z29080]; BsteX: Bacillus stearothermophilus 21 Xyn [D28121]; BovaI: Bacteroides ovatus V975 XylI [U04957]; BfibA: Butyrivibrio fibrisolvens 49 XynA [P23551]; BfibB: Butyrivibrio fibrisolvens H17C XynB [P26223]; CsacA: Caldocellum saccharolyticum XynA [P23556]; Csac4: Caldocellum saccharolyticum ORF4 [P23557]; Cfimi: Cellulomonas fimi Cex [P07986]; CsteB: Clostridium stercorarium F9 XynB [D12504]; CtheZ:Clostridium thermocellum XynZ [P10478]; CtheX: Clostridium thermocellum XynX [144776];DtheA: Dictyoglomus thermophilum Rt46B.1 [973983]; EnidC: Emericella (Aspergillus) nidulans XlnC [1050888];ErumX: Eubacterium ruminantium Xln [974180]; PchrP: Penicillium chrysogenum Q176 XylP [M98458]; PfluA: Pseudomonas fluorescens cellulosa XynA [P14768]; PfluB: Pseudomonas fluorescens cellulosa XynB [P23030]; RflaA: Ruminococcus flavefaciens XynA [P29126]; TsacA: Thermoanaerobacter saccharolyticum B6A-RI XynA [P36917]; TaurX: Thermoascus aurantiacus Xyn [P23360]; TmarA: Thermotoga maritima XynA [559960]; TneaA: Thermogota neapolitana XynA [603892]; TbacA: Thermophilic bacterium Rt8.B4A XynA [P40944]*

the effect of the substrate protection for each enzyme, a small gain in half-life of 21% was observed for the wild-type enzyme, whereas that for H86W increased by 525%. This result indicates that the presence of a bulky, hydrophobic residue in place of H86 is more detrimental to the stability of the enzyme if this residue is exposed to the solvent. On the other hand, the replacement of H86 by a lysine (H86K) had an opposite effect. In absence of substrate, this mutant showed a slight gain in stability while in presence of xylan, no improvement of thermostability but a decrease in the half-life by 27% was observed.

Table IV. Kinetic and biochemical parameters of wild-type and mutant xylanases

Enzyme	Specific activity[a] (IU/mg)	K_M (mg/mL)	k_{cat} (sec^{-1})	Temperature optimum[b] (°C)	pH optimum[b]	½ life (60°C) without[c] (min)	½ life (60°C) with[d] (min)
Wild type	176	0.22	170	65	6.5	140	170
H86K	205	0.23	158	65	6.5	166	120
H86W	205	0.60	197	60	6.5	16	100
H207R	1.3	---	---	60	5.5-6.5	11	90

[a]:Activity was determined with 1% xylan in 50 mM citrate buffer pH 6.0 at 60°C for 10 minutes using 300 ng of pure enzyme (mean of three assays).
[b]: Optimum temperatures and pH were determined using 1% xylan in 50 mM citrate buffer, pH 6.0, for 10 minutes at various temperatures and pH.
[c]: Thermostabilities without substrate were measured by incubating the enzyme preparations at 60°C. At given time, aliquots were withdrawn and the remaining activity was measured at 60°C with 1% xylan in 50 mM citrate buffer, pH 6.0 for 10 minutes.
[d]:Thermostabilities with substrate (7.2 mg/mL birchwood xylan) were measured by incubating the enzyme preparations at 60°C. At given time, aliquots were withdrawn and the amount of released reducing sugar determined.

H86 is hydrogen-bonded indirectly to E128, the acid/base catalyst, through a water molecule (Figure 3). This residue is located in the crevice of the $(\alpha/\beta)_8$ barrel and consequently exposed to the solvent. Replacement of this residue by tryptophan (H86W) would expose to the solvent a hydrophobic side chain creating repulsive interactions which could decrease stability in absence of the substrate. Moreover, a neighbouring tryptophan at position 85 would certainly create a steric hindrance, destabilising further the local structure. However, from the kinetic parameters K_M and k_{cat} it can be seen that H86W could accommodate the substrate. Only the stabilization of its ground state (indicated by the variation of K_M value) is affected by the replacement of H86 by tryptophan. Stabilization of the transition intermediate (indicated by the k_{cat} value) leading to hydrolysis of the polymeric substrate is not affected. In the case of replacement of H86 by lysine (H86K), the presence of a charged side chain in the active site did not affect the stability of the enzyme, but seems to provide a slight improvement. In conclusion, replacement of H86 by tryptophan and lysine did not affect the catalytic properties of the XlnA. However, it is still possible that in both cases, the indirect hydrogen bond with E128 through a water molecule is conserved and therefore, in order to evaluate the importance of this particular hydrogen bond, other mutations will be required.

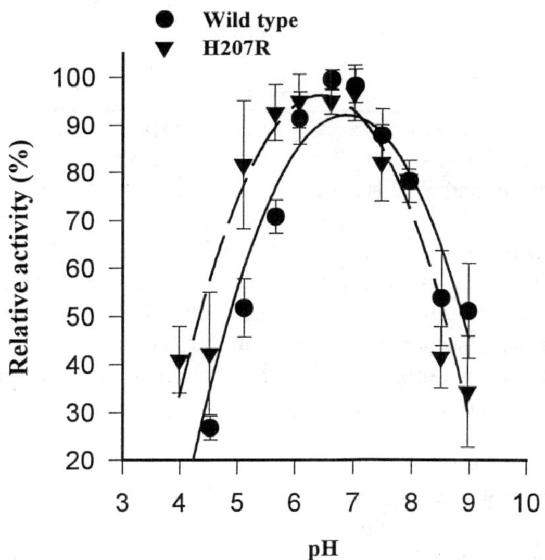

Figure 6. *pH profiles of wild-type XlnA and mutant H207R*

H207 interacts directly with the nucleophile E236 by a hydrogen bond through its Nє2 atom. Replacement of this residue by an arginine (H207R) led to drastic diminution of the catalytic capacity of the enzyme and loss of stability as shown by the decrease of half-life at 60°C (Table IV). Despite the low residual activity, it was possible to determine the pH profile of the mutant enzyme. As shown in Figure 6, modification of H207 by arginine produced an enzyme with a different pH profile than that of the wild-type. H207R retained activity at lower pH than the wild-type enzyme (e.g. 60% at pH 4.5 for H207R compared to 25% for wild-type). Based on the three-dimensional structure of XlnA, one could speculate that replacement of H207 with arginine disturbs the local surroundings of the nucleophile, which would account for the decrease in activity. R207 could still be hydrogen-bonded with E236 and the modified pH profile would be a consequence of the interaction between R207 and E236. Since arginine has a higher pK_a than histidine, it would retain its proton stronger at a lower pH than would a histidine residue. This allows the nucleophile E236 to maintain its ionized state at this lower pH. To verify this hypothesis, once again further mutations and analysis will be necessary. These results show the potential of modifying the pH profile of the enzyme by changing the local interactions with the nucleophile. Similar experiments are underway with the acid/base catalyst E128.

Concluding Remarks

Site-directed mutagenesis of highly conserved residues among xylanases from family 10 was used to probe and define the role of these residues in *S. lividans* XlnA. The utilization of a phagemid yielding expression of the *xlnA* gene into *E. coli* allowed rapid screening of the numerous mutants produced. Thus residues affecting either catalytic efficiency, hydrolysis pattern, thermostability and pH profile or many of these properties were identified. The three-dimensional structure of the catalytic domain of XlnA was then used to explain at the molecular level the changes observed upon the mutation introduced. The structure confirmed the role of E128 and E236 as catalytic residues of the enzyme. Residues interacting with these catalytic residues were further investigated. Among them, it was shown that H207 monitor the ionization state of the nucleophile E236 and modification of this residue will affect the pH profile of the enzyme.

Literature Cited

1. Shareck, F.; Roy, C.; Yaguchi, M.; Morosoli, R.; Kluepfel, D. *Gene*, **1991,** *107*, 75.
2. Kluepfel, D.; Shareck, F.; Senior, D.J.; Bernier, R.L.; Morosoli, R. In *Industrial microorganism: basic and applied molecular genetics*; Baltz, R.H.; Hegeman, G.D.; Skatrud, P.L., Eds.; American Society for Microbiology: Washington, 1993; pp.137-142.
3. Henrissat, B.; Bairoch, A. *Biochem . J.* **1993,** *293*, 781.
4. Henrissat, B.; Claeyssens, M.; Tomme, P.; Lemesle, L.; Mornon, J.P. *Gene* **1989,** *81*, 83.
5. Henrissat, B.; *Biochem. J.* **1991,** *280*, 309.
6. Kunkel, T., *Proc. Natl. Acad. Sci. USA.* **1985,** *74*, 488.
7. Moreau, A.; Roberge, M.; Manin, C.; Shareck, F.; Kluepfel, D.; Morosoli, R. *Biochem. J.* **1994,** *302*, 291.

8. Katsube, Y.; Hata, Y.; Yamaguchi, H.; Moriyama, H.; Shinmyo, A.; Okada, H. In *Protein engineering: Protein design in basic research, medecine, and industry*; Ikehara, M., Ed.; Japan Scientific Societies Press: Tokyo, 1990; pp 91-96.
9. Derewenda, U.; Swenson, L.; Green, R.; Wei, Y.; Morosoli, R.; Shareck, F.; Kluepfel, D.; Derewenda, Z. S. *J. Biol. Chem.* **1994,** *269,* 20811.
10. Dupont, C.; Kluepfel, D.; Morosoli, R. In *Lysozymes: Model enzymes in biochemistry and biology*; Jollès, P. Ed.; Birkhäusser Verlag: Basel, 1996; chapter 20; in press
11. White, A.; Withers, S.; Gilkes, N.R.; Rose, D.R. *Biochemistry* **1994,** *33,* 12546.
12. Harris, G.W.; Jenkins, J.A.; Connerton, I.; Cummings, N.; Leggio, L.L.; Scott, M.; Hazlewood, G.P.; Laurie, J.I.; Gilbert, H.J.; Pickersgill, R.W. *Structure* **1994,** *2,* 1107.
13. Dominguez, R.; Souchon, H.; Spinelli, S.; Dauter, Z.; Wilson, K.S.; Chauvaux, S.; Béguin, P.; Alzari, P.M. *Nature Struct. Biol.* **1995,** *2,* 569.
14. Campbell, R.L.; Rose, D.R.; Wakarchuk, W.W.; To, R.; Sung, W.; Yaguchi, M. In *Proceedings of the second TRICEL symposium on Trichoderma reesei cellulases and other hydrolases*; Suominen, P.; Reinikainen, T., Eds.; Foundation for biotechnical and industrial fermentation research: Helsinski, 1993, Vol. 8; pp 63-72.
15. Wakarchuk, W.W.; Campbell, R.L.; Sung, W.L.; Davoodi, J.; Yaguchi, M. *Protein Sci.* **1994,** *3,* 467.
16. Törrönen, A.; Harkki, A.; Rouvinen, J. *EMBO J.* **1994,** *13,* 2493.
17. Törrönen, A.; Rouvinen, J. *Biochemistry* **1995,** *34,* 847.
18. Sinnot, M. L. *Chem. Rev.* **1990,** *90,* 1171.
19. McCarter, J.D.; Withers, S.G. *Curr. Opin. Struct. Biol.* **1994,** *4,* 885.
20. Moreau, A.; Shareck, F.; Kluepfel, D.; Morosoli, R. *Eur. J. Biochem.* **1994,** *219,* 261.
21. Moreau, A.; Shareck, F.; Kluepfel, D.;Morosoli, R. *Enzyme Microb. Technol.*, **1994,** *16,* 420.

Lignin Degrading Enzymes

Chapter 10

Laccase-Producing White-Rot Fungus Lacking Lignin Peroxidase and Manganese Peroxidase

Role of Laccase in Lignin Biodegradation

Claudia Eggert, Ulrike Temp, and Karl-Erik L. Eriksson

Center for Biological Resources Recovery, Department of Biochemistry and Molecular Biology, Life Sciences Building, University of Georgia, Athens, GA 30602–7229

This paper presents an overview of phenoloxidases involved in lignin degradation with a particular focus on laccase. To investigate the role of laccase in lignin degradation, *Pycnoporus cinnabarinus* was identified as an excellent organisms for further study. It produces only one isoelectric form of laccase and yet degrades lignin fast and extensively. We report here on the purification and characterization of the *P. cinnabarinus* laccase. Laccases are generally considered to have a too low redox potential to attack non-phenolic lignin substrates. This is also true for the laccase from *P. cinnabarinus*. However, to make up for this deficiency, *P. cinnabarinus* was found to produce a redox mediator, 3-hydroxy anthranilic acid (3-HAA). Laccase plus the redox mediator, was demonstrated to degrade a non-phenolic lignin model dimer.

Lignins constitute the second most abundant group of biopolymers in the biosphere; thus, their biodegradation occupies an important position in the global carbon cycle. Studies of lignin biodegradation are also of great importance for possible biotechnological applications since lignin polymers are a major obstacle to the efficient utilization of lignocellulosic materials in a wide range of industrial processes *(1)*. The turnover of lignin biomass occurs primarily through the action of white-rot fungi, consequently, this ecological group has received a considerable amount of research attention, as have the enzymatic mechanisms employed by these fungi for the breakdown of lignin.

Most of our understanding of the enzymology of lignin biodegradation stems from studies of a single species of white-rot fungus, *Phanerochaete chrysosporium.* In *P. chrysosporium,* lignin degradation occurs only after secondary metabolism has been triggered by starvation for nitrogen, sulfur or phosphorus *(2)*. Prodigious production of extracellular phenoloxidases is a characteristic of this ligninolytic metabolism, and was noted even in very early studies of the differences between

0097–6156/96/0655–0130$15.25/0
© 1996 American Chemical Society

white-rot and other wood-rotting fungi *(3)*. Ligninolytic activity by *P. chrysosporium* is closely correlated with secretion of two peroxidases, lignin peroxidase (LiP) and manganese peroxidase (MnP) *(4, 5, 6)*. These phenoloxidases are thought to initiate attack on the lignin polymer by a single-electron oxidation step, either directly or through the action of low molecular weight mediators, followed by chemical bond rearrangements that occur during the resultant free radical reactions. Biochemically, this is a very unusual process, and no doubt the random nature of these reactions has contributed greatly to the difficulties encountered to date in elucidating the exact mechanisms underlying this important biological process.

Since their discovery, *P. chrysosporium* LiP and MnP have been extensively characterized with respect to their biochemistry. These enzymes, as well as similar enzymes found in many other white-rot fungi, generally occur as a multitude of isoenzymes encoded by multiple genes. The molecular genetics of these ligninolytic peroxidases have been widely studied, and to date, several cDNA and genomic clones of LiP, as well as MnP, have been cloned and sequenced *(7, 8, 9, 10)*. Additionally, the X-ray crystal structure of LiP has recently been elucidated *(11)*. These studies have, no doubt, lead to a better understanding of the regulation and structure of the lignin-degrading system produced by *P. chrysosporium.* Yet, with all these advances, it has proven surprisingly difficult to demonstrate extensive ligninolysis using isolated LiP or MnP *(12, 13)*. In fact, several investigators have reported polymerization, rather than depolymerization, of lignin preparations by LiP *in vitro (14, 15)*.

To make matters more confusing, an increasing number of studies have indicated that the value of *P. chrysosporium* as a model organism for lignin degradation might be limited for the majority of species within the ecological group of white-rot fungi. Widely distributed among white-rot fungi, is the production during ligninolytic metabolism of an additional extracellular phenoloxidase, i.e. laccase. Laccase (*p*-diphenol:0_2 oxidoreductase, EC 1.10.3.2) is a blue copper protein that contains four copper atoms per polypeptide chain and is capable of catalyzing the four-electron transfer reaction necessary to fully reduce dioxygen to water. Although by definition *p*-diphenols serve as electron-donors for laccases, these enzymes have a broad substrate specificity that also includes *ortho-* and *meta*-substituted polyphenols, methoxy-substituted phenols, aryl diamines, and a considerable variety of other natural and synthetic substrates *(16)*. Laccases have been identified in fungi, insects, higher plants, and very recently in bacteria *(16)*. In addition to their involvement in lignin degradation, fungal laccases appear to play important roles in several aspects of fungal metabolism, ranging from the polymerization of melanin precursors in differentiated cell walls (appressoria and rhizomorphs, for example) *(17, 18)* to the detoxification of antimicrobial phytoalexins produced by plants *(19, 20)*. A laccase has even been proposed to function as an alternative terminal oxidase for electron transport in *Podospora anserina (21)*. Many fungi produce multiple laccase isoforms under the appropriate inductive conditions; however, different laccase isoenzymes appear to be specific for particular metabolic, developmental or environmental situations.

Recent studies suggest that in white-rot fungi the combination of laccase with either LiP and/or MnP is a much more common combination of phenoloxidases than the LiP/MnP pattern found in *P. chrysosporium*. It was generally believed that *P. chrysosporium* lacks laccase. Only very recently the fungus was found to produce a laccase-like activity *(22)*. The authors were able to detect phenoloxidase activity when the fungus was grown on cellulose as the sole carbon source. However, unlike typical laccase-producing white-rot fungi, laccase production by *P. chrysosporium* seems to be repressed by glucose and the enzyme is overall secreted in relatively low amounts. Considering the many functions laccase is involved in, a more detailed characterization of the laccase-like activity seems to be required to draw firm conclusions about the eventual involvement of the *P. chrysosporium* laccase in ligninolytic functions. Indeed, the estimated 2000-odd species of white-rot fungi actually appear to be much more diverse in their patterns of ligninolytic phenoloxidases (laccases and peroxidases) than we have previously appreciated (Table I). Additionally, production of laccase in the company of either LiP or MnP has been shown to be sufficient for the complete degradation of lignin by those fungi producing this pattern of phenoloxidases *(23, 24, 25)*. Since *P. chrysosporium,* the model white-rot fungus of choice, was believed not to produce laccase, the role of laccase in lignin degradation has been less intensely studied and, thus, it is not known whether this enzyme is critical for those white-rot fungi that do produce it. Obviously, laccase is not a prerequisite for lignin degradation by all white-rot fungi, but by not producing a laccase under ligninolytic conditions, *P. chrysosporium* seems to be more the exception than the rule.

The possibilities for elucidating the mechanisms for lignin degradation through analysis of phenoloxidase-less mutants of white-rot fungi were appreciated very early on; however, most of such mutants identified to date have proven to be pleiotropic with respect to secondary metabolic functions *(26, 27)*. A study of phenoloxidase-less mutants by Ander and Eriksson *(28)* lent early support to the possibility that laccase could play an important role in lignin degradation. That study demonstrated that a strain of phenoloxidase-less (most likely a LiP^-/MnP^- double-mutant) *Sporotrichum pulverulentum* (anamorph of *P. chrysosporium)* had a greatly reduced ability to degrade lignin. However, lignin degradation was restored when cultures of the mutant were supplemented by addition of a laccase isolated from *Coriolus (Trametes) versicolor* or by production of phenoloxidase-positive revertants. Although *P. chrysosporium* does not produce laccase under ligninolytic conditions, it was concluded that phenoloxidases, specifically laccases, were critical for ligninolysis. Needless to say, that original study was carried out before the discovery of the two specific peroxidases, LiP and MnP, and at a time when laccase was the only type of extracellular phenoloxidase known to be produced by white-rot fungi. The fact that the study by Ander and Eriksson *(28)* is still one of the most cited concerning the importance of laccases for lignin degradation -- see, for example, a recent comprehensive review of fungal laccases by Thurston *(29)* -- only reflects the minimal effort expended to date in studying the enzymes produced by other species of white-rot fungi during lignin degradation.

Table I. Distribution of Ligninolytic Phenoloxidases in White-Rot Fungi

Organism	LiP	MnP	Lac	Reference
Coriolopsis occidentalis	+	+	?	Nerud et al. *(76)*
Phlebia brevispora	+	+	+	Rüttimann et al. *(24)*
Phlebia radiata	+	+	+	Vares et al. *(77)*
Pleurotus ostreatus	+	+	+	Waldner et al.*(78)*; Beckeret al. *(38)*
Pleurotus sapidus	+	+	+	Orth et al. *(79)*
Trametes gibbosa	+	+	+	Nerud et al. *(80)*
Trametes hirsutua	+	+	+	Nerud et al. *(80)*
Trametes versicolor	+	+	+	Waldner et al. *(78)*; Johansson et al. *(59)*; Jonsson et al. *(81)*
Phanerochaete chrysosporium	+	+	+[1]	see Eriksson et al. *(1)*
Perenniporia medulla-panis	-	+	-	Orth et al. *(79)*
Trametes cingulata	-	+	-	Orth et al. *(79)*
Phanerochaete sordida	-	+	-	Rüttimann-Johnson et al. *(82)*
Bjerkandera sp.	-	+	+	de Jong et al. *(25)*
Ceriporiopsis subvermispora	-	+	+	Rüttimann et al. *(76)*
Cyathus stercoreus	-	+	+	Orth et al. *(79)*
Daedaleopsis confragosa	-	+	+	de Jong et al. *(25)*
(*Coriolus pruinosum*)	+	+	?	Waldner et al. *(78)*
Dichomitus squalens	-	+	+	Perie & Gold *(83)*; Nerud et al. *(80)*; Orth et al. *(79)*
Ganoderma volesiocum	-	+	+	Nerud et al. *(80)*
Ganoderma colossum	-	+	+	Horvarth et al. *(84)*
Ganoderma lucidum	-	+	+	Orth et al. *(79)*
Grifola frondosa	-	+	+	Orth et al. *(79)*; Kimura et al. *(85)*
Lentinus (Lentinula) edodes	-	+	+	Leatham *(86)*
	-	+	+	Orth et al. *(79)*
Panus tigrinus	-	+	+	Maltseva et al. *(87)*
Pleurotus eryngii	-	?	+	Martinez et al. *(88)*
	+	+	-	Orth et al. *(79)*
Pleurotus pulmonarius	-	?	+	Masaphy et al. *(89)*
	+	+	-	Orth et al. *(79)*
Rigidoporus lignosus	-	+	+	Galliano et al. *(32)*
Stereum hirsutum	-	+	+	de Jong et al. (*25*)
Stereum spp.	-	+	+	de Jong et al. (*25*)
Trametes villosa	-	+	+	de Jong et al. (*25*)
Pycnoporus cinnabarinus	-	-	+	Nerud et al. (*80*); Eggert et al. (*67*)
Junghuhnia separabilima	+[2]	-	+	Vares et al. (*90*)
Phlebia tremellosa (*Merulius tremellosus*)	+	?	+	Vares et al. (*77*)
Bjerkandera adusta	+	-	-	Huoponen et al. (*91*)
(*Polyporus adustus*)	-	+[3]	+	de Jong et al. (*25*)
	+	+	?	Waldner et al. (*78*)
Coriolus consors	+	?	?	Kimura et al. (*85*)

[1]Srinivasan et al., *(22)* [2]novel heme peroxidase [3]Mn-inhibited peroxidase
[?]information not given

There can be no doubt that laccases can play an important role in lignin degradation by some, if not the majority of white-rot fungi. Sound biochemical support for this comes from a series of studies showing that laccase can take part in many of the reactions required for ligninolysis *(1, 13)*. Like MnP, laccase can catalyze the alkyl-phenyl and Cα-Cβ cleavages of phenolic dimers which are used as model lignin substructures *(30)*, and it can catalyze the demethoxylation of several lignin model compounds *(31)*. In an *in vitro* system, the laccase and MnP from *Rigidoporus lignosus,* a fungus that lacks LiP, acted synergistically to breakdown lignin *(32)*. Archibald and Roy *(33)* demonstrated that a system containing a purified laccase from *T. versicolor* and various monophenols could produce Mn(III) chelates from Mn(II) in the presence of a suitable organic acid. That Mn(III), produced through the actions of enzymes from ligninolytic cultures, can be chelated by certain organic acids so as to form complexes capable of oxidizing lignin has been shown *(34)*, but the oxidation of Mn(II) to Mn(III) had previously only been thought to be catalyzed by MnP *(35)*. Thus, in some ligninolytic systems, laccase appears to fulfill the role played by MnP in *P. chrysosporium* cultures, while in other systems it may act more like the *P. chrysosporium* LiP.

One troubling point in efforts to assign to laccase a substitute role for either LiP or MnP has been that the redox potentials of the laccases studied so far have not been thought sufficiently high to remove electrons from the non-phenolic aromatic substrates that must be oxidized during lignin degradation. In fact, LiP has so far been the only ligninolytic phenoloxidase found, all by itself, capable of oxidizing such non-phenolic compounds *in vitro.* However, Bao et al., *(36)* have recently shown that MnP, in the presence of unsaturated lipids, can also oxidize non-phenolic lignin structures. Kersten et al. *(37)* compared the abilities of LiP from *P. chrysosporium* and laccase from *T. versicolor* to oxidize members of a homologous series of methoxybenzenes with different redox potentials. The twelve methoxybenzene congeners had known half-wave potentials that differed by as much as ≈1 Volt. LiP oxidized the ten compounds with the lowest half wave potentials, whereas laccase oxidized only the one with the lowest potential, i.e. 1,2,4,5-tetramethoxybenzene. However, Bourbonnais and Paice *(38)* showed that two artificial laccase substrates, ABTS (2,2'-azinobis-[3-ethylbenzthiazoline-6-sulfonate]) and Remazol Blue, could act as redox-mediators which enabled laccase to oxidize non-phenolic lignin model compounds. The initial results of Bourbonnais and Paice *(39)* were extended when they later demonstrated the partial delignification and demethylation of kraft pulp by *T. versicolor* laccase in the presence of ABTS *(40)*. The biotechnological possibilities suggested by this finding are apparently beginning to be exploited as Call and Mücke *(41)* have recently reported that a laccase has been used in combination with an unidentified (proprietary) redox-mediator to delignify wood pulp in a pilot plant-scale process. These claims have significantly increased interest in the possible use of laccase for one step in the bleaching of wood pulp in pulp and paper processing. This suggests that laccases may be capable of catalyzing all of the necessary oxidative processes required for complete degradation of lignin *in vivo* if white-rot fungi also produce some compound as a physiological redox-mediator. Kawai et al. *(42)* reported that syringaldehyde might function as such a physiological redox-mediator

for laccase-catalyzed lignin degradation; however, these results have proven difficult to reproduce *(43)*. Thus, (at this point) the existence, not to mention the identity, of physiological redox-mediators produced by white-rot fungi *in vivo* remained to be demonstrated.

Although laccases, as a class of enzymes, oxidize a wide variety of substrates, their ability to oxidize compounds generally considered to be laccase substrates varies with the enzyme source *(44, 45)*. Some of this discrimination very likely arise from incompatibility between the redox potentials of the enzyme and the respective substrate *(46, 47)*. The redox potentials so far determined for plant and fungal laccases, as well as some other blue copper oxidases, vary considerably *(48, 49)* and with the accumulation of amino acid sequence data for comparison with other copper-containing redox enzymes, certain patterns in the structure-function relationships governing redox potential are beginning to emerge. For example, crystallographic analyses of azurins *(50, 51)*, plastocyanin *(52)*, and other small blue proteins harboring Type-I copper centers *(53)* have shown that the cupric ions are coordinated in a trigonal bipyramidal fashion *(54)*. The Cu^{2+} resides in a plane with strong His, His, and Cys ligands, and also has a weak axial interaction with the sulfur in a Met residue *(55)*. The remaining axial interaction appears to derive from the backbone carbonyl of a Gly residue. This general structure is conserved in the blue copper oxidases with the exception that the axial Met residue is replaced by Leu in most fungal laccases *(56)*. Karlsson et al. *(57)* used site-directed mutagenesis to convert the axial Met in *Pseudomonas* azurin to Leu and showed that the redox potential of the Type-1 copper increased by 70 mV. Messerschmidt and Huber *(56)* subsequently suggested that the Met→Leu conversion in the fungal Type-1 copper center might be responsible for the large difference in redox potentials between plant (+394 mV) and fungal (+780 mV) laccases *(48)*, even though no sequence information for plant laccases was available to say that they retained the axial Met residue. We expect that further studies of laccase gene sequences will help in developing a better understanding of the structure-function relationships that govern substrate specificities and functions of laccases in different biological systems.

Although, the studies described above offer strong support for the involvement of laccase in lignin degradation, there are certainly observations that suggest otherwise. In an elegant study with *Coriolus (Trametes) versicolor* (one of the same species used to demonstrate the significance of laccase for ligninolysis *(33, 39, 40)*, Evans *(58)* demonstrated that degradation of milled-wood lignin was unaffected when laccase activity was inhibited by addition of a laccase-specific antibody. That study did not, however, allow for a generalized conclusion since *T. versicolor* produces LiP, as well as MnP, in ligninolytic cultures *(59)*. The presence of one or the other of these enzymes might have easily masked any lost contribution attributable to laccase. Such an interpretation is supported to some extent by the fact that Evans *(58)* also observed partial depolymerization of the same milled-wood lignin by the purified *T. versicolor* laccase. There are numerous studies in which the question of polymerization/depolymerization of lignin *in vitro* by laccase has been investigated. The majority of these studies, performed mainly with the enzyme isolated from *T. versicolor*, show that laccase has a tendency to polymerize, rather than depolymerize, a variety of

lignin substrates *(60, 61, 1)*. However, such an activity does not eliminate the possibility that laccases contribute to lignin breakdown, particularly since purified LiP and MnP each show a similar tendency to also polymerize the same substrates *in vitro*. It should be noted that this points to the likelihood that accessory enzymes, such as cellobiose dehydrogenases, may be required in ligninolytic enzyme systems to prevent repolymerization of phenoxyradicals created by phenoloxidases *(62)*.

We interpret the foregoing evidence to strongly support an important role for laccases in the degradation of lignin by at least some species of white-rot fungi. This being the case, there have been surprisingly few detailed studies of the biochemical properties of laccases purified from ligninolytic cultures of different white-rot fungi, particularly with respect to their redox potentials and the possible existence of redox-mediators for the degradation of non-phenolic lignin components. In contrast, the laccases which function to polymerize aromatic precursors in fungal spore walls have been examined in fine detail using biochemical and molecular biological methods *(62, 64, 65, 66)*. Several gene sequences for laccases from white-rot fungi have been placed in GenBank. However, none of these species lack LiP under ligninolytic conditions. If we are to gain a full understanding of how white-rot fungi employ laccases in the degradation of lignin, it is imperative that we examine a species in which laccase is the predominant phenoloxidase produced in ligninolytic culture from which LiPs are absent.

Through an extensive screening program to identify white-rot fungi that produce large amounts of laccase, we have found that *Pycnoporus cinnabarinus* PB, an isolate obtained from decaying pine wood in Queensland, Australia, seems to be an ideal candidate for in-depth studies to examine the significance of laccase in the degradation of lignin. This fungus combines several attributes that make it an attractive model for basic studies of laccase.

Crucial for the choice of *P. cinnabarinus* as a model organism is the unique combination of phenoloxidases produced by this fungus under ligninolytic conditions. *P. cinnabarinus* produces only one isoelectric form of laccase. In addition, an as yet unidentified phenoloxidase, a peroxidase whose substrate specificities do not fit with those of either LiP or MnP, is secreted into the medium.

Purification and Characterization of the Laccase from *Pycnoporus cinnabarinus*. (for experimental details see *(67)*)

A spore suspension of *P. cinnabarinus* was used for inoculation of the medium (for culture conditions see Eggert et al., *(67)*). Extracellular laccase activity in shake flasks was detectable after 48 hours and reached a maximum of 1.2 U/ml on day 7 in a medium with a C:N ratio of about 15 (3 g of glucose per liter; 2.4 mM $(NH_4)_2\ SO_4$) (Figure 1A). Although the laccase production in the cultures was only to a lesser extent repressed by high nitrogen levels, the laccase/peroxidase activity ratio varied with the nitrogen concentration ranging between 48 (2.4 mM $(NH_4)_2\ SO_4$ and 90 (24 mM $(NH_4)_2\ SO_4$). No LiP activity was detectable under conditions known to stimulate production of that enzyme, such as N-starvation or addition of veratryl alcohol. Moreover, the peroxidase activity could not be ascribed to a MnP type

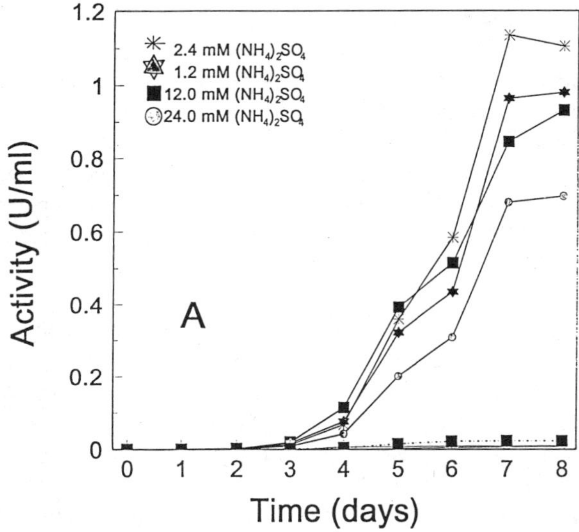

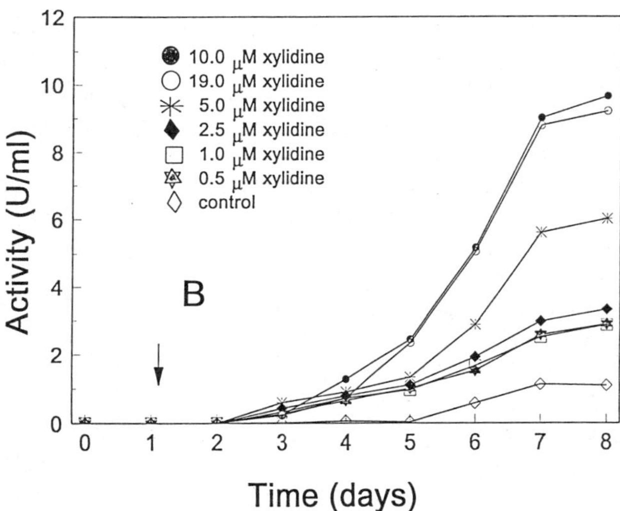

Figure 1: A) Influence of nitrogen concentration on extracellular laccase (solid lines) and peroxidase (dashed lines) production by *P. cinnabarinus* grown in shake flask cultures (5 g of glucose per liter as carbon source). (B) Stimulation of laccase production in cultures of *P. cinnabarinus* [2,4 mM $(NH_4)_2SO_4$; 5 g of glucose per liter] by addition of various amounts of 2,5-xylidine. The arrow indicates the time of 2,5-xylidine addition. Values represent the averages from three independent experiments, with a sample mean deviation of less than ± 7% of the reported values.
(Reproduced with permission of reference 67. Copyright 1996 American Society for Microbiology.)

enzyme either. The observed peroxidase activity may therefore emanate from a new type of peroxidase produced by a white-rot fungus.

To enhance laccase production, several possible laccase inducers were tested. 2,5-Xylidine, a known laccase inducer, *(68)* showed the most pronounced effect with optimal production reached at concentrations between 10 and 19 μM. When added after 24 hours to shaken cultures (Figure 1B) this amount of laccase inducer enhanced laccase activity about 9 fold, i.e., to approximately 9.6 U/ml, compared to the basal level obtained in 2.4 mM ammonium sulfate cultures without inducer.

To produce enough laccase for characterization of the enzyme as well as for its use in pulp bleaching experiments, production was scaled up to a 100 liter fermentor. Fermentor cultivation in the presence of 2,5-xylidine increased laccase activity 2 fold or to about 18 U/ml compared with shake flask xylidine-cultures. Also, maximum enzyme activity was reached on day 5, three days earlier than in shake flasks, when laccase constituted 70% of the total extracellular protein. The enzyme was purified to homogeneity using three steps, i.e. ion-exchange and gel-filtration. With this purification scheme the enzyme yield was about 47% and the purification factor 36 *(67)*.

P. cinnabarinus laccase is a single polypeptide with a molecular mass of approximately 76,500 Da as determined by SDS-PAGE (Figure 2) or 81,000 Da if calibrated gel filtration chromatography was used. This molecular mass is comparable to other fungal laccases *(1)*. Isoelectric focusing showed one single isoform of laccase with an isoelectric point of 3.7 *(67)*.

To study the nature of the catalytic center of the laccase, the enzyme was characterized spectroscopically. A typical laccase spectrum was obtained both in the UV/visible region as well as with EPR technique. The results confirmed the presence of 4 Cu(II) ions typical for an intact laccase active center.

The enzyme was found to be glycosylated containing about 9% of total carbohydrate. Mannose was found to be the predominant glycosyl residue (> 70%). Typical fungal laccases contain about 8-20% carbohydrate of mostly unknown composition *(29)*.

Temperature stability of the *P. cinnabarinus* laccase is shown in Figure 3 which demonstrates high stability below 50°C. At 70°C, the half-life of the enzyme was about 60 minutes, whereas at 80°C, the laccase was completely inhibited in 40 minutes. The pH activity profile for the *P. cinnabarinus* laccase is presented in Figure 4. Optimal activity is at pH 4 which is well within the range found for other fungal laccases isolated from ligninolytic cultures.

The substrate specificity of *P. cinnabarinus* laccase was compared with the laccase from *Coriolus hirsutus*, producing both LiP and MnP, to find out if a laccase from a white-rot fungus such as *P. cinnabarinus,* lacking an enzyme with the high redox potential of LiP, might secrete a laccase with different catalytic properties. However, comparison of the two laccases did not indicate any major differences in this respect. Also, the sensitivity of *P. cinnabarinus* laccase towards several putative laccase inhibitors was very similar to that seen with the laccase from *C. hirsutus*.

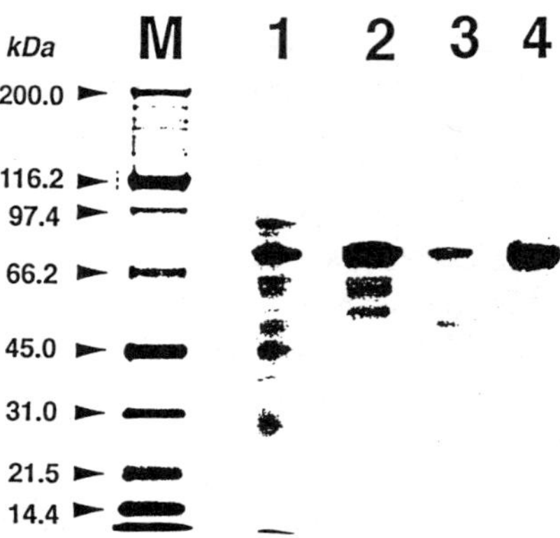

Figure 2: Electrophoresis of *P. cinnabarinus* laccase after different purification steps. Protein was stained with Coomassie Brilliant Blue R-250. SDS-PAGE (10% T; 0.1% SDS). M: protein marker, 1: concentrated culture supernatant (track loaded with approx. 25 μg protein), 2: DEAE-M column eluate (25 μg), 3: Butyl-Toyopearl column eluate (10 μg), 4: S-400 gel filtration eluate (25 μg). (Reproduced with permission of reference 67. Copyright 1996 American Society for Microbiology.)

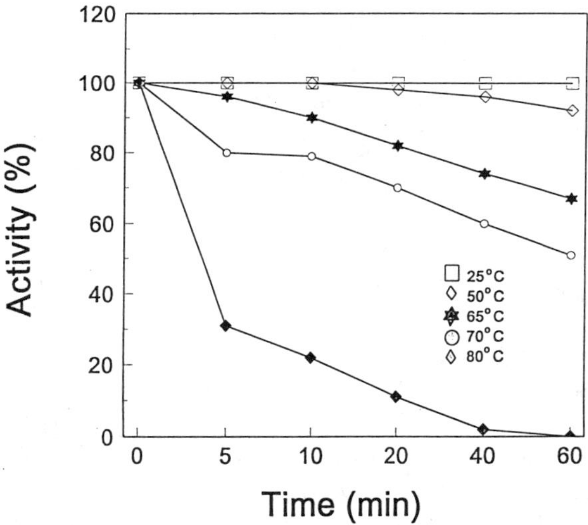

Figure 3: Activity of purified *P. cinnabarinus* laccase after preincubation at different times and temperatures. One hundred percent activity refers to 24 U/ml, using 5 mM guaiacol (50 mM sodium tartrate, pH 4.0) as the substrates. (Reproduced with permission of reference 67. Copyright 1996 American Society for Microbiology.)

Both enzymes were completely inhibited by 0.1 mM Na-azide, the most effective inhibitor.

N-terminal amino acids sequence analysis of the purified laccase showed only a single polypeptide sequence (Table II). We have obtained a gene fragment which encodes a protein matching the N-terminal protein sequence data and have recently isolated the corresponding full-length laccase cDNA clone.

Comparison of the N-terminal protein sequence of the *P. cinnabarinus* laccase with those of other fungal laccases showed the closest similarity to laccase II from *T. versicolor* (86%). About 76% similarity was found with the laccase from *C. hirsutus*, the basidiomycete PM1, and laccase IIIc from *T. versicolor*, 70% similarity to *C. subvermispora* laccase, and 64% similarity to *Phlebia radiata* laccase. All of these fungi are basidiomycetes white-rot fungi. In contrast, the N-terminal sequences of laccases isolated from non-wood-rotting fungi such as *Neurospora crassa*, the yeast-like fungus *Cryptococcus neoformans* were significantly different, with similarities of 18 and 0%, respectively.

In summary, *P. cinnabarinus* has been found to produce an unusual set of extracellular phenoloxidases consisting of a single isoform of laccase and a peroxidase that is neither LiP nor MnP. *P. cinnabarinus* represents a common type of white-rot fungi, devoid of LiP but in possession of laccase in combination with a peroxidase of lower redox potential. Although the fungus appears to lack the enzymes having the high oxidation potential thought to be necessary for the depolymerization of the non-phenolic structures of lignin, it very efficiently degrades lignin in wood *(14)*. Furthermore, the lack of LiP does not seem to require compensation by a laccase having an unusually high redox potential. Our comparison of the laccases from *P. cinnabarinus* and *C. hirsutus*, a fungus that also secretes LiP and MnP, did not show any significant differences in their oxidation capacity for various substrates. We have taken these results as support for the possible presence of a physiological enzyme-mediator system, allowing for the oxidation of non-phenolic lignin structures by laccase *(39, 69)*.

In Search for a Physiological Redox-Mediator.
(for experimental details see ref. *(70, 71)*)

When *P. cinnabarinus* was grown at 30°C in shake flask cultures with glucose (3g/l) as the sole carbon source, over time a color change in the initially colorless medium was noted. After 3 days, the culture medium turned pale yellow and later dark red. The cell-free culture solution showed an increased absorbance with time, around 450 nm. This red pigment and laccase activity were found to develop and accumulate in parallel in liquid culture solutions of the fungus. The adsorption maximum at 450 nm, for the red color was in good agreement with the spectrum reported previously for cinnabarinic acid (CA), *(72, 73)*. Since the *o*-aminophenol, 3-hydroxyantranilic acid (3-HAA), had been demonstrated to be a precursor for CA in other biological systems, excretion of this kynurenine metabolite in *P. cinnabarinus* cultures was measured. TLC analysis of the culture samples revealed a peak having a retention time identical with authentic 3-HAA. After 3-4 days of cultivation, 3-HAA was no longer

Table II: Comparison of the N-Terminal Amino Acid Sequence of *P. cinnabarinus* laccase with other fungal laccases

Microorganism		N-Terminal Amino Acid Sequence
P. cinnabarinus		A I G P V A D L T L T N A A V S P D G F S
T. (Coriolus) versicolor[a]	I	A I G P V A S L V V A N A P V S P D G F L
	II	G I G P V A D L T I T D A A V S P D G F S
	III	G I G P V A D L T I T D A E V S P D G L S
Basidiomycete PM 1[b]		S I G P V A D L T I S N G A V S P
C. hirsutus[c]		A I G P T A D L T I S N A E V S P D G F A
C. subvermispora[d]		A I G P V T D L E I T D A F V S P D G P
P. radiata[e]		S I G P V T D F H I V N A A V S P
A. bisporus[f]		D T - K T F N F D L V N T R L A -
N. crassa[g]		G G G G G C N S P T N R Q C W S P
C. neoformans[h]		X K T D E S P E A V S D N Y M P K

a) Bourbonnais et al. *(74)*; b) Coll et al. *(92)*; c) Kojima et al. *(93)*; d) Fukushima and Kirk *(94)*; e) Saloheimo et al. *(95)*; f) Perry et al. *(96)*; g) Germann et al. *(97)*; h) Williamson *(98)*.

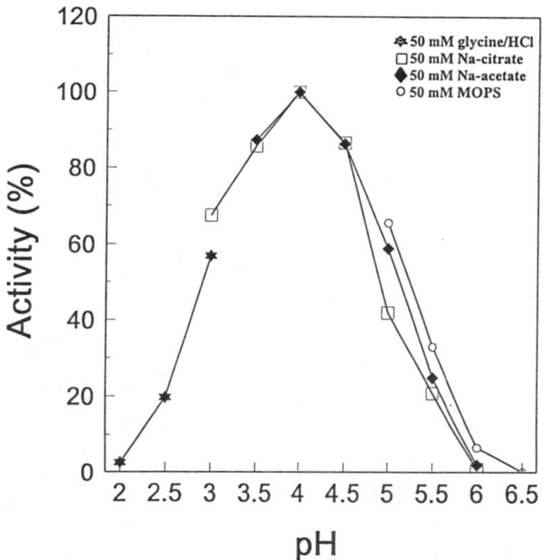

Figure 4: pH optimum curve of purified *P. cinnabarinus* laccase. One hundred percent activity refers to 24 U/ml, using 5 mM guaiacol (50 mM sodium tartrate, pH 4.0) as the substrate. (Reproduced with permission of reference 67. Copyright 1996 American Society for Microbiology.)

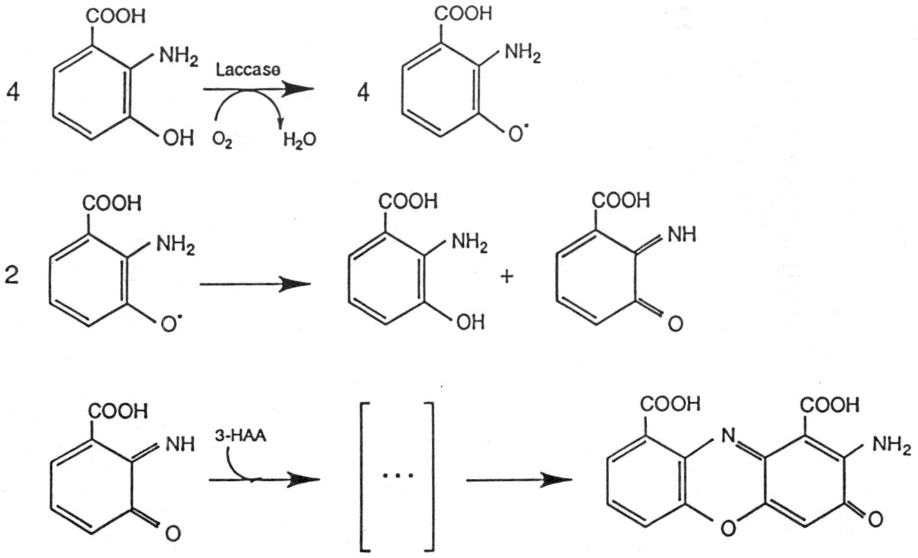

Figure 5: Hypothetical reaction mechanism for laccase mediated conversion of 3-hydroxyanthranilate into cinnabarinic acid. (Reprinted from reference 70. Copyright 1995, with kind permission from Elsevier Science - NL, Sara Burgerhartstraat 25, 1055 KV Amsterdam, The Netherlands)

detectable, most likely due to conversion of 3-HAA to CA. *In vitro* studies with purified laccase clearly demonstrated the ability of laccase to catalyze the formation of CA by oxidative coupling of 3-HAA. The K_m value obtained for 3-HAA was in the same range as other common laccase substrates such as guaiacol. A hypothetical reaction mechanism for laccase mediated conversion of 3-HAA into cinnabarinic acid is presented in Figure 5.

The earlier mentioned finding by Bourbonnais and Paice, *(39, 74)* that the synthetic laccase substrate, ABTS can act as a redox mediator enabling laccase to indirectly oxidize non-phenolic lignin model compounds and also to partially delignify wood pulp, was suggested to result from the interaction of highly reactive intermediates with the model compound. There was a potential for a similar reaction to proceed from the 3-HAA oxidation product.

Synthetic substrates representing lignin substructures are commonly used to assess ligninolytic activities *(1)*. Five such model compounds (Table III, I-V), representing both phenolic and non-phenolic structures commonly found in lignin, were used to test the oxidative ability of the *P. cinnabarinus* laccase and to compare this activity with that of LiP. The results verified the high redox potential of LiP but did not indicate a similarly high redox potential for the *P. cinnabarinus* laccase (Table III). Typical for fungal laccases only the lignin model compounds I-IV, all containing phenolic hydroxyl groups were oxidized by the *P. cinnabarinus* enzyme while the non-phenolic lignin model dimer, veratryl glycerol-β-guaicyl ether (α-carbonyl form) (V), was not. This clearly suggested that *P. cinnabarinus,* to mineralize lignin using only laccase, must use an indirect mode of action for the attack of non-phenolic portions of the polymer. To test the hypothesis that 3-HAA was that missing link, i.e., a physiologically produced redox mediator, lignin model compound V was incubated with the *P. cinnabarinus* laccase and 3-HAA for 48 hours. After this time 58% of the model compound V was oxidized. The decrease in compound V coincided with the formation of 2 peaks, identified as veratric acid (VA) and guaiacol (G) from the cleavage of compound V (Figure 6A). A major peak corresponding to cinnabarinic acid (CA) as well as several minor peaks representing oxidation products of 3-HAA were also resolved by HPLC. These minor peaks were also observed when laccase was incubated with 3-HAA in the absence of compound V (Figure 6B). In the absence of laccase, 3-HAA and the lignin dimer did not react for at least 72 hours (Figure 6C).

To assess whether the ratio of mediator to model compound was important for optimizing degradation of compound V with laccase varying amounts of 3-HAA were examined. An unexpectedly sharp optimal concentration ratio of about 15:1 was determined for this mediator/model compound pair (Figure 7). The mechanisms underlying the laccase/mediator reactions are, as yet, unclear. The oxidation products of 3-HAA show striking similarities to the synthetic mediators recently developed for pulping processes. However, unlike the fungal mediator, 3-HAA, the synthetic mediators are heterocyclic compounds belonging to the general classes of phenoxazinones, phenoxythiazones *(75)* or phenoxybenzothiazoles *(69)*. However, not all phenoxazinones can act as redox mediators. We could thus show that cinnabarinic acid could not serve in this function. We therefore anticipate that

Table III: Oxidation of five dimeric lignin model compounds by lignin peroxidase, laccase, and laccase plus redox mediator, 3-hydroxyanthranilate. Structures of lignin model compounds: *I:* 1-(4-hydroxy-3-methoxyphenyl)-2-(4-hydroxy-3,5-dimethoxyphenyl)-1,3-propanediol; *II:* 1,2-bis(4-hydroxy-3-methoxyphenyl)-1,3-propanediol; *III:* guaiacyl glycerol-β-guaiacyl ether (α-alcoholic form); *IV:* guaiacyl glycerol-β-guaiacyl-ether (α-carbonyl form); *V:* veratryl glycerol-β-guaiacyl ether (α-carbonyl form); +: oxidation*; -: no oxidation; n.d.: not determined.

	I 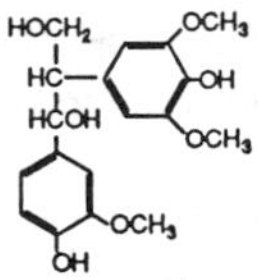	II 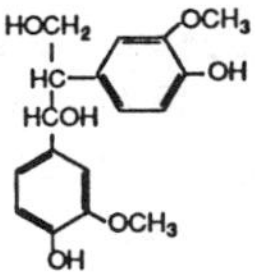	III 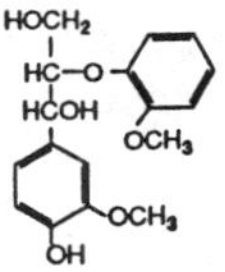	IV 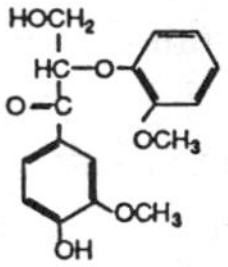	V 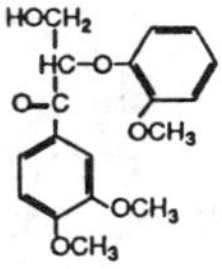
LiP*	++++	+++	+++	+++	++
Laccase*	++++	+++	++	+	-
Laccase + 3-HAA†	n.d.	n.d.	n.d.	n.d.	+†

*Oxidations were monitored photospectroscopically at 30°C in 50 mM sodium tartrate buffer with 3 μg *P. cinnabarinus* laccase (pH 4.0) and 2.5 μg *P. chrysosporium* lignin peroxidase (LiP) (pH 3.5), respectively. For LiP reactions hydrogen peroxide was included at a final concentration of 100 μM. To these reaction mixtures, each model compound was added to a final concentration of 125 μM.

†Reaction products were analyzed after 48h incubation at 25°C by reversed phase chromatography.

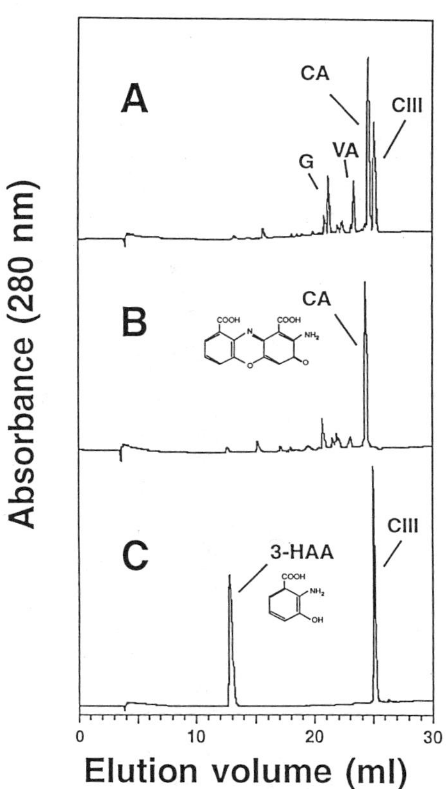

Figure 6: HPLC-analysis of degradation products obtained by oxidation of model compound V (CV) in the presence of laccase and 3-hydroxyanthranilate (3-HAA). (**A**) Oxidation and cleavage of compound V by laccase plus 3-HAA. Two peaks corresponding to guaiacol (G) and veratric acid (VA) are produced. (**B**) Oxidation of 3-HAA by laccase; (**C**) in a mixture, 3-HAA and lignin model compound V are stable for 72 h. (Reprinted from reference 71, Copyright 1995, with kind permission from Elsevier Science - NL, Sara Burgerhartstraat 25, 1055 KV Amsterdam, The Netherlands)

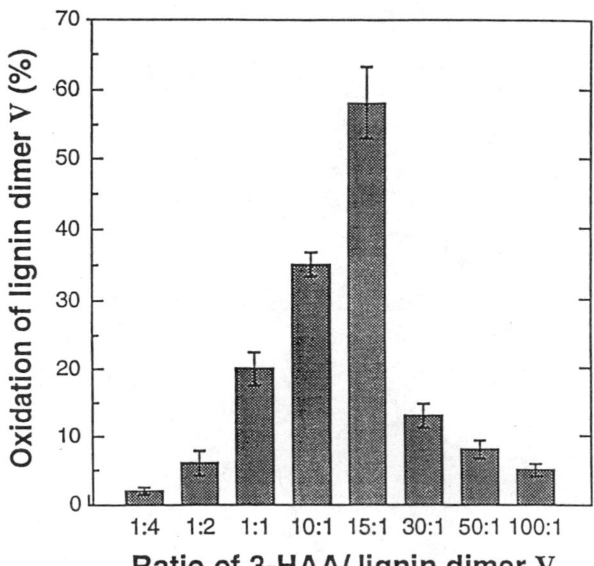

Figure 7: Effect of various ratios of laccase and 3-hydroxyanthranilate on oxidation of model compound V. Concentrations of mediator and model compound V were varied between 15 μM and 250 μM for 3-HAA and 125 μM to 1.5 mM for 3-HAA final concentrations in 1 ml of 50 mM 50 mM sodium tartrate (pH 4.0). Reaction products were analyzed by HPLC after 48 h incubation at 25 °C *(71)*. Error bars show sample mean deviations for at least three independent experiments.

studies of naturally occurring mediators will help to uncover compounds that serve more efficiently as redox mediators in biotechnological processes such as delignification of wood pulp.

Literature Cited

1. Eriksson, K.-E. L.; Blanchette, R. A.; Ander, P.; *Microbial and Enzymatic Degradation of Wood and Wood Components;* Springer Verlag: Berlin, **1990**; pp 416.
2. Keyser, P.; Kirk, T. K.; Zeikus, J. G. *Bacteriol.* **1978**, *135*, pp. 790-797.
3. Bavendamm, W. *Z. Pflanzenkr. Pflanzenschutz* **1928**, *38*, pp 257-276.
4. Tien, M.; Kirk, T. K. *Science* **1983**, *221,* pp 661-663.
5. Glenn, J. K.; Morgan, M. A.; Mayfield, M. B.; Kuwahara, M.; Gold, M. H. *Biochem. Biophys. Res. Comm.* **1983**, *114*, pp 1077-1083.
6. Kuwahara, M. J.; Glenn, M. K.; Morgan, M. A.; Gold, M. H. *FEBS Lett.* **1984**, *169*, pp 247-250.
7. Tien, M.; Tu, C.-P. D. *Nature (London)* **1987**, *326*, pp 520-523.
8. Brown, J. A.; Sims, P. F. G.; Raeder, U.; Broda, P. *Gene* **1988**, *73*, pp 77-85.
9. Pribnow, D.; Mayfield, M. B.; Nipper, V. J.; Brown, J. A.; Gold, M. H. *J. Biol. Chem.* **1989**, *264,* pp 5036-5040.
10. Gold, M. H.; Alic, M. *Microbiol. Rev.* **1993**, *57(3*), pp 605-622.
11. Edwards, L.; Raag, R.;Wariishi, H.; Gold, M. H.; Poulos, T. L. *Proc. Natl. Acad. Sci. USA* **1993**, *90*, pp 750-754.
12. Hammel, K. E.; Moen, M. A. *Enzyme Microb. Technol.* **1991**, *13*, pp15-18.
13. Higuchi, T. *Wood Sci Technol.* **1990**, *24*, pp 23-63.
14. Hatakka, A.; Uusi-Rauva, A. K. *Eur. J. Appl. Microbiol. Biotechnol.* **1983**, *17*, pp 235-242.
15. Odier, E.; Mozuch, M. D.; Kalyanaraman, B.; Kirk, T. K. *Biochimie* **1988**, *70*, pp 847-852.
16. Dean, J. F. D.; Eriksson, K.-E. L. *Holzforschung* **1994**, *48*, pp 21-33.
17. Worrell, J. J.; Chet, I.; Hüttermann, A. *J. Gen. Microbiol.* **1986**, *117*, pp 327-338.
18. Rehmann, A. U.; Thurston, C. F. *J. Gen. Microbiol.* **1992**, *138*, pp 1251-1257.
19. Bar-Nun, N.; Mayer, A. M. *Phytochemistry* **1989**, *28*, pp 1369-1371.
20. Choi, G. H.; Larson, T. G.; Nuss, D. L. *Molec. Plant-Microbe Interact.* **1992** *5*, pp 119-128.
21. Frese, D.; Stahl, U. *Mech. Aging Develop.* **1992**, *65*, pp 277-288.
22. Srinivasan, C.; D'Souza, T. M.; Boominathan, K.; Reddy, C. A. *Appl. Environ. Microbiol* **1995**, *61*, pp 4274-4277.
23. Sarkanen, S.; Razal, R. A.; Piccariello, T.; Yamamoto, E.; Lewis, N. G. *J. Biol. Chem.* **1991,** *266*, pp 3636-3643.
24. Rüttimann, C.; Schwember, E.; Salas, D.; Cullen, D.; Vicuna, R. *Biotechnol. Appl. Biochem.* **1992**, *16*, pp 64-76.
25. de Jong, E.; de Vries, F. P.; Field, J. A.; Zwan, R. P. v. d.; deBont, J. A. M. *Mycol. Res.* **1992**, *96*, pp 1098-1104

26. Gold, M. H.; Mayfield, M. B.; Cheng, T. M.; Krisnangkura, K.; Shimada, M.; Enoki, A.; Glenn, J. K. *Arch. Microbiol.* **1982** *132*, pp 115-122.
27. Liwicki, R.; Paterson, A.; MacDonald, M. J.; Broda, P. *J. Bacteriol.* **1985** *2*, pp 641-644.
28. Ander, P.; Eriksson, K.-E. *Arch. Microbiol.* **1976**, *109*, pp 1-8.
29. Thurston, C. F. *Microbiology* **1994**, *140*, pp 19-16.
30. Ishihara, T. In *Lignin degradation: Microbiology, Chemistry And Potential Applications*; Kirk, T. K., Higuchi, T., Chang H.-m., Eds.; CRC Press, Boca Raton, FL, 1980; pp17-31.
31. Morohoshi, N. In *Enzymes in Biomass Conversion*; Leatham, G. F., Himmel, M. H., Eds.; ACS Symposium Series: Washington, D.C., 1991, Vol. 460, pp.207-224.
32. Galliano, H.; Gas, G.; Seris, J. L.; Boudet, A. M. *Enzyme Microb. Technol.* **1991**, *13*, pp 478-482.
33. Archibald, F.; Roy, B. *Appl. Environ. Microbiol.* **1992**, *58*, pp 1496-1499.
34. Tuor, U.; Wariishi, H.; Schoemaker, H. E.; Gold, M. H. *Biochemistry* **1992**, *31*, pp 498-4995.
35. Wariishi, H.; Dunford, H. B.; MacDonald, I. D.; Gold, M. H. *J. Biol. Chem.* **1989**, *264*, pp 3335-3340.
36. Bao, W. L.; Iukushima, Y.; Jensen, K. A.; Moen, M.A.; Hammel, K.E. *FEBS Lett.* **1994**, *354*, pp 297-300.
37. Kersten, P. J.; Kalyanaraman, B.; Hammel, K. E.; Reinhammar, B.; Kirk, T. K. *Biochem. J.* **1990**, *268*, pp 475-480.
38. Becker, H. G.; Sinitsyn, A. P. *Biotechnol. Lett.* **1993**, *15(3)*, pp 289-294.
39. Bourbonnais, R.; Paice, M. G. *FEBS Lett.* **1990**, *267*, pp 99-102.
40. Bourbonnais, R.; Paice, M. G. *Appl. Microbiol. Biotechnol.* **1992**, *36*, pp 823-827.
41. Call, H. P.; Mücke, I. In *TAPPI Pulping Conference Proceedings, San Diego, CA;* TAPPI Conference Proceedings: San Diego, CA, 1994, pp 83.
42. Kawai, U. S. T.; Higuchi T. *Wood Res.* **1989**, *70,* pp 10-16.
43. Muheim, A.; Fiechter, A.; Harvey, P. H.; Schoemaker, H. E. *Holzforschung* **1992**, *46*, pp 121-126.
44. DeVries, O. M. H.; Kooistra, W. H. C. F.; Wessels, G. H. *J. Gen. Microbiol.* **1986**, *132*, pp 2817-2826.
45. Sterjiades, R.; Dean, J. F. D.; Gamble, G.; Himmelsbach, D. S.; Eriksson, K.-E. L. *Planta* **1993**, *190*, pp 75-87.
46. Reinhammar, B. *Biochim. Biophys. Acta* **1972**, *275*, pp 245-259.
47. Reinhammar, B.; Malmström, B. G. In *Copper Proteins (Metal Ions in Biology);* Spiro, T. G., Ed.; John Wiley & Son: New York, NY, 1981; pp 109-149.
48. Reinhammar, B. In *Copper Proteins and Copper Enzymes* Lontie, R., Ed.; CRC Press, Inc., **1984**; pp 1-35.
49. Xu, F.; Shin, W.; Brown, S. H.; Wahleithner, J. A.; Sundaram, U.M.; Solomon, E.I. *Biochim. Biophys. Acta* **1996**, *1292,* pp 303-311.
50. Baker, E. N. *J. Mol. Biol.* **1988**, *203*, pp 1071-1095.

51. Nar, H.; Messerschmidt, A.; Huber, R.; van de Kamp, M.; Canters, G.W. *J. Mol. Biol.* **1991b**, *221*, pp 765-772.
52. Guss, J. M.; Freeman, H. C. *J. Mol. Biol.* **1983**, *169*, pp 521-563.
53. Adman, E. T.; Turley, S.; Bramson, R.; Petratos, K.; Banner, D.; Tsernoglou, D.; Beppu, T.; Watanabe, H. *J. Bol. Chem.* **1989,** *264*, pp 87-99
54. Nar, H.; Messerschmidt, A.; Huber, R.; van de Kamp, M.; Canters, G.W. *J. Mol. Biol.* **1991a**, *218*, pp 427-447.
55. Solomon, E. I.; Baldwin, M. J.; Lowery, M. D. *Chem. Rev.* **1992**, *92*, pp 521-542.
56. Messerschmidt, A.; Huber, R. *Eur. J. Biochem.* **1990**, *187*, pp 341-352.
57. Karlsson, B. G.; Aasa, R.; Malmström, B.; Lundberg, L. G. *FEBS Lett.* **1989**, *253*, pp 99-102.
58. Evans, C. S. *FEMS Microbiol. Lett.* **1985**, *27*, pp 339-343.
59. Johansson, T.; Welinder, K. G.; Nyman, P. O. *Arch. Biochem Biophys.* **1993**, *300*, pp 57-62.
60. Kaplan, D. L. *Phytochemistry* **1979**, *18*, pp. 1917-1919.
61. Leonowicz, A.; Szklarz, G.; Wojtas-Wasilewska, M. *Phytochemistry* **1985**, *24*, pp 393-396.
62. Eriksson, K.-E. L.; Habu, N.; Samejima, M. *Enzyme Microb. Technol.* **1993**, *15*, pp 1002-1008.
63. Clutterbuck, A. J. *J. Gen. Microbiol.* **1972**, *70*, pp 423-435.
64. Aramayo, R.; Adams, T. H.; Timberlake, W. E. *Genetics* **1989**, *122*, pp.65-71.
65. Aramayo, R.; Timberlake, W. E. *Nucleic Acid Res.* **1990**, *18*, pp 3415.
66. Tanaka, C.; Tajima, S.; Furusawa, I.; Tsuda, M. *Mycol. Res.* **1992**, *96*, pp 959-964.
67. Eggert, C.; Temp, U.; Eriksson, K.-E.L. *Appl. Environ. Microbiol.* **1996,** 62, pp 1151-1158.
68. Bollag, J. M.; Leonowicz, A. *Appl. Environ. Microbiol.* **1984**, 48, pp 849-854.
69. Mücke, M. (December **1994**) World patent application WO 94/29510.
70. Eggert, C.; Temp, U.; Dean, J. F. D.; Eriksson, K.-E. L. *FEBS Lett.* **1995**, *376*, pp 202-206.
71. Eggert, C.; Temp, U.; Dean, J. F. D.; Eriksson, K.-E. L. *FEBS Lett.* 1995, In Press
72. Christen, S.; Southwell-Keely, P.; Stocker, R. *Biochemistry* **1992**, *31*, pp 8090-8097.
73. Toussaint, O.; Lerch, K. *Biochemistry* **1987**, *26*, pp 8568-8571.
74. Bourbonnais, R.; Paice, M. G.; Reid, I. A.; Lanthier, P.; Yaguchi., M. *Appl. Envion. Microbiol.* **1995**, *61*, pp 1876-1880.
75. Novo Nordisk A/S (January **1995**) World patent application WO 95/01426.
76. Nerud, F.; Misurcova, Z. *Biotechnol. Lett.* **1989**, *11*, pp 427-432.
77. Vares, T.; Niemenmaa, O.; Hatakka, A. *Environ. Microbiol.* **1994**, *60*, pp 569-575.
78. Waldner, R.; Leisola, M. S. A.; Fiechter, A. *Appl. Microbiol. Biotechnol.* **1988**, *29*, pp 400-407.

79. Orth, A. B.; Royse, D. J.; Tien, M. *Appl. Environ. Microbiol.* **1993**, *59*, pp 4017-4023.
80. Nerud, F.; Zouchova, Z.; Misurcova, Z. *Biotechnol. Lett.* **1991**, *13(9)*, pp 657-660.
81. Jonsson, M.; Pettersson, E.; Reinhammar, B. *Acta Chem. Scand.*, **1968**, *22*, p 2135.
82. Rüttimann-Johnson, C.; Cullen, D.; Lamar, R. *Appl. Environ. Microbiol.* **1994**, *60(2)*, pp 599-605.
83. Perie, F. H.; Gold, M. H. *Appl. Environ. Microbiol.* **1991**, *57*, pp 2240-2245.
84. Horvarth, E. M.; Srebotnik, E.; Messner, K. In *FEMS Symposium: Lignin Biodegradation and Transformation;* Duarte, J. C., Ferreira, M. C., Ander, P, Eds.; Forbitec Editions: Lisboa, Portugal, 1993, pp 163-164.
85. Kimura, Y.; Asada, Y.; Kuwahara, M. *Appl. Microbiol. Biotechnol.* **1990**, *32*, pp 436-443.
86. Leatham, G. F. *Appl. Microbiol. Biotechnol.* **1986**, *24*, pp 51-58.
87. Maltseva, O. V.; Niku-Paavola, M.-L.; Leontievsky, A. A.; Myasoedova, N. M.; Golovleva, L. A. *Biotechnol. Appl. Biochem.* **1991**, *13*, pp 291-302.
88. Martinez, M. J.; Munoz, C.; Guillen, F.; Martinez, A. T. *Appl. Microbiol. Biotechnol.*, **1994**, *41*, pp 500-504.
89. Masaphy, S.; Levanon, D. *Appl. Microbiol. Biotechnol.* **1992**, *36*, pp 828-832.
90. Vares, T.; Lundell, T.; Hatakka, A. I. *FEMS Microbiol. Lett.* **1992**, *99*, pp 53-58.
91. Huoponen, K.; Ollikka, P.; Kalin, M.; Walther, I.; Mantsala, P.; Reiser, J. *Gene* **1990**, *89*, pp145-150.
92. Coll, P. M.; Fernandez-Abalos, J. M.; Villanueva, J. R.; Santamaria, R.; Perez, P. *Appl. Environ. Microbiol.*, **1993**, *59*, pp 2607-2613.
93. Kojima, Y.; Tskuda, Y.; Kawai, Y.; Tsukamoto, A.; Sugiura, J.; Sakaino, M.; Kita, Y. *J. Bacteriol.* **1990**, *265*, pp 15224-15230.
94. Fukushima, Y.; Kirk, T. K. *Appl. Environ. Microbiol.* **1995**, *61*, pp 872-876.
95. Saloheimo, M.; Niku-Paavola, M.-L.; Knowles, J. K. C. *J. Gen. Microbiol.* **1991**, *137*, pp 1537-1544.
96. Perry, C. R.; Matcham, S. E.; Wood, D. A.; Thurston, D. F. *J. Gen. Microbiol,* **1993**, *139*, pp 171-178.
97. Germann, U. A.; Müller, G.; Hunziker, P. E.; Lerch, K. *J. Biol. Chem.* **1988**, *263*, pp 885-896
98. Williamson, P.R. *J. Bacteriol.* **1994**, *176*, pp 656-664.

Chapter 11

Enzymology of Kraft Pulp Bleaching by *Trametes versicolor*

M. G. Paice[1], F. S. Archibald[1], R. Bourbonnais[1], L. Jurasek[1], I. D. Reid[1], T. Charles[2], and T. Dumonceaux[2]

[1]Pulp and Paper Research Institute of Canada, 570 boulevard Saint Jean, Pointe-Claire, Québec H9R 3J9, Canada
[2]Department of Natural Resource Sciences, Macdonald Campus, McGill University, Sainte Anne de Bellevue, Québec H9X 1C0, Canada

Trametes versicolor and other white-rot fungi can lower the residual lignin content (kappa number) and increase brightness of kraft pulps without diminishing the pulps' strength or yield. Experiments with ^{14}C-labelled lignin indicate that the residual lignin is solubilized but not extensively mineralized by *T. versicolor*. Laccase, manganese peroxidase (MnP) and cellobiose dehydrogenase (CDH) are produced by the fungus during bleaching, but based on molecular modelling studies, they are unable to freely access residual lignin in the kraft pulp fibre wall. Two isozymes of laccase were purified and compared: both enzymes required a mediator such as ABTS for pulp delignification, and under optimum conditions could produce up to 55% lignin removal. However, several lines of evidence indicate that MnP is the key enzyme required for fungal bleaching. CDH has several potential roles in delignification, including generation of complexing agents and Mn(II) for MnP.

Kraft pulps are brown, and must be bleached to acquire the high and stable brightness desired in fine writing and printing papers. Bleaching is achieved by removing the residual lignin from the pulp. Traditionally this was done with Cl_2, but the production of organochlorine by-products has made chlorine bleaching an unpopular technology. Alternative bleaching chemicals, such as oxygen, ozone, and peroxide, are being adopted, but they are more expensive, or more likely to damage the strength of the pulp, than chlorine was. Biological delignification with fungi and their enzymes is an alternative approach which has shown promise, but has not yet attracted as much development effort as have the bleaching chemicals.

Fungal Bleaching

The lignin-degrading fungi are called white rots because of their characteristic bleaching effect as they decay wood. The first attempt to bleach kraft pulps with these fungi was

0097–6156/96/0655–0151$15.00/0
© 1996 American Chemical Society

made by Kirk and Yang with *Phanerochaete chrysosporium* (*1*). That fungus lowered the lignin content (kappa number) of the pulp, but also attacked the cellulose and significantly decreased the pulp strength. Subsequently, other white-rot fungi have been found to more selectively delignify kraft pulps. Paice et al. (*2*) found that *Trametes versicolor* markedly increased the brightness of dilute suspensions of hardwood kraft pulp, and lowered their lignin content. Japanese scientists have extensively screened for fungi that bleach kraft pulps; Nishida isolated an unidentified strain, called IZU-154, that delignifies and brightens kraft pulps under solid-state fermentation conditions (*3-6*). Kondo has selected an isolate, YK-624, of *Phanerochaete sordida* which is also very effective (*7-9*).

The highest brightness that has been achieved by fungal treatment is 80%, after a two-stage treatment of hardwood pulp with YK-624 (*9*). Usually the pulp brightness after fungal treatment is 50-60%, but it can be increased to 80-90% by post-treatment with chlorine dioxide (*2*) or peroxide (*5,10*). Fungal treatment often measurably decreases the viscosity and zero-span tensile strength of pulps (*2,11*), indicating that some cellulose depolymerization is taking place. Whether this attack on cellulose is hydrolysis by low levels of cellulase, or oxidation by radical by-products of lignin degradation, has not yet been clarified. In any case, the fungal treatment seems to improve inter-fiber bonding, since the sheet strength properties of the fungally delignified pulps are equal to or better than those of chemically bleached pulps (*2,5,6,11*).

Because the residual lignin is immobilized inside the pulp fiber walls, its removal must involve diffusible agents released by the fungal hyphae. Experiments in which the fungus was separated from the pulp by membranes have confirmed that delignification does not require direct contact between fungus and fiber (*7,12*). Extensive bleaching does require the presence of living fungus, however. Apparently one or more components of the delignifying system are labile or consumed, and need to be replenished constantly. One obvious candidate for this factor is H_2O_2, but just supplying H_2O_2 is not sufficient to replace the fungus. As shown in Figure 1, a filtrate from a bleaching culture of *T. versicolor* supplemented with glucose and glucose oxidase to provide H_2O_2 could demethylate kraft pulp almost as extensively as and faster than an intact culture. The methanol is released from methoxyl groups on lignin aromatic rings with a free phenolic hydroxyl by manganese peroxidase (*13*); its appearance indicates that conditions were favourable for MnP activity. But the brightening effect of the culture fluid was short-lived and much less than that of the whole culture. This shows that some other factor, not yet identified, may be required for optimal activity of the *Trametes* bleaching system. The enzymes laccase and manganese peroxidase, found in pulp bleaching cultures, can effectively delignify pulp under conditions optimized for their action (see below). Their measured activities in the cultures, however, do not account for the delignification and brightening produced by the fungus. This suggests that other enzymes contribute to pulp bleaching.

The residual lignin in kraft pulps is heavily modified during pulping; the alkyl-aryl ethers characteristic of lignin are depleted and diphenylmethane linkages that do not occur in native lignins are formed (for a review see Ref.*14*). Little is known about the chemical changes that white-rot fungi produce to remove this modified lignin from the

fiber walls. Demethylation of the phenolic rings occurs early in pulp delignification by *T. versicolor*, and seems to lead to an increase in the alkali extractability of the demethylated lignin (*15*). We have prepared kraft pulps containing ^{14}C-labelled lignin. Some of the lignin carbon is mineralized, but more of it is solubilized by the fungus (*16*). Detailed comparison of the effects of laccase and manganese peroxidase and intact fungal cultures on this labelled pulp should provide more clues to the nature of the enzymes involved in pulp delignification.

Molecular Architecture of Kraft Pulp Fibres

Enzymatic bleaching of kraft fibres take place in the complex three-dimensional environment of the fibre wall. Therefore it is not only the chemical reactivity of lignin and other components of the wall that determine the course of the biological bleaching, but also the configuration of the cell wall components. The three-dimensional structure of the cell wall is difficult to study, and as a result our knowledge is rather limited. In order to gain a better understanding of the structure, we have applied molecular modelling to construct a representative sample of the wall.

A model of a small piece of the secondary wall structure was produced using the method of coarse lignin model building developed previously (*17*). This structure was then subjected to simulated kraft pulping as described by Archibald et al. (*14*). The simulation was designed to provide a simultaneous partial removal of lignin and hemicellulose. The residual lignin was much richer in condensed structures than the native lignin, due to the preferential breakage of non-condensed bonds during the simulated kraft pulping. Thus a model of unbleached pulp fibre structure was obtained (Figure 2A). Although the removal of some fibre components leads to an increased porosity, the diffusion of relatively large redox enzyme molecules would still be strongly hindered. Fortunately neither manganese peroxidase nor laccase has to contact residual lignin in the kraft fibre directly, but rather can effect the delignification via low molecular weight intermediates such as, for example, ABTS (with laccase) or complexed manganese (III) ions (with manganese peroxidase). The sizes of these mediators are such that their diffusion deep into the kraft fibre walls appears feasible. These mediators can then either reduce the size of the residual lignin, and thus accelerate leaching, or achieve a similar effect by increasing solubility through oxidation or other modifications. The sizes of the residual lignin molecules, as shown in Figure 2A, appear to be such that only a small change in size would be required to allow their leaching from the cell wall.

The present molecular model has imperfections but none of them appear insurmountable. Our current focus is on the refinement of the atomic coordinates in order to obtain a model that could be subjected to further study using computational chemistry methods. An example of the refinement is shown in Figure 2, where a small part of the coarse model of lignin (Figure 2A) has been translated into atomic coordinates (Figure 2B). Part of a lignin macromolecule is shown in the vicinity of manganese peroxidase (18) together with hydrogen peroxide and manganic oxalate, both of which are required for the enzyme function.

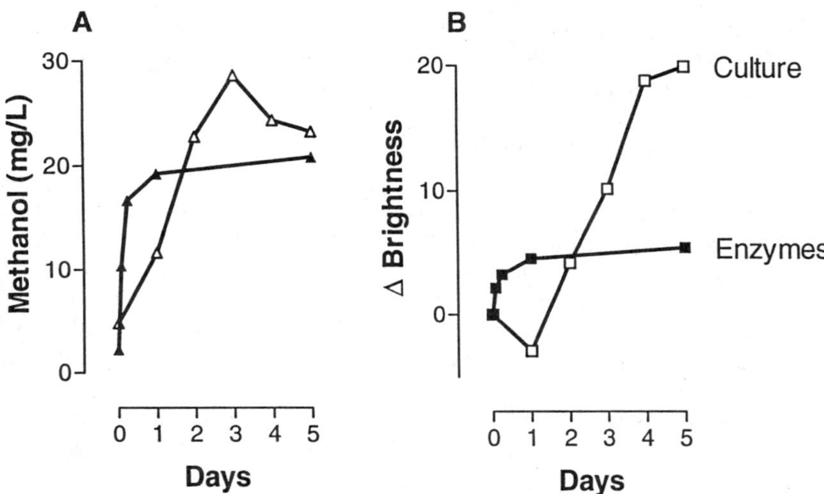

Figure 1. A. Release of methanol and B. brightening of hardwood kraft pulp by a culture of *T. versicolor* (△, □) and by culture fluid supplemented with glucose oxidase and glucose (▲, ■). Redrawn from Paice et al. (*13*).

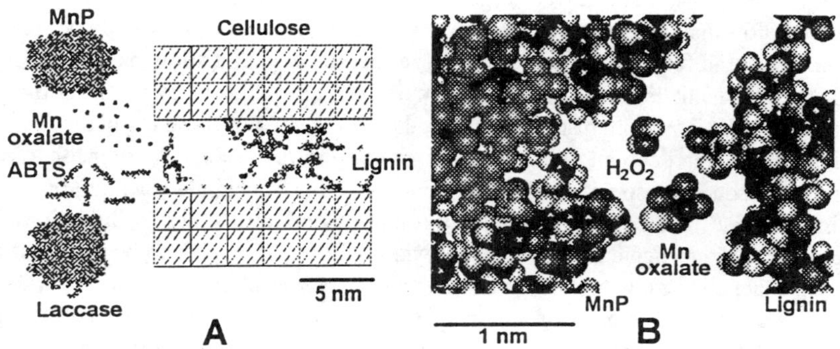

Figure 2. A coarse molecular model (A) of a segment of the secondary wall of an unbleached kraft fibre is shown together with two redox enzyme molecules and their mediators. A refined model (B) shows a detail of manganese peroxidase structure, adjacent to a part of a lignin molecule produced by refinement of the coarse model. The manganese peroxidase molecular model is that of Sundaramoorthy *et al.* (*18*).

Lignin Oxidation and Pulp Delignification by Laccase and Mediators

Laccase, a four-copper phenoloxidase (*19*), is produced abundantly by *T. versicolor* during kraft pulp bleaching (*2*). Laccase performs one-electron oxidation of many aromatic substrates including polyphenols, methoxy-substituted monophenols, and aromatic amines (*20*). In lignin, only phenolic subunits are attacked by laccase, giving oxygen-centered radicals that can subsequently polymerize or depolymerize (*21-23*). However, we have shown that the substrate range of laccase can be extended to non-phenolic subunits of lignin by inclusion of a mediator such as 2,2'-azinobis-(3-ethylbenzthiazoline-6-sulfonate) (ABTS) (*24*). This enzyme-mediator system was also shown to effectively demethylate and delignify kraft pulp (*25*). Demethylation of most free phenolic methoxyl groups in lignin was shown to precede the delignification reaction (*26*). It seems likely that the oxidized ABTS must function as a diffusible electron carrier, because laccase is a large molecule (molecular weight around 70,000) and therefore cannot enter the secondary wall to contact the lignin substrate directly (Figure 2A). Recently, over 50% delignification of kraft pulps has been reported with laccase and another mediator, 1-hydroxybenzotriazole (*27*), or by repeated treatment with laccase/ABTS followed by alkaline extraction (*28*). These observations have provoked considerable interest in enzyme-catalyzed oxidative bleaching of kraft pulps.

Two laccase isozymes were purified from a 2,5-xylidine induced culture of *T. versicolor* (*29*). Laccase I and II are both glycoproteins with MWs around 70,000. Their N-terminal amino acid sequences indicate that laccase I and II are sufficiently different to likely originate from different genes. The laccase II sequence was found to be highly similar to the published sequence of laccase III-c from *Coriolus versicolor* (*30*), which was reported earlier to depolymerize certain lignin preparations (*31*). Laccase I and II were tested alone and in the presence of mediators, for their capacity to delignify kraft pulps. Neither laccase I nor II was able to delignify kraft pulp without the presence of a mediator, whereas both enzymes were equally effective in delignifying and demethylating pulp when they were coupled to ABTS. The reactivity of the two laccase isozymes towards various monomeric and polymeric substrates was found to be very similar, except with kraft lignin and a polymeric dye where the relative rate of oxidation was higher with laccase II. Both laccases were also tested for their capacity to depolymerize ^{14}C-labelled kraft lignin (*29*). We found by gel permeation chromatography that when either of the two laccases was applied alone, the average molecular weight of the radio-labelled kraft lignin was increased. However, the addition of ABTS at the beginning of the incubation completely prevented polymerization of kraft lignin by both laccases. When ABTS was added following a preincubation of the lignin with laccases, the laccase-condensed lignin was depolymerized within a relatively short time. It was previously reported that laccase catalyzes the coupling of ABTS with various phenolic derivatives to produce coloured compounds (*32*). The coupling of ABTS with phenolics in lignin is likely to form hydrophilic lignin-ABTS complexes. This and the proven cleavage of $C\alpha$-$C\beta$ bonds of non-phenolic sites in lignin are two modes of action that can explain the delignifying activity of the laccase-ABTS system.

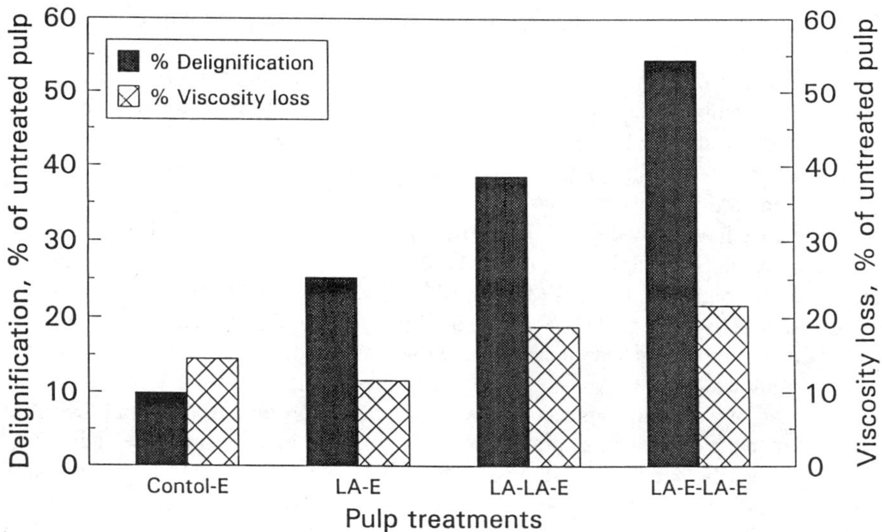

Figure 3. Sequential bleaching of oxygen-delignified softwood kraft pulp by laccase/ABTS (LA) couple and alkaline extraction (E). The initial pulp kappa number was 17.1, and viscosity 25.5 mPa.s.

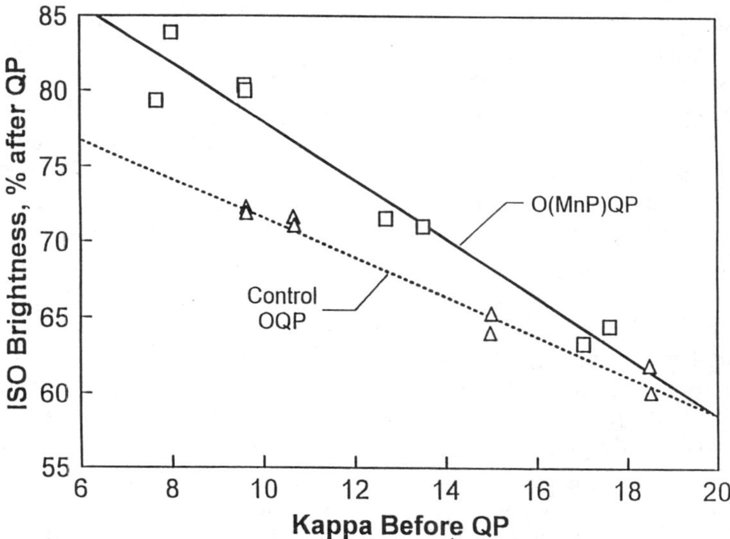

Figure 4. Sequential bleaching of oxygen-delignified kraft pulp by MnP, chelation (Q) and alkaline hydrogen peroxide (P) gives a higher brightness pulp than controls without MnP. Originally published in ref. 10. Copyright TAPPI 1995.

The laccase-ABTS system was optimized to delignify chemical pulps under conditions (time, temperature, consistency, oxygen pressure) compatible with current bleaching technology (*28*). Pulps were treated at 10% consistency for two hours at 60°C, pH 5, 300 kPa of O_2 pressure, with 1% of ABTS and 5U/g of laccase on pulp, followed by an alkaline extraction. Under these conditions, the extent of delignification varied from 25 to 40% with various kraft pulps, and was over 50% with a sulfite pulp. By repeating the treatment with laccase-ABTS and alkaline extraction, the kappa number of a softwood kraft pulp was decreased by 55% (Figure 3). ABTS was developed for analytical purposes and the current price is too high for industrial pulp bleaching. Further work is required to find a cost-effective mediator. We plan to investigate the reaction mechanism of laccase and mediators on lignin, with the objective of bringing this concept to a commercial application.

Manganese Peroxidase Plays a Key Role in Fungal Bleaching

Since its discovery in *Phanerochaete chrysosporium* in 1984 (*33*), manganese peroxidase (MnP) has been found to be secreted by many white-rot fungi during wood delignification (*34*). *T.versicolor* produces a number of MnP isozymes under N-limited conditions (*35-37*). The enzyme oxidizes Mn(II) to Mn(III), and chelated Mn(III) is the proximal oxidant for lignin oxidation. Initially, the enzyme was thought to play only a limited role in lignin depolymerization, oxidizing only phenolic subunits, while lignin peroxidase was regarded as the key enzyme (*38*). Subsequently, MnP was found to depolymerize DHP lignin (*39*), to oxidize non-phenolic model compounds (*40*), and to decolorise dissolved, coloured lignin in bleached kraft mill effluent (*41*). In 1993, we found that MnP was produced by *T. versicolor* pulp bleaching cultures during growth (*13*), and that isolated enzyme supplied with sufficient hydrogen peroxide and chelator could delignify kraft pulp. Subsequently, we applied the enzyme to various pulps (*10*), and found that totally chlorine free pulp could be produced with a 10-point brightness gain when MnP was included prior to conventional chemical chelation and peroxide brightening (Figure 4). Furthermore, Kondo et al (*42*) found that at low pulp consistency, the addition of Tween-80 or unsaturated fatty acids improved pulp brightness attained with MnP, and that high brightness gains could be achieved by applying multiple sequences of MnP and alkaline extraction. Thus MnP can undoubtedly oxidize residual lignin in kraft pulp.

A question remains as to the importance of MnP in kraft pulp bleaching by *T. versicolor*: can this enzyme alone achieve the high brightness gains observed with the fungal system? Several lines of evidence indicate that MnP is a necessary activity for fungal bleaching. Firstly, the maximum secreted activity of MnP by fungal cultures occurs at the maximum rate of bleaching (*13*). Secondly, the extent of fungal bleaching is decreased when hydrogen peroxide or Mn(II) are limited in fungal cultures (*13*). Thirdly, during bleaching the fungus secretes oxalate and glyoxalate, potential chelators for Mn(III) (*43*). As well as this circumstantial evidence, experiments with *T. versicolor* oxidase-less mutants have yielded more conclusive proof (*44*). Thus monokaryon strains of the fungus were mutagenized with uv light, and screened on guaiacol plates to give

```
60                  70                  80
ALGGAMNNDLLLVAWANGNQIVSS
                LLLVAMPNGDWIVSS
```

Figure 5. Alignment of cyanogen bromide cleavage-derived peptide sequence from *T. versicolor* CDH 4.2 (lower line (*52*)) with CDH from *P. chrysosporium* (upper line (*50,51*)). The methionine residue at position 65 of the *P. chrysosporium* protein is hypothesized to be involved in complexing the heme prosthetic group (*50*).

laccase- and MnP-negative mutants. One MnP-negative mutant, named M49, did not brighten or delignify kraft pulp, but this ability was partially restored by addition of exogenous MnP. Similar experiments with laccase addition to laccase-negative mutants were unsuccessful in restoring bleaching. Taken together, the above observations indicate an important role for MnP in fungal bleaching. Nevertheless, as argued previously and in the next section, there remains the possibility that other enzymes, such as cellobiose dehydrogenase or enzymes of unknown function, are required for extensive delignification *in vivo*.

As shown in Figure 2, the proximal oxidant for pulp bleaching by MnP is chelated Mn(III). The redox potential, stability and net charge on this complex will all influence the efficiency of residual lignin oxidation in the secondary wall. Since the interior of kraft pulp fibres are negatively charged, the distribution of ions will be governed by the Donnan Equilibrium (*45*), and a high concentration of Mn(III) complex inside the fibre will be found when the complex has a net positive charge. Complicating the issue when oxalate is the chelator is the competitive reduction of Mn(III) by oxalate (*46*), and the possible formation of oxalate radicals. Oxalate is ubiquitous in kraft mill process waters, tending to form scale deposits of calcium oxalate. Oxidation of oxalate by Mn(III) would be a useful byproduct of MnP bleaching.

The Role of Reducing Enzymes

Are there fungal enzymes whose reductive capabilities are important in lignin degradation and biological bleaching? Westermark and Eriksson (*47,48*), first proposed this after detecting a secreted enzyme that, in cellulose-containing *P. chrysosporium* and *T. versicolor* cultures, decolorizes the brown oxidation product of 2-methoxyphenol (guaiacol) (*48*). This enzyme, isoforms of which are referred to as cellobiose dehydrogenase (CDH) or cellobiose:quinone oxidoreductase (CBQ), carries out the 1-electron reduction of a wide variety of quinones, semiquinones, organic free radicals, metal salts and dyes by oxidizing cellobiose to cellobionolactone.

Two isoforms of this reducing enzyme are found in cultures of *T. versicolor*. In keeping with Henriksson (*49*) and other recent publications, this review will call the *T. versicolor* heme-flavin adenine dinucleotide (FAD)-cofactored glycoprotein CDH 4.2 (based on its isoelectric point) and the FAD-only CDH glycoprotein or proteolysis fragment CDH 6.4. We have found that *T. versicolor* CDH is induced and secreted into the culture medium in response to the presence of cellulose, cellobiose, pulp, or kraft black liquor lignin (Indulin) (*50,51*).

The heme-flavin CDH of *P. chrysosporium* is the only CDH for which an amino acid sequence has been published (*52,53*). The enzyme has 755-770 amino acids with a predicted molecular weight of about 80 kDa. Despite the differences between the molecular weights of CDH from *P. chrysosporium* and *T. versicolor* (80 kDa vs 58 kDa), an 18 amino acid sequence from *T. versicolor* CDH 4.2 showed 79% identity with a region of the *P. chrysosporium* sequence (Figure 5) (*52,53*), suggesting a strong similarity between the CDH proteins of these two fungi.

Substrates of *T. versicolor* CDH. Tests in which a wide range of mono-, di-, and polysaccharides, including all those commonly present in wood, were tried as electron donors to CDH 4.2 and CDH 6.4, showed only cellobiose, lactose, and cellulose to be oxidized (51,54). This utilization pattern suggested that *T. versicolor* CDH recognizes and utilizes only β 1→ 4 linked glucose residues.

In sharp contrast to this specificity for electron sources, *T. versicolor* CDH 4.2 donates electrons to a wide range of substances including semiquinones, ortho- and para-quinones, azo dyes, kraft lignins, benzthiazolines, a variety of organic free radicals, and Cu (II), Fe(III), MnO_2, and Mn(III) complexes. The metal reductions have important implications for delignification; MnO_2 reduction because it produces Mn(III) complexes directly and provides the Mn(II) necessary for MnP to function (*51*), and Fe(III) reduction because it provides, in the presence of H_2O_2, the Fe(II) required for the "Fenton" reaction which produces the potent, cellulose-depolymerizing hydroxyl radical. The reduction of Cu(II) provides Cu(I) which may also react with H_2O_2 to produce hydroxyl radicals.

CDH can also reduce MnP- or laccase-generated organic free radicals of many phenolics, as well as substances such as veratryl alcohol, chlorpromazine, and ABTS (*50,54*). CDH 4.2 showed considerable selectivity among simple methylated ortho-quinones, the position of the methylation varying their reducibility from near zero to rapid (*54*). In general, para-quinones, especially the larger ones, were not good CDH substrates. In contrast to the high selectivity among ortho-quinone electron acceptors, a very diverse group of organic radicals (e.g. chlorpromazine, the veratric and guaiacyl and other phenoxy radicals, and ABTS•+) were readily reduced by CDH 4.2 (*54*). In *P. chrysosporium* it has been reported that the heme-cofactored CDH (analogous to *T. versicolor* CDH 4.2) can reduce oxygen to hydrogen peroxide, albeit very slowly (*55*). *T. versicolor* CDH 4.2 does not reduce O_2, even under pure O_2, at a detectable rate. The *T. versicolor* analyses used a Clark oxygen cell, high oxygen levels, long incubation times and catalase, thus avoiding artefacts found when using peroxidase-coupled assays (*56*).

CDH Counteracts Polymerization of Phenolics by Laccase and Peroxidase. Oxidation of guaiacol (2-methoxyphenol) as well as sinapyl and coniferyl alcohols by laccases and peroxidases usually leads to the formation of brown, insoluble polymerizates. These reactions (*48*) and their products have been proposed as useful models of lignin formation and lignin respectively (*57*). Growing cultures of *T. versicolor* and *P. chrysosporium* have been shown to oxidize and polymerize guaiacol, and then, over time, to decolorize it (*44,47,48*). The reductive activity of purified *T. versicolor* CDH 4.2 has been shown to prevent the polymerization and darkening of guaiacol and kraft lignin (Indulin) by reforming reduced substrates oxidizable by laccase (*54*). Oxygen consumption and laccase activity therefore continue unabated as long as there is cellobiose present, as the CDH constantly regenerates laccase-oxidizable substrate. Both quinones and phenoxy and other organic radical intermediates are reduced, as seen by UV-visible and electron paramagnetic resonance (EPR) spectroscopy (*54*). Likewise, the MnP-catalyzed oxidation of the phenolphthalein derivative phenol red can be completely blocked by CDH 4.2 (*54*). Thus, CDH 4.2 can interact with MnP, LiP, and laccase

enzymes by reducing the organic free radical species which are their initial oxidation products back to the oxidizing enzymes' substrates. The fact that *T. versicolor* secretes MnP, laccase, and CDH concurrently during delignification indicates that CDH and the oxidative enzymes could set up futile cycles in which certain substrates are endlessly redox cycling, with a net oxidation of carbohydrate to sugar acids and reduction of peroxide or oxygen to water.

Biological Functions of CDH. Based on our current knowledge of its capabilities, the following are possible biological roles of *T. versicolor* CDH: 1) Promotes MnP activity by producing the cellobionic acid necessary for cellobionic acid-Mn(III) complexes. 2) CDH could prevent lignin repolymerization after an oxidative enzyme attack, and generate fresh laccase and peroxidase substrates. CDH could also prevent the redeposition of lignin on exposed cellulose surfaces. This would improve the accessibility of cellulose to the fungus. 3) CDH could catalyze hyphal invasion and cellulose depolymerization through hydroxyl radical generation via CDH 4.2-mediated Fe(III) → Fe(II) reduction and Cu(II)-Cu(I) reduction in the presence of H_2O_2 (Fenton reaction). 4) Promotes MnP activity through the solubilization of MnO_2 and the production of Mn(II) and Mn(III). CDH 4.2 can also produce Mn(III) directly and "recycle" Mn oxidized to insoluble MnO_2. 5) Promotion of the net degradation of lignin through redox cycling of lignin phenolics until they become unreducible fragments. The delignifying actions of CDH and the oxidative enzymes may be concomitant or sequential. 6) Reduction of toxic quinones such as methoxyquinones (formed by peroxidases and laccases) to the corresponding phenols.

Prospects for Industrial Pulp Bleaching by Enzymes and Fungi

There are not yet any commercial applications of biopulping or biobleaching with fungi, despite intensive research and development, first at STFI in Stockholm, and more recently at the Forest Products Lab. in Wisconsin, at various locations in Japan, and at the Pulp and Paper Research Institute of Canada. The long times required for fungal delignification and the cost of preventing contamination with unwanted microbes have been prohibitive. On the other hand, the Cartapip process for removing wood extractives with the fungus *Ophiostoma* has been successfully applied at mill scale. The ability of *Ophiostoma* to compete with other microbes during chip colonization and the shorter times required for extractives metabolism than for delignification account for the success of the Cartapip process. Even if highly competitive isolates of white-rot fungi and the environmental conditions that favour their establishment in kraft pulp were found, we would face the problem that these fungi have evolved to consume all the components of wood, not to selectively remove the lignin. In the short to medium term, applying enzymes isolated from these fungi gives us more control to optimize the rate and selectivity of biological delignification. In the longer term, genetic engineering of a fungus to produce a designed mixture of enzymes and their co-substrates *in situ* may become more cost-effective than producing and applying the enzymes in separate steps.

Experience with xylanase prebleaching has already demonstrated the feasibility of large-scale application of microbial enzymes in kraft pulp mills. However, due to the limited chemical savings achieved with xylanase, only six out of 40 Canadian kraft mills are currently applying the enzyme. Both laccase and MnP can achieve more substantial delignifying action than xylanase, but there are obstacles to be overcome before either enzyme can be used in a cost-effective manner. There is currently no large-scale commercial source for either enzyme, so the bulk cost remains to be established. For laccase, so far we do not have a mediator sufficiently effective and inexpensive to be commercially viable. On the positive side, the enzyme is fairly thermotolerant, operates at medium pulp consistency within a wide range of acidic pHs, and requires only a low oxygen pressure. For MnP application, most pulps contain sufficient manganese to permit the enzyme to function efficiently, so no mediator addition is required. However, the requirement of a low but constant supply of hydrogen peroxide is a major obstacle which prevents operating at pulp consistences above about two percent. The MnP produced by *T. versicolor* is also intolerant of temperatures above 30°C. Finally, for cellobiose dehydrogenase, a significant delignifying effect on kraft pulp remains to be demonstrated, and production of the enzyme in *T. versicolor* is significantly lower than laccase or MnP under the conditions tried to date.

Acknowledgments

This work was financially supported by Genencor International, Industry Canada, the Natural Sciences and Engineering Research Council (CRD program), and by the maintaining member companies of the Pulp and Paper Research Institute of Canada.

Literature Cited

1. Kirk, T. K.; Yang, H. H. *Biotechnol. Lett.* **1979**, *1*, 347-352.
2. Paice, M. G.; Jurasek, L.; Ho, C.; Bourbonnais, R.; Archibald, F. *Tappi J.* **1989**, *72*(5), 217-221.
3. Nishida, T.; Kashino, Y.; Mimura, A.; Takahara, Y. *Mokuzai Gakkaishi* **1988**, *34*, 530-536.
4. Fujita, K.; Kondo, R.; Sakai, K.; Kashino, Y.; Nishida, T.; Takahara, Y. *Tappi J.* **1991**, *74*(11), 123-127.
5. Murata, S.; Kondo, R.; Sakai, K.; Kashino, Y.; Nishida, T.; Takahara, Y. *Tappi J.* **1992**, *75*(12), 91-94.
6. Fujita, K.; Kondo, R.; Sakai, K.; Kashino, Y.; Nishida, T.; Takahara, Y. *Tappi J.* **1993**, *76*(1), 81-84.
7. Kondo, R.; Kurashiki, K.; Sakai, K. *Appl. Environ. Microbiol.* **1994**, *60*, 921-926.
8. Hirai, H.; Kondo, R.; Sakai, K. *Mokuzai Gakkaishi* **1994**, *40*, 980-986.
9. Tsuchikawa, K.; Kondo, R.; Sakai, K. *Jpn. Tappi J.* **1995**, *49*, 1332-1338.
10. Paice, M. G.; Bourbonnais, R.; Reid, I. D. *Tappi J.* **1995**, *78*(9), 161-169.
11. Reid, I. D.; Paice, M. G.; Ho, C.; Jurasek, L. *Tappi J.* **1990**, *73*(8), 149-153.

12. Archibald, F. S. *Holzforschung* **1992**, *46*, 305-310.
13. Paice, M. G.; Reid, I. D.; Bourbonnais, R.; Archibald, F. S.; Jurasek, L. *Appl. Environ. Microbiol.* **1993**, *59*, 260-265.
14. Archibald, F. S.; Bourbonnais, R.; Jurasek, L.; Paice, M. G.; Reid, I. D. *J. Biotechnol.* **1996**, in press.
15. Reid, I. D.; Paice, M. G. *FEMS Microbiology Reviews* **1994**, *13*, 369-376.
16. Reid, I. D.; Paice, M. G. In *Frontiers in Industrial Mycology*; Leatham, G. F., Ed.; Chapman & Hall: New York, 1992; Chapter 8, pp. 112-126.
17. Jurasek, L. *J. Pulp Pap. Sci.* **1995,** *21(8)*, J274-J279.
18. Sundaramoorthy, M.; Kishi, K.; Gold, M. H.; Poulos, T. L. *J. Biol. Chem.* **1994**, *269*,32759-32767.
19. Reinhammar, B. In *Copper proteins and copper enzymes*; Lontie, R. Ed.; CRC Press: Boca Raton, Florida, **1984**, Vol.31; pp. 1-35.
20. Thurston, C.F. *Microbiology.* **1994,** *140*, 19-26.
21. Higuchi, T. In *Plant cell wall polymers: biogenesis and biodegradation*; Lewis, N.G.; Paice, M.G., Eds.; ACS Symposium Series 399; **1989**, 482-502.
22. Kawai, S.; Umezawa, T.; Higuchi, T. *Arch. Biochem. Biophys.* **1988,** *262*, 99-110.
23. Ishihara, T. *In lignin biodegradation: Microbiology, chemistry and potential applications*; Kirk, T.K.; Higuchi, T.; Chang, H.M. Eds.; CRC Press: Boca Raton, Florida, **1980,** Vol. 2; pp. 17-31.
24. Bourbonnais, R.; Paice, M.G. *FEBS Lett.* **1990**, *267*, 99-102.
25. Bourbonnais, R.; Paice, M.G. *Appl. Microbiol. Biotechnol.* **1992,** *36*, 823-827.
26. Bourbonnais, R.; Paice, M.G.; Reid, I.D. In *Biotechnology in Pulp and Paper Industry*, Kuwahara, M.; Shimada, M., Eds.; Uni Publishers Co. Ltd.: Tokyo, Japan, **1992**, pp 181-186.
27. Call, H.P. *PCT. World patent application*: WO 94/29510, **1994.**
28. Bourbonnais R.; Paice M.G. *Tappi J.*, **1996**, *79*, in press.
29. Bourbonnais, R.; Paice, M.G.; Reid, I.D.; Lanthier, P.; Yaguchi, M. *Appl. Environ. Microbiol.* **1995,** *61,* 1876-1880.
30. Iimura, Y.; Takenouchi, K.; Nakamura, M.; Kawai, S.; Katayama, Y.; Morohoshi, N. In *Biotechnolgy in Pulp and Paper Industry*; Kuwahara, M.; Shimada, M., Eds.; Uni Publishers Co. Ltd.: Tokyo, Japan, **1992**, pp. 427-431.
31. Morohoshi, N.; Fujita, K.; Wariishi, H.; Haraguchi, T. *Mokuzai Gakkaishi.* **1987,** *33*, 310-315.
32. Matsumura, E.; Yamamoto, E.; Numata, A.; Kawano, T.; Shin, T.; Murao, S. *Agric. Biol. Chem.* **1986,** *50*, 1355-1357.
33. Kuwahara, M.; Glenn, J.K.; Morgan M.A.; Gold, M.H. *FEBS Letters* **1984**, *169*, 247-250.
34. Hatakka, A. *FEMS Microbiol. Rev.* **1994**, *13*, 125-135.
35. Johansson, T.; Nyman, P.O. *Arch. Biochem Biophys.* **1984**, *300*, 49-56.
36. Johansson, T.; Welinder, K.G.; Nyman, P.O. *Arch. Biochem. Biophys.* **1984**, *300*, 57-62.

37. Jonsson, L.; Becker, H.G.; Nyman, P.O. *Biochim Biophys. Acta.* **1994**, *1207*, 255-259.
38. Kirk, T.K.; Farrell, R.L. *Annual. Rev. Microbiol.* **1987**, *41*, 465-505.
39. Wariishi, H.; Valli, K.; Gold, M.H. *Biochem. Biophys. Res. Commun.* **1991**, *176*, 269-275.
40. Bao, W.L.; Fukushima, Y.; Jensen, K.A.; Moen, M.A. *FEBS Letters* **1994**, *354*, 4359-4363 41. Michel, F.C.; Dass, S.B.; Grulke; E.A.; Reddy, C.A. *Appl. Environ. Microbiol.* **1991**, *57*, 2368-2375.
42. Kondo, R.; Harazona, K.; Sakai, K. *Appl. Environ. Microbiol.* **1994**, *60*, 4359-4363.
43. Roy, B.P.; Archibald, F.S. *Appl. Environ. Microbiol.* **1993**, *59*, 1855-1863 .
44. Addleman, K.; Dumonceaux, T.; Paice, M.G.; Bourbonnais, R.; Archibald, F.S. *Appl. Environ. Microbiol.* **1995**, *61*, 3687-3694.
45. Scallan, A.M., Katz, S., Argyropoulos, D.S.In *Cellulose and Wood - Chemistry and Technology*; C. Schuerch, Ed,. John Wiley, New York. **1989**, pp 1457-1471.
46. Harazona, K.; Kondo, R.; Sakai, K. *Appl. Environ. Microbiol.* **1996**, *62*, 913-917.
47. Westermark, U.; Eriksson, K.-E. *Acta Chem. Scand.* **1974**, *B 28,*204-208.
48. Westermark, U.; Eriksson, K.-E. *Acta Chem. Scand.* **1974**, *B28,* 209-214.
49. Henriksson, G. *Structure, Function, and Applications of Cellobiose Dehydrogenase from Phanerochaete chrysosporium;* Ph.D. Thesis, Uppsala University, Uppsala, Sweden.
50. Roy, B.P.; Archibald, F. *Appl. Environ. Micriobiol.* **1993**, *59,*1855-1863.
51. Roy, B.P.; Paice, M.G.; Archibald, F.S. *J. Biol. Chem.* **1994**, *31,*19745-19750.
52. Raices, M.; Paifer, E.; Cremata, J.; Montesino, R.; Stahlberg, J.; Divine, C.; Szabo, I.J.; Henriksson, G.; Pettersson, G. *FEBS Lett.* **1995**, *368,*233-238.
53. Li, B.; Nagalla, S.R.; Renganathan, V. *Appl. Environ. Microbiol.* **1996**, *62,*1329-1335.
54. Roy, B.P. *The Role of Reductive Enzymes in Trametes versicolor-mediated Kraft Pulp Bleaching;* Ph.D. Thesis, McGill, Montreal, Canada. **1994**.
55. Ayers, A.R.; Ayers, S.B.; Eriksson, K.-E.L. *Eur. J. Biochem.* **1978**, *90,*171-178.
56. Ander, P. *FEMS Microbiol. Rev.* **1994**, *13,*297-312.
57. Crawford, R.L.; Robinson, L.E.; Foster, R.D. *Appl. Environ. Microbiol.* **1981**, *41,*1112-1116.

ENZYME PRODUCTION AND CHARACTERIZATION

Chapter 12

Short-Fiber Formation During Cellulose Degradation by *Trichoderma pseudokoningii* S38

Jie Liu and Peiji Gao

State Key Laboratory of Microbial Technology, Institute of Microbiology, Shandong University, Jinan 250100, China

A low-molecular weight component, which could cause depolymerization of native cellulose to form insoluble short fibres without production of soluble reducing sugar, was separated from the cell-free culture of cellulolytic fungal strain *Trichoderma pseudokoningii S38*. It is termed Short Fibre Generating Factor (SFGF) according to its function. Its essential features, as well as the process of short fibre formation, which is the intermediate state of cellulose degradation, were studied.

Much attention has been given to study the mechanisms of biological degradation of cellulose. Due to the structural complexity of native cellulose materials, the degradation of cellulose is a quite complicated process. Insoluble crystalline microfibrils are highly resistant to enzymatic hydrolysis. Cellulase systems contain a multiplicity of enzyme components showing a marked synergism against crystalline cellulose (*1,2*). But the degrading efficiency of native cellulose is still quite low, which partly lies in the uncertainty of degradation mechanism. The appearance of short, insoluble fibres during the cellulose degradation was regarded as the initial stage or the first step, which was not accompanied by the formation of soluble reducing sugar, but which made the substrate more accessible to endoglucanase attack. Evidence (*3,4*) suggested that besides the hydrolytic mode, oxidative degradation is also a part of the degradation process. Although the former mode has the main effect on cellulose degradation, the latter helps to accelerate the process and make use of cellulosic materials. Comparatively speaking, hydrolytic degradation is better understood than oxidative degradation. The present paper shows the isolation of a kind of non-enzymatic component in filmentous fungi and its functions during cellulose degradation. It is termed short fibre generating factor (SFGF).

0097–6156/96/0655–0166$15.00/0
© 1996 American Chemical Society

Isolation of SFGF

Media and Culture Conditions. The cellulolytic fungal strain *Trichoderma pseudokoningii S38* (*5*) was maintained at 30 ℃ for 6 days in a solid-layer fermentation system, consisting of (W:W%) corn stalk powder 80, wheat bran 20, and water 250 supplemented with Mandels' mineral salts (*6*).

Assay of Cellulase Activity. Cellulase activity (filter paper activity, FPA) was measured by incubating enzyme solution with 50 mg filter paper (Whatman No.1) in a 1.5 ml acetate buffer (100 mM, pH 4.8) system at 50 ℃ for one hour. The activity was expressed as equivalent of reducing sugar produced, which was assayed by DNS (3,5-dinitrosalicylic acid) reagent (*7*). And one unit (U) was defined as the amount (μ mol) of glucose produced by enzymes per minute under given condition.

Assay of short fibre formation. 50 mg filter paper taken as substrate and the sample solution supplemented with phosphate buffer (20 mM, pH 6.0) upto 10 ml, were incubated in a 50 ml flask at 45 ℃, 30 rpm for one day. The turbidity of the suspension was determined using an integrating sphere attachment (Spectrophotometer UV-VIS-240, Shimadzu). The increase ratio of turbidity was defined as following:

$$T\% = (Ts-Tc) / Tc$$

Ts,Tc represented the turbidity of sample and control respectively.

Separation of SFGF from Crude Enzyme. The crude cell-free culture filtrate (crude enzyme) was collected by centrifugation from acetate buffer extracts (50 mM, pH 4.8) of solid-layer culture. The crude enzyme had a relatively high ability to degrade cellulose, including both reducing-sugar production (1.2 U cellulase activity per milliliter) and short-fibre formation. The components with molecular weights above 10,000 Da, such as cellobiohydrolase (CBH), endo-glucanase (EG) and β -glucosidase (β G), were removed from the crude enzyme broth by ultrafiltration (Amicon, PM-10 with cut-off of 10,000 Da, Pharmarcia). The permeate from ultrafiltration had little cellulase activity (CBH, EG, β G) that could be neglected, but it could decompose native cellulose to produce short fibres, which were still insoluble.

After concentrated by ultrafiltration (Amicon, YM-1 with cut-off of 1,000 Da), the filtrate was applied in turn to columns of Sephadex LH-20, DEAE-Sephadex A-25, Bio-gel P-4, and HPLC (Supelcosil, LC-18-DB), eluted with appropriate buffers indicated respectively in Section A, B, C and D of Figure 1. In each purification step, the component having the ability of generating short fibre (shown in Figure 1) was collected and prepared for the next step. In the step of HPLC, a single peak was obtained.

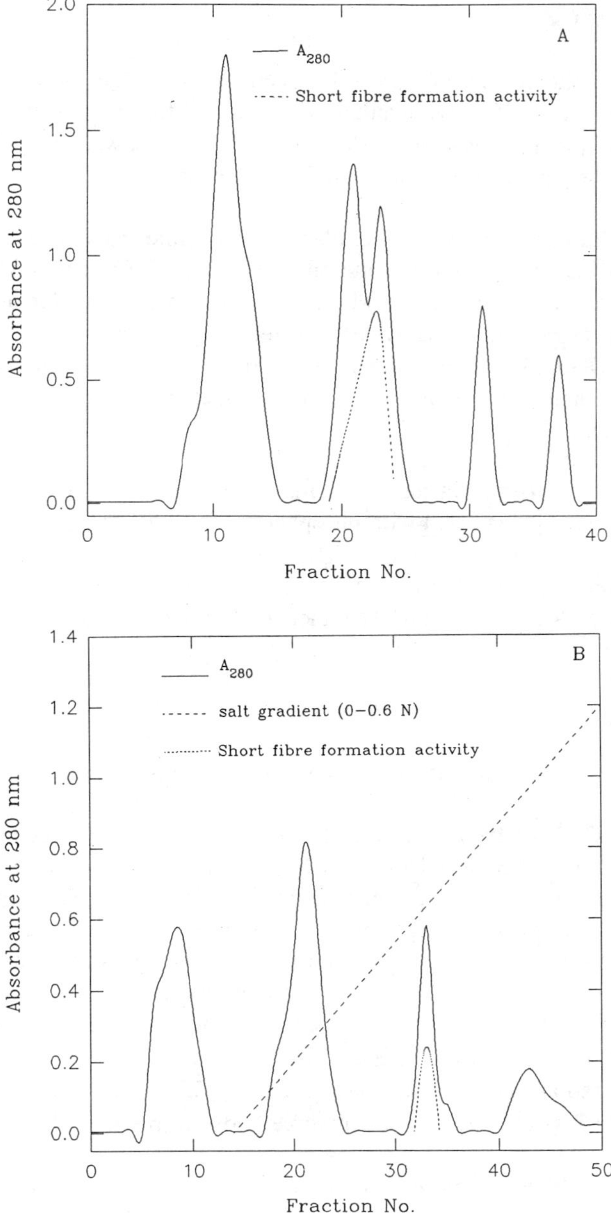

Figure 1. Isolation processes of SFGF

A. Separation of SFGF on a Sephadex LH-20 column eluted with 20mM, pH 6.0 phosphate buffer.

B. Separation of SFGF on a Sephadex A-25 column eluted with 20 mM, pH 6.0 phosphate buffer and NaCl gradient from 0 to 0.6.

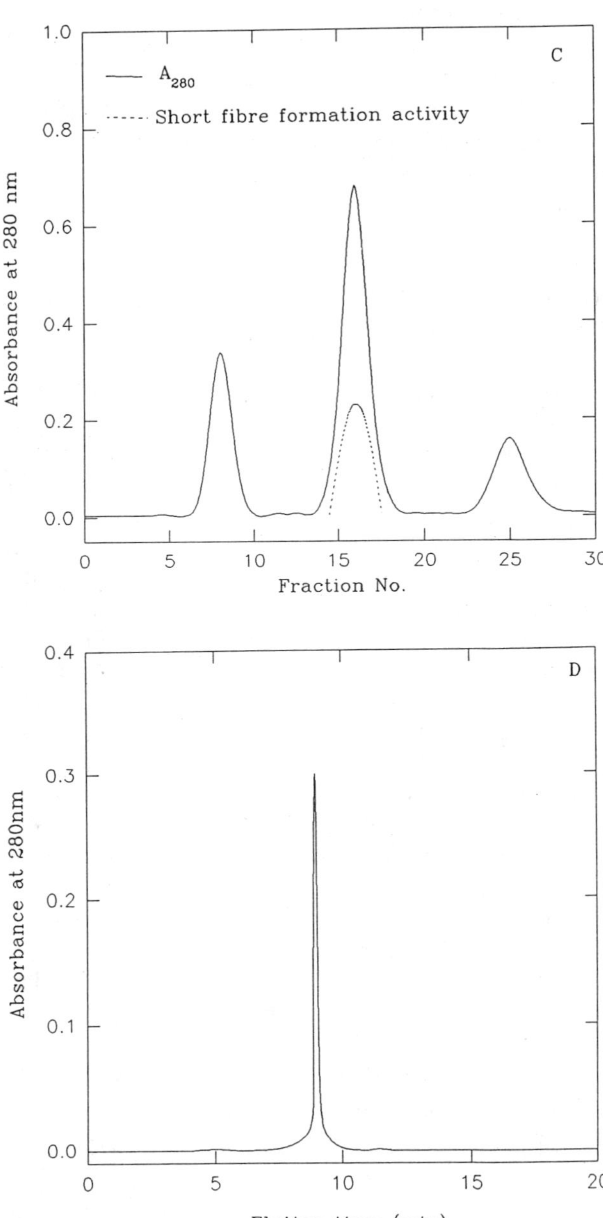

Figure 1. *Continued*

C. Separation of SFGF on a Bio-gel P-4 column eluted with distilled water.

D. Separation of SFGF on a HPLC (LC-18-DB) eluted with methanol and water in the volume proportion of 7:3.

Functions of SFGF During Cellulose Degradation

The fraction of SFGF was responsible for generating short fibres from native cellulose and processed cellulose. The reaction took place at 45 ℃, pH 6.0 under shaking aerobic condition (120 rpm) in a 50 ml flask. The relevant 20 mM, pH 6.0 phosphate buffer was used as the control without exception. Because of the long reaction time, 0.02% azide sodium was added to each sample to avoid interference from microbes.

Action of SFGF on Dewaxed Cotton Fibre. The reaction system containned 50 mg dewaxed cotton fibre (purchased as commercial reagent) and 10 ml SFGF solution.

After three days, the dewaxed cotton fibre was still in a bundle. But fine fibres suspended in the solution were easily observed with the naked eye. The turbidity of the suspension was measured as the method mentioned above. 1.5 ml supernate was taken to be determined by DNS reagent (*7*). No reducing sugar was produced. And no obvious weight loss could be detected..

Figure 2. showed the relationship between reaction time and the amount of short fibres formed, as determined by the turbidity of the suspension. As the reaction period proceeded, the amount of short fibres increased. But after six days, no further increase was observed. This suggested that the formation of short fibres was the intermediate stage in the whole process of cellulose degradation: the insoluble native fibre was degraded into short fibres, which acted as the substrate of cellulases in the following steps to produce reducing sugar. When there were enough short fibres formed, in other words, no further hydrolytic process followed, the formation of short fibre was "inhibited". In this experiment, the cellulases were removed from the filtrate, and the short fibres produced by SFGF could not be degraded further into reducing sugar, so the process of short fibre formation was blocked.

The same result was obtained when filter paper and paper pulp were used as the substrates. The degree of polymerization (DP) of paper pulp was detected using the TAPPI standard viscosity assay method before and after the reaction with SFGF. The value of DP decreased by 40-50% after three days.

Action of SFGF on Powdered Materials. α -cellulose (Sigma), reprecipited cellulose (cellulose reprecipited from H_3PO_4) (*8*) and microcrystalline cellulose (Sigma Cell Type50) were used. α -cellulose was taken for example.

Thirty mg of α -cellulose powder between 150 and 200 mesh was mixed with 10 ml SFGF solution under the same conditions as discribed above. A partial of the supernatant solution was used to estimate reducing sugars. The rest was added with distilled water up to 100 ml and the turbidity was determined. The turbidity of suspension attacked by SFGF for three days increased by 30-40% compared with the control, which suggested that SFGF caused the amount of cellulose particles in the reaction solution to increase. No reducing sugar could be detected in this case, either.

Action of SFGF on Soluble Cellulose. Carboxylmethylcellulose (CMC-Na, Sigma) was used as a soluble substrate. When the solution of SFGF and 1% CMC-Na were incubated together in a 1:1 ratio, the viscosity of the mixture (determined by Ubbelohde Viscosimeter) decreased by 50% within two hours, compared with 0.5% CMC-Na alone. No reducing sugar was detected.

Synergistic Degradation of Cellulose by SFGF and Cellulase Complex.

Synergism in the cellulose degradation was examined by assaying for reducing sugar produced from filter paper by the cellulase complex from *Trichoderma pseudokoningii S-38* and SFGF, alone and in combination (Figure. 3). When SFGF and the cellulase complex were mixed together, synergistic saccharfication was observed. The result indicated that the reducing sugar production in combination was about 40% more than that of cellulase complex alone.

On the other hand, after washed with distilled water, the dewaxed cotton fibre attacked by SFGF was used as a substrate to estimate cellulolytic activity instead of commercial dewaxed cotton fibre. The cellulase complex showed nearly 25% "higher activity" than the normal activity. These indicated that SFGF made the substrate more easy to be degraded by cellulases.

Essential Features of SFGF

SFGF absorbed light at 280 nm and it also had positive result in biuret reaction (*9*), which indicated its peptide nature. Amino acids composition was analyzed after acid hydrolysis using 835 HITACHI Amino Acid Analyzer. Table I showed the relative mole percentage of main amino acids. Its molecular weight was estimated to be approximately 2000 Da according to the results from gel filtration chromatography, which was available for separation of SFGF.

Table I Amino Acids Composition of SFGF Analyzed after Acid Hydrolysis

Amino Acid	Composition (mol %)
Aspartic acid	13.9
Threonine	7.5
Serine	6.2
Glutamic acid	15.0
Proline	6.2
Glycine	9.1
Alanine	12.6
Cysteine	4.5
Valine	8.2
Leucine	7.1
Phenylalanine	4.3
Lysine	4.8

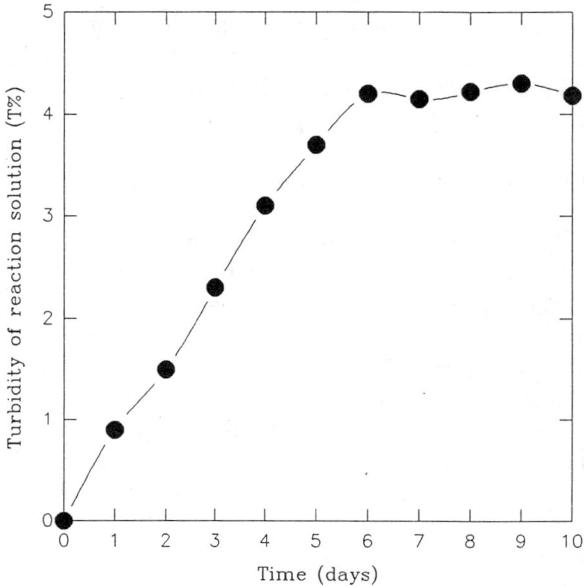

Figure 2. The relationship between the reaction time and the short fibre formation.

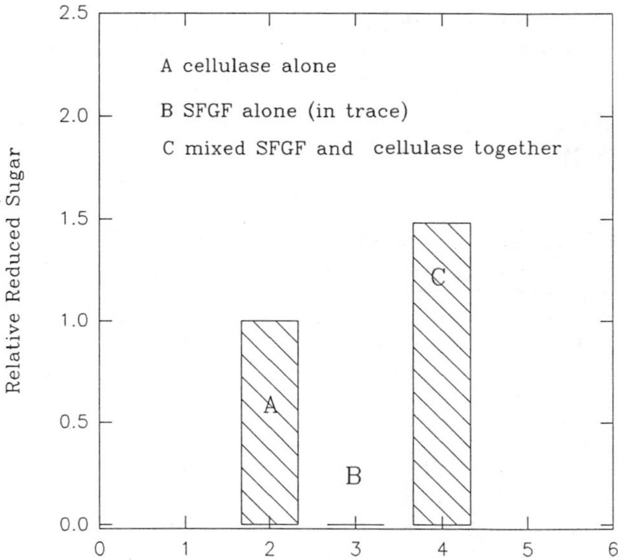

Figure 3. Synergistic hydrolysis of filter paper by mixed SFGF and cellulose complex.

The soluble SFGF could react with CAS reagent (Chrome azurol S / iron (III) / hexa-decyltrimethylammonium bromide System, Sigma) as a method of detecting the existence of siderophore (*10*). As previously reported, iron plays an important role in the biological oxidation-reduction process (*11*). The positive result mentioned above suggested SFGF was able to chelate Fe^{3+} and might function as siderophore. So SFGF might be involved in some kind of oxidation-reduction reaction.

The oxidative capability of SFGF was determined using 2-keto-4-thiomethyl-butyric acid (KTBA, Sigma) (*12*), which releases ethylene upon one-electron oxidation. Reactions were carried out in a stoppered tube filled with oxygen, and the ethylene was detected by gas chromatagraphy (GC-8A, Shimadzu) assay. About 0.85×10^{-8} mol ethylene was released from KTBA per hour by SFGF. This indicated that radical might be the product of oxidative reaction and had effect on the decomposition of cellulose.

In addition, the optimal working temperature and pH of SFGF were found to be 45 ℃ and pH 6.0 respectively. These conditions were a little different from those of cellulases (50 ℃, pH4.8) (*13*). Oxygen was necessary for short fibre formation. No short fibre was generated as the reaction took place under anaerobic conditions. When pure oxygen was supplied into the reaction system, the functional effect of SFGF increased by 45%. Moreover, adding ferric ion and H_2O_2 to the reaction system was conductive to the function of SFGF.

SFGF, as a kind of non-enzymatic peptide, plays an important role in the initial decomposition of cellulose. Its characterization could be regarded as evidence for the existence of an oxidative mechanism in cellulose degradation.

Conclusion

1. Short cellulosic fibres were formed during the biological degradation of cellulose. This process was thought to be the intermediate state, and might be the rate-limiting step in cellulose degradation. Moreover short fibre formation was not limited to the fungi of *Trichoderma pseudokoningii S38*, but could apply to all the cellulolytic fungi, including filamentous fungi and basidiomycetes, some of which were also used to detect short fibre formation and the similar results were obtained (dissertation of Liu, J., Shandong University, China, unpublished data).
2. A low-molecular weight component SFGF was isolated from the crude enzyme, responsible for short fibre formation. As to its characterization and functional mode, there were great differences between SFGF and cellulases. That means this kind of functional peptide plays its role in some non-enzymatic mode. Most likely it is involved in oxidative degradation. Further study is needed to clarify its mechanism during cellulose degradation.
3. The investigation suggested SFGF had synergistic action with cellulases. As well known, synergism in cellulose degradation has only been demonstrated by recombination experiments of the purified endo- and exo-glucanases (*14*). The existence of SFGF, causing short fibre formation, helped to make the substrate more susceptible to the hydrolytic enzymes, which led to the synergism.

Acknowledgment

This work was supported by Natural Science Foundation of China (NSFC).

Literature Cited

1. Beguin, P.; Aubert, J. P. *FEMS Microbiol. Rev.* **1994**, *13*, 25.
2. Coughlan, M.P. In *Microbial Enzymes and Biotechnology;* Fogarty, W. M.; Kelly, C. T., Eds; 2nd Edition, Elsevier Science Publishers: Essex, England, 1990; 1-29.
3. Vaheri, M. P. *J. Appl. Biochem.* **1982**, *4*, 356.
4. Griffin, H.; Dintzis, F. R.; Krull, L.; Baker, F. L. *Biotech & Bioeng.* **1984**, *26*, 296.
5. Ma, D. B.; Gao P. J.; Wang Z. N. *Enzyme Microb. Technol.* **1990**, *12*, 631.
6. Mandels, M.; Weber, J. *Adv. Chem. Ser.* **1969**, *95*, 391.
7. Miller, G.L. *Anal. Chem.* **1959**, *31(3)*, 426.
8. Halliwell, G. *Biochem., J.* **1965**, *95*, 270.
9. Ellman, G.L. *Anal. Biochem.* **1962**, *3*, 40.
10. Schwyn B.; Neilands, J. B. *Anal. Biochem.* **1987**, *160*, 47.
11. Brown, D.H.; Smith W.E. In *Enzyme Chemistry: impact and application;* Suckling, C. J., Ed; Chapman and Hall: New York, NY,1984; Vol.2,167-171.
12. Enoki, A.; Tanaka, H.; Higashiosaka, G. F. *Wood Sci Technol.* **1989**; *23,* 1.
13. Wood, T.M.; Mahalingeshwara, K. In *Method in Enzymology*; Wood, W.A.; Kellogg, S.T., Eds; Part A Cellulose and Hemicellulose; Academic Press: San Diego, CA, 1988; Vol. 160; 87-112.
14. Henrissat B.; etal *Bio/technology.* **1985**, *3*, 722.

Chapter 13

Production of Cellulase and Xylanase with Selected Filamentous Fungi by Solid Substrate Fermentation

G. Szakacs[1] and R. P. Tengerdy[2]

[1]Technical University of Budapest, Department of Agricultural Chemical Technology, HU–1521 Budapest, Gellert ter 4, Hungary
[2]Colorado State University, Department of Microbiology, Fort Collins, CO 80523–1677

Lignocellulolytic fungi were tested for growth and enzyme production on solid substrates such as extracted sweet sorghum pulp and wheat straw. A *Gliocladium* sp. was best adapted to sweet sorghum pulp with about 6.0 Filter Paper Unit (FPU)/Dry Weight (DW) cellulase activity, 14.0 IU/g DW beta glucosidase activity, 1900 IU/g DW xylanase activity and 1500 Endoglucanase Unit (EGU)/g DW endoglucanase activity. On wheat straw a *Trichoderma hamatum* strain produced about 7000 IU/g DW xylanase activity and a *Penicillium* strain 34 IU/g DW beta glucosidase activity, surpassing *Trichoderma reesei* Rut C30 in these categories (10.0 IU/g DW beta glucosidase and 3600 IU/g DW xylanase). All selected wild strains surpassed *Trichoderma reesei* QM 6a, the parent of Rut C30 in enzyme production. The *Gliocladium* and *Trichoderma hamatum* strains grew considerably faster on both substrates than other tested fungi.

Lignocellulolytic enzymes may be produced economically and efficiently by solid substrate fermentation (SSF) (*1,2,3*). The advantage of SSF is operational simplicity and economy in a water restricted environment, resulting in high volumetric productivity, high product concentration, and possibility for use of the product with little or no downstream processing.

Crude SSF enzymes may be used directly in agrobiotechnological applications such as ensiling (*4*), feed additives, retting, soil additives, and in biotechnology industries such as the paper industry and the biofuel industry (*3*). *In situ* SSF enzyme production may be incorporated into such biotechnologies with great saving in process cost (*5,6*).

The purpose of this work was finding an efficient and economical enzyme source for the bioprocessing of sweet sorghum for ethanol production. We have reported earlier the successful application of enzyme assisted ensiling (ENLAC) for increasing plant cell wall permeability, resulting in much higher yield in sugar extraction (*7*). Although the process is economical even with commercial enzymes, a cost reduction of about 100

0097–6156/96/0655–0175$15.00/0
© 1996 American Chemical Society

times may be achieved by using *in situ* enzymes, produced by solid substrate fermentation (SSF) on recycled extracted sweet sorghum pulp (*4*). The process scheme is shown in Figure 1. The advantages of the *in situ* process is that it uses a pretreated, near sterile, no-cost recyclable substrate, and produces enzymes that can be used without downstream processing in ENLAC.

In this paper we report the cell wall degrading enzyme production of selected filamentous fungi by SSF that may be used as *in situ* enzyme source in sweet sorghum processing, and/or in other similar processes. The enzyme productivities of selected fungi were compared on extracted sweet sorghum pulp and wheat straw, a commonly used reference substrate in SSF.

Materials and Methods

Cultures. *Trichoderma hamatum* TUB F-105 (ATCC 62392) was isolated from decaying reed in Hungary by G. Szakacs, and identified by L. Vajna, Budapest; *Trichoderma* sp. TUB F-426 was isolated from soil in Queensland, Australia by G. Szakacs; *Trichoderma* sp. TUB F-482 was isolated from forest soil in Florida by G. Szakacs; *Trichoderma* sp. TUB F-486 was isolated from decaying wood, near Lake Placid, N.Y.; *Gliocladium sp.* TUB F-498 was isolated from soil in Germany; *Penicillium aculeatum* NRRL 2129, *Pencillium funiculosum* NRRL 1132 and *Penicillium funiculosum* NRRL 3647 strains were kindly donated from NRRL collection (Peoria, Illinois). *Trichoderma reesei*, Rut C30 and QM 6a were obtained from the American Type Culture Collection (ATCC).

Substrates. Extracted sweet sorghum pulp was obtained from the bioprocessing of sweet sorghum by ENLAC, followed by countercurrent diffusion for sugar extraction (4). The extracted and dried pulp contains about 25% available carbon in the form of cellulose and hemicellulose, and about 0.6% nitrogen in the form of protein. Wheat straw, supplemented with 10% wheat bran contains about 30% available carbon and 0.9% nitrogen. The dried substrates were milled to about 1 cm particle size and supplemented with the following salt solution: NH_4NO_3, 5.0g/l; KH_2PO_4, 5.0 g/l; $MgSO_4.7H_2O$, 1.0 g/l; NaCl, 1.0 g/l; $MnSO_4$, 1.6 mg/l; $ZnSO_4.7H_2O$, 3.4 mg/l; $CoCl_2.6H_2O$, 2.0 mg/l; $FeSO_4.7H_2O$, 5.0 mg/l. For 1 g dry substrate 3 ml salt solution was added to give a moisture content of 75% and approximate C:N ratio of 25 for sweet sorghum and 18 for wheat straw/bran. The initial pH before sterilization was 4.8.

Solid Substrate Fermentation. SSF was performed in 250 ml double plastic cups, the inner cup having perforations for air excess, the outer one providing protection from contamination. The prepared substrate was sterilized at 121°C for 30 min, then inoculated with the spore suspension of the test fungi to a final concentration of 10^7 colony forming units (CFU)/g DW, and loosely packed into the inner cup (75 g/cup). The spore suspension was prepared by washing spores from the surface of 10 days old sporulating potato dextrose agar plate cultures of the respective fungi with 0.1% Tween 80 containing water. The inoculated cups were incubated at 28°C in 99% relative humidity chamber for 12 days.

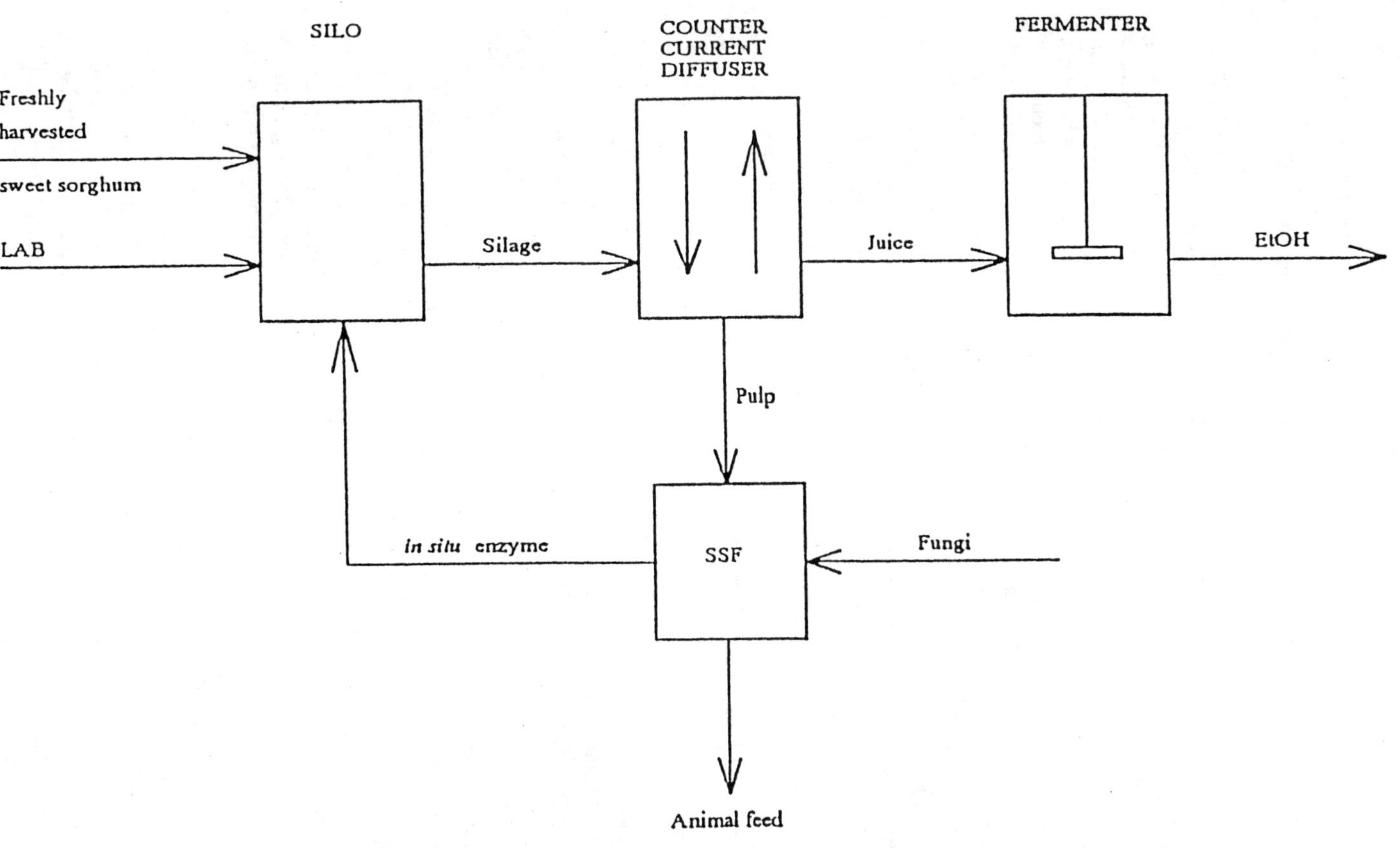

Figure 1. Scheme of integrated bioprocessing of sweet sorghum for ethanol production.

Larger scale SSF with strain F-498 was performed in aluminum trays (780 x 510 x 80 mm), 2000 g wet substrate per tray in a layer thickness of about 5 cm. SSF was performed as above for 6 days at 28°C in a 99% relative humidity chamber. During incubation the trays were covered with plastic wrap and ventilated daily for a few minutes.

Analytical Procedures. Fungal growth was determined by measuring the insoluble N (protein) content of the fermentum by a modified Kjeldahl procedure (*8*). The N value was multiplied by 6.25 to get the protein content, and by 2.7 to get the estimated biomass content, assuming that 37% of the fungal mycelium is protein (average value observed for 2-3 days growth of *T. reesei* Rut C30). Soluble protein, indicative of secretion capability of a fungus, was measured by the bicinchoninic acid method (*9*). Enzyme activities were determined from the culture extract of SSF samples: 5 g DW fermented substrate was extracted with 95 ml water, containing 0.1% Tween 80, by shaking for 60 min at room temperature. From the centrifuged extract, filter paper activity (FPA) for cellulase was determined by standard IUPAC method (*10*). Beta-1,4-endoglucanase activity (EG) was determined according to Bailey and Nevalainen (*11*). Xylanase activity was assayed according to Bailey et al. (*12*). Beta-glucosidase activity (BG) was determined following Kubicek (*13*). Each variable was tested in four reps (four cups), and each assay was done in duplicate. The means of these tests are shown in the results. The enzyme activities were measured daily from day 2, and the peak enzyme activities are reported on respective days.

Results and Discussion

The comparison of the peak lignocellulolytic enzyme activities of the selected fungi are shown in Table I. All selected strains had higher enzyme activities than the reference wild strain *T. reesei* QM 6a, from which the commercially successful mutant Rut C30 has been developed. Some strains had better beta-glucosidase and xylanase activity than Rut C30. Most strains grew better and produced more enzymes on the richer wheat straw/bran medium, but one isolate, F-498, performed better on extracted sweet sorghum pulp. For industrial applications a short fermentation time is at premium, therefore, fast enzyme production is an important selection criterion. The kinetics of cellulase production on sweet sorghum pulp is shown in Figure 2. The initial rate of cellulase production of strains F-498 and F-105 was almost identical to that of Rut C30. Owing to the rapid enzyme production and the favorable cellulase-beta glucosidase-xylanase ratio in the produced enzyme complex, accompanied by a rapid growth rate and lack of strong sporulation, strain F-498 was selected for *in situ* enzyme production in the sweet sorghum bioprocessing scheme shown in Figure 1.

The enzyme production in large tray SSF was satisfactory and reproducible (Table II). The volumetric productivity in this *in situ* enzyme preparation was 1.35 FPU/ml for cellulase and 570 IU/ml for xylanase, surpassing the volumetric productivities achievable with this strain in submerged fermentation (SF) (1.0 FPU/ml for cellulase and 380 IU/ml for xylanase) (Szakacs, G. and Tengerdy, R.P., Techn. U. Budapest, unpublished data). The volumetric productivities were calculated on the basis of 30 g/100 ml packing density in SSF.

Table I. Lignocellulolytic enzyme production in solid substrate fermentation (SSF)

		Enzyme activities			
Strain	Substrate	FPA	BG	Xylanase	EG
Trichoderma reesei QM 6a	S	2.68 (8)	1.43 (8)	1500 (8)	420 (8)
	W	2.92 (8)	1.82 (8)	2090 (6)	580 (8)
Trichoderma reesei Rut C30	S	9.87 (10)	8.70 (10)	2020 (10)	2410 (12)
	W	10.60 (10)	10.30 (10)	3580 (10)	4120 (10)
Trichoderma hamatum TUB F-105	S	4.87 (6)	6.07 (8)	3470 (4)	960 (4)
	W	6.12 (4)	11.41 (6)	7070 (4)	1580 (4)
Gliocladium sp. TUB F-498	S	5.67 (6)	14.00 (10)	1890 (4)	1510 (6)
	W	1.53 (6)	10.21 (6)	830 (6)	130 (6)
Trichoderma sp. TUB F-426	S	5.02 (8)	6.06 (8)	3050 (8)	1180 (8)
	W				
Penicillium aculeatum NRRL 2129	S	3.87 (12)	18.02 (12)	2480 (12)	1170 (8)
	W	5.92 (10)	33.75 (10)	3060 (8)	1530 (10)
Penicillium funiculosum NRRL 1132	S	4.02 (8)	22.83 (8)	2480 (12)	890 (8)
	W	2.10 (10)	15.30 (10)	260 (10)	330 (10)
Trichoderma sp. TUB F-482	S	3.71 (12)	12.72 (12)	2000 (12)	730 (12)
	W	3.80 (10)	14.52 (10)	1170 (10)	1150 (10)
Trichoderma sp. TUB F-486	S	3.82 (10)	5.01 (10)	1430 (10)	610 (10)
	W	4.60 (10)	10.43 (10)	1670 (10)	840 (10)

Conditions: Fermentation in perforated plastic cups, 75 g inoculated substrate/250 ml cup; inoculum 10^7 spores/g DW; moisture content 75%, t=28°C. The peak enzyme activities are shown on days in parenthesis on two substrates: S = extracted sweet sorghum pulp; W: wheat straw + wheat bran (9+1); FPA = filter paper activity, FPU/g DW; BG = beta glucosidase activity, IU/g DW; xylanase activity, IU/g DW; EG = 1,4-beta-endoglucanase activity, EGU/g DW. The data shown are averages from four parallel cups.

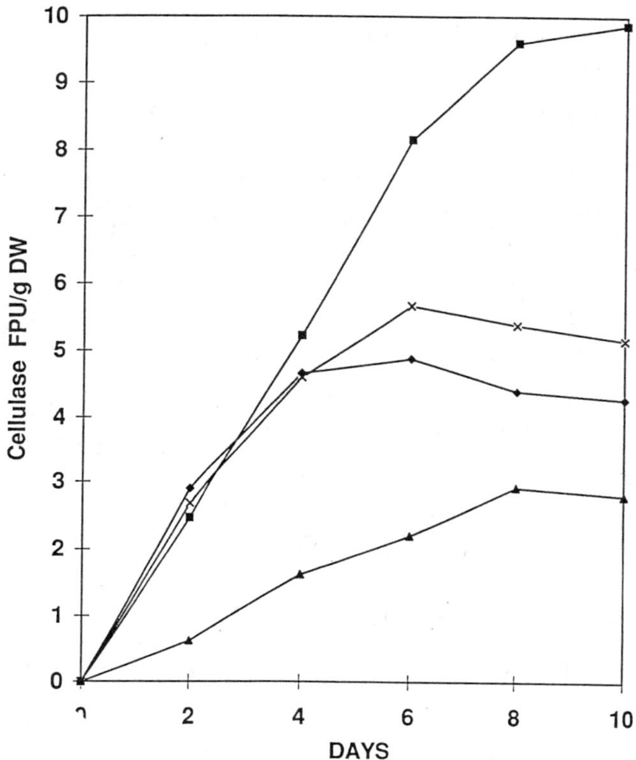

Figure 2. The kinetics of cellulase production on sweet sorghum pulp in solid substrate fermentation. ▲ *T. reesei* QM 6a; ■ *T. reesei* Rut C30; X *Gliocladium* sp. F-498; ♦ *T. hamatum* F-105.

Table II. Enzyme production by *Gliocladium* sp. F-498 in large tray SSF

		Enzyme activities			
Run No.	Final pH	FPA	BG	Xylanase	EG
1	6.0	4.2	13.4	2370	1570
2	5.9	4.1	7.1	1680	1510
3	6.0	4.0	7.5	2560	1150
4	6.1	4.7	11.7	3090	1530

Conditions: 2000g inoculated extracted sweet sorghum pulp in 780x510x80 mm aluminum tray, 5 cm layer thickness, inoculum 10^7 spores/g DW, moisture content 75%, t=28°C, fermentation time 6 days, FPA = filter paper activity, FPU/g DW; BG = beta glucosidase activity, IU/g DW; xylanase activity, IU/g DW; EG = 1,4-beta-endoglucanase activity, EGU/g DW.

The conditions of SSF particularly enhanced the production of xylanase compared with SF: for Rut C30 in SF the volumetric productivity was 200 IU/ml, in SSF 1075 IU/ml (Szakacs, G. and Tengerdy, R.P., Techn. U. Budapest, unpublished data). The most rapid and best xylanase producer was strain F-105 as shown in Figure 3. The peak production, 7070 IU/g DW, corresponds to a volumetric productivity of 2140 IU/ml.

In conclusion, *in situ* enzymes produced by SSF are in sufficiently concentrated form to be applied economically in biotechnology processes. These results show that one of the most important criterium for strain selection in SSF is the substrate specificity under SSF conditions. The wild strains selected under these conditions all produced higher levels of enzymes than *T. reesei* QM 6a, the parent of the highly successful Rut C30 mutant selected under SF conditions. These strains probably have a greater potential for genetic improvement aimed specifically for enhancing enzyme production by SSF.

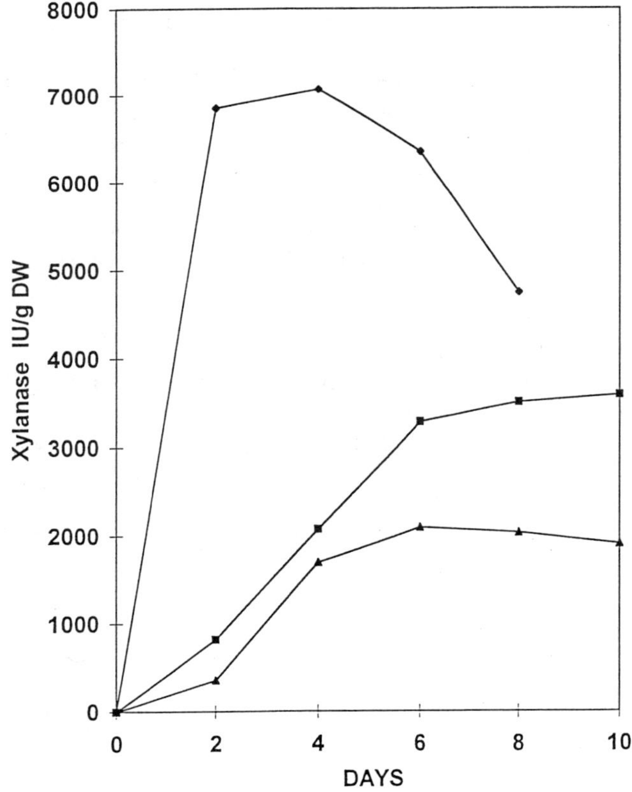

Figure 3. The kinetics of xylanase production on wheat straw in solid substrate fermentation. ▲ *T. reesei* QM 6a; ■ *T. reesei* Rut C30; ♦ *T. hamatum* F-105.

Acknowledgement

This research was supported by a USA-Hungary Science Cooperation Grant NSF INT 8722686, by a USA-Hungary Joint Board Grant No. 307/92, and by a grant from the Hungarian Academy of Sciences, OTKA T-017201.

Literature Cited

1. Lonsane, B.K.; Ghildyal, N.P. In "*Solid State Cultivation*", H.W. Doelle, D.A. Mitchell, C.E. Rolz, Eds.; Elsevier, N.Y. **1992**, pp. 191-209.
2. Persson, I.; Tjerneld, F.; Hahn-Hagerdal, B. *Proc. Biochem.* **1991**, *26*, 65-74.
3. Tengerdy, R.P.; *Journal of Scientific and Industrial Research*, **1996**, *55*, 313-316.
4. Tengerdy, R.P.; Szakacs, G.; Sipocz, J. *Appl. Biochem. Biotechn.* **1996**, *57/58*, 563-569.
5. Pandey, A. *Proc. Biochem.* **1992**, *27*, 109-117.
6. Tengerdy, R.P. In "*Solid State Cultivation*", H.W. Doelle, D.A. Mitchell, C.E. Rolz, Eds.; Elsevier,N.Y. **1992**, pp. 269-282.
7. Tengerdy, R.P.; Schmidt, J.; Nyeste, L.; Castillo, M.R.; Szakacs, G.; Pecs, M.; Sipocz, J. In "*Biomass for Energy, Environment, Agriculture and Industry*", P. Chartier, A. Beenackers, G. Grossi, Eds.; Pergamon Press, Oxford, U.K. **1995**, Vol. 2, pp. 1455-1459.
8. Hach Inc. Manual. In "*Procedures Manual: Systems for Food, Feed and Beverage Analysis*", Hach Inc., Loveland, CO, **1990**, pp. 15-21.
9. Smith, P.K.; Krohn, R.I.; Hermanson, G.T.; Mallia, A.K.; Gardner, F.H.; Provenzano, M.D.; Fujimoto, E.K.; Goeke, N.M.; Olson, B.J.; Klenk, D.C. *Anal. Biochem.* **1985**, *150*, 76-85.
10. Ghose, T.K. *Pure and Appl. Chem.* **1987**, *59*, 257-268.
11. Bailey, M.J.; Nevalainen, K.M.H. *Enzyme Microb. Technol.* **1981**, *3*, 153-157.
12. Bailey, M.J.; Biely, P.; Poutanen, K. *J. Biotechnol.* **1992**, *23*, 257-270.
13. Kubicek, C.P. *Arch. Microbiol.* **1982**, *132*, 349-354.

Chapter 14

MnP Isoenzymes Produced by Two *Pleurotus* Species in Liquid Culture and During Wheat-Straw Solid-State Fermentation

M. J. Martinez, B. Böckle, S. Camarero, F. Guillén, and A. T. Martinez

Centro de Investigaciones Biológicas, Consejo Superior de Investigaciones Cientificas, Velázquez 144, E–28006 Madrid, Spain

Pleurotus eryngii and *P. pulmonarius* are being investigated for straw biopulping. They produce up to seven MnP isoenzymes which have been purified and characterized from solid-state fermentation (SSF) and liquid cultures. *Pleurotus pulmonarius* produces under both conditions a unique MnP, with similar pI but different molecular mass. However, *P. eryngii* secretes five MnP isoenzymes: three in SSF, differing in both molecular mass and pI, and two in liquid culture, differing in pI. The isoenzymes from each fungus differ in their N-terminal sequences. A noteworthy characteristic of six of them is their Mn-independent activity on 2,6-dimethoxyphenol and veratryl alcohol. Among them, those produced under liquid or SSF conditions share several characteristics, including two N-terminal consensus sequences. Finally, the isoenzyme with the highest pI, among those purified from *P. eryngii* SSF, is more related to *Phanerochaete chrysosporium* MnP.

Fungi from the genus *Pleurotus* are being investigated for the biological delignification ("biopulping") of non-woody materials (*1*). Using ^{14}C-lignin wheat-straw it has been shown with *Pleurotus pulmonarius* that lignin mineralization is enhanced by manganese. This suggests the involvement of Mn^{3+} chelates formed by the action of manganese-oxidizing peroxidases (MnP) (*2*). In addition to MnP, laccase and aryl-alcohol oxidase (AAO) are secreted by *Pleurotus* species during straw solid-state fermentation (SSF) (*3*). A lignin peroxidase (LiP) similar to that described in *Phanerochaete chrysosporium* (conidial state *Sporotrichum pruinosum*) and other ligninolytic fungi has not been detected in *Pleurotus* species. It has been demonstrated that laccase from *Pleurotus eryngii* can oxidize phenolic as well as non-phenolic aromatic compounds in the presence of mediators (*4*), but evidence on its natural role are still lacking. However, the structural analysis of lignin during straw SSF showed a preferential attack to phenolic units, which can be oxidized by laccase or MnP (*5*).

Lignin mineralization by *Pleurotus* species seems to be related also to the production of H_2O_2, which is necessary for the action of MnP and involved in the generation of hydroxyl free radical. In these fungi it has been found that extracellular H_2O_2 is generated by the redox cycling of *de novo* synthesized anisylic compounds (*6-8*). Mycelium-associated alcohol and aldehyde dehydrogenases and

0097–6156/96/0655–0183$15.00/0
© 1996 American Chemical Society

extracellular AAO participate in this process. The latter enzyme, an oxidase of polyunsaturated primary alcohols characteristic of all *Pleurotus* species investigated (*1*), has been fully characterized in *P. sajor-caju* (*9*), *P. eryngii* (*10,11*) and *P. ostreatus* (*12*).

MnP was first described in cultures of *P. chrysosporium* growing on glucose-ammonium tartrate medium (*13*). The production of four different MnP isoenzymes was reported in this N-limited medium (*14*). The fungi from the genus *Pleurotus* produce laccase and AAO in glucose-ammonium tartrate media, but no detectable levels of MnP are formed (*11*). However, MnP attained relatively high levels during straw degradation by five *Pleurotus* species (2-5 U g^{-1} of initial substrate) under SSF conditions, compared with those produced by *P. chrysosporium* and *Trametes versicolor* (0.2-0.7 U g^{-1}) (*2,3*). Moreover, high MnP levels (up to 1 U ml^{-1}) were obtained in liquid cultures of *P. eryngii* and *P. pulmonarius* using peptone as the N source (*15,16*). Mn-peroxidases seem to be involved in degradation of lignin *via* Mn^{3+} chelates (*17,18*). However, these enzymes as well as laccases, could participate also in the generation of active oxygen species, through quinone redox cycling. The operation of this system (which produces hydroxyl free radical *via* the Haber-Weis reaction) has been demonstrated in *Pleurotus* (*19*) and its relevance in lignin degradation and involvement of MnP is currently under investigation. In the present study, the different MnP isoenzymes produced by *P. eryngii* and *P. pulmonarius* in liquid culture and during wheat-straw SSF were characterized in order to establish a basis to understand their contribution to lignin degradation processes.

Materials and Methods

Enzyme Purification. MnP isoenzymes from *P. eryngii* CBS 613.91 (IJFM A169) and *P. pulmonarius* CBS 507.85 (strains maintained in 2% malt extract agar) were purified from liquid and SSF cultures grown at 28 °C. The liquid cultures were incubated for 6 days in 1-liter flasks (shaken at 180 rev min^{-1}) with 200 ml of the following medium: 2% glucose, 0.5% peptone, 0.2% yeast extract, 0.1% KH_2PO_4 and 0.05% $MgSO_4 \cdot 7\ H_2O$ (pH 5.5) (*20*). For preparation of SSF inoculum, stationary cultures in 1-liter flasks containing 100 ml of the medium described elsewhere (*11*) were used. After development of a fungal mat (20 days), the mycelia were recovered, homogenized, and incubated at 170 rev min^{-1} for four days in 1-liter flasks containing 200 ml of the same medium. The pellets from each flask were washed, suspended in 300 ml of water, and used to inoculate the straw in one bottle. The SSF cultures were grown for 15 days in a rotary fermenter including six 2-liter bottles, each of them contained 150 g straw (5 cm), 150 ml of water and 300 ml of inoculum. They were incubated with continuous aeration (166 ml min^{-1} wet air per flask) and intermittent agitation of 1 rev min^{-1} (*21*).

The filtrates from liquid media and the water extracts from the straw treated under SSF conditions (obtained by adding 3 liters of water to each flask, agitating 1 h at 200 rev min^{-1}, and vacuum filtering through paper and 0.8 μm filter) were concentrated by ultrafiltration (5-kDa cut-off membrane) and dialyzed against 10 mM sodium tartrate, pH 4.5. The concentrate, approximately 140 ml, was loaded onto a Biorad Q-cartridge using the above buffer (1 ml min^{-1}) and retained fractions were eluted with 1M NaCl. Fractions with MnP activity, estimated by 2,6-dimethoxyphenol (DMP) oxidation as described below, were concentrated and 1 ml samples applied onto a Sephacryl S-200 HR column, using the same buffer (0.8 ml min^{-1}). Fractions containing MnP activity were pooled, concentrated, dialyzed against 10 mM sodium tartrate, pH 5, and 1 ml samples were applied to a Mono-Q anion-exchange column. The MnP isoenzymes were eluted using 0 to 0.25 M NaCl gradient in the latter buffer from 2 to 32 min at 0.8 ml min^{-1}.

Enzymatic Activities. MnP activity was estimated directly by the formation of Mn^{3+}-tartrate complex (ε_{238} 6 500 M^{-1} cm^{-1}) from 0.1 mM $MnSO_4$ in 0.1 M sodium tartrate (pH 5) using 0.1 mM H_2O_2, or indirectly by the Mn-mediated oxidation of 0.1 mM DMP to coerulignone under the same conditions (two moles of DMP form one mole of coerulignone, ε_{469} of 55 000 M^{-1} cm^{-1} in the range of 0-5 µM). Mn-independent peroxidase activities on 2 mM veratryl alcohol (VA) (LiP activity) - measured by the veratraldehyde formed, ε_{310} 9 300 M^{-1} cm^{-1} - and 0.1 mM DMP were also estimated, using 0.1 mM H_2O_2, in 0.1 M sodium tartrate buffer at pH 3 and 5, respectively. AAO activity was determined as the veratraldehyde formed from 5 mM VA in 0.1 M phosphate buffer, pH 6 (*11*). Laccase activity was measured with 5 mM 2,2'-azinobis-(3-ethylbenzothiazoline-6-sulphonate) (ABTS) in 0.1 M acetate buffer, pH 5 (ε_{436} of ABTS cation radical 29 300 M^{-1} cm^{-1}). One unit of enzymatic activity was defined as the amount of enzyme transforming 1 µmol of substrate per minute.

Enzyme Characterization. Protein concentration was determined with the Bradford reagent. IEF was performed in 5% polyacrylamide gels with a thickness of 1 mm and a pH range from 2.5 to 5.5. SDS-PAGE was carried out in 12% polyacrylamide gels, including low molecular-mass standards from Biorad. Protein bands in PAGE and IEF gels were stained with $AgNO_3$ and Coomassie Blue R-250, respectively. Proteins were deglycosylated using Endo-H from Boehringer Manheim. Molecular-mass estimations were also carried out by Superose-12 chromatography (using Biorad standards), and by MALDI/TOF-MS (matrix-assisted laser-desorption ionization/time of fly-mass spectrometry) in sinapic acid matrix, using a Brucker equipment. The N-terminal sequences were obtained by automated Edman degradation of 20 µg of protein in an Applied Biosystems 477A pulsed-liquid protein sequencer.

Kinetic Studies. The kinetic constants for H_2O_2 and Mn^{2+}, corresponding to the Mn^{2+}-peroxidase activity of the different MnP isoenzymes, were calculated by the formation of Mn^{3+}-tartrate (pH 5). Mn-mediated peroxidase activity on phenols was also investigated and the kinetic constants for H_2O_2, Mn^{2+} and DMP obtained for coerulignone formation (pH 5). These enzymes also exhibited Mn-independent peroxidase activities on DMP and VA and the kinetic constants were calculated, respectively, from coerulignone and veratraldehyde formation at pH 3. Finally, the kinetic constants for NADH (ε_{340} 6 220 M^{-1} cm^{-1}) were obtained corresponding to NADH oxidase activity (the optimum pH was around 3.5 but the activity was measured at pH 5 to minimize interferences due to substrate autoxidation). All reactions were carried out in 0.1 M sodium tartrate buffer. Superoxide anion radical generation during MnP oxidation of 0.3 mM NADH was evidenced by reduction of 0.4 mM Nitro Blue Tetrazolium (490 nm) and 30 µM cytochrome-C (550 nm). O_2 consumption due to oxidase activity on NADH and other substrates (using different pH and buffers) and O_2 release due to Mn^{2+}-peroxidase activity (in the presence of different substrates) were measured with an Clark-type O_2 electrode.

Results

Isoenzyme Purification. The same purification scheme was used for the different MnP isoenzymes produced by *P. eryngii* and *P. pulmonarius* in liquid and SSF cultures. The different steps of MnP purification from straw treated with *P. eryngii* under SSF conditions are presented in Fig. 1. During Q-cartridge chromatography MnP was not retained but a portion of the protein, most of the yellow pigment synthesized by the fungus and a part of the brown-colored compounds formed during straw degradation bound to the gel. During Sephacryl S-200 chromatography of the MnP-containing fraction, the MnP activity appeared as a sharp peak. This peak presented a symmetric shape during MnP purification from

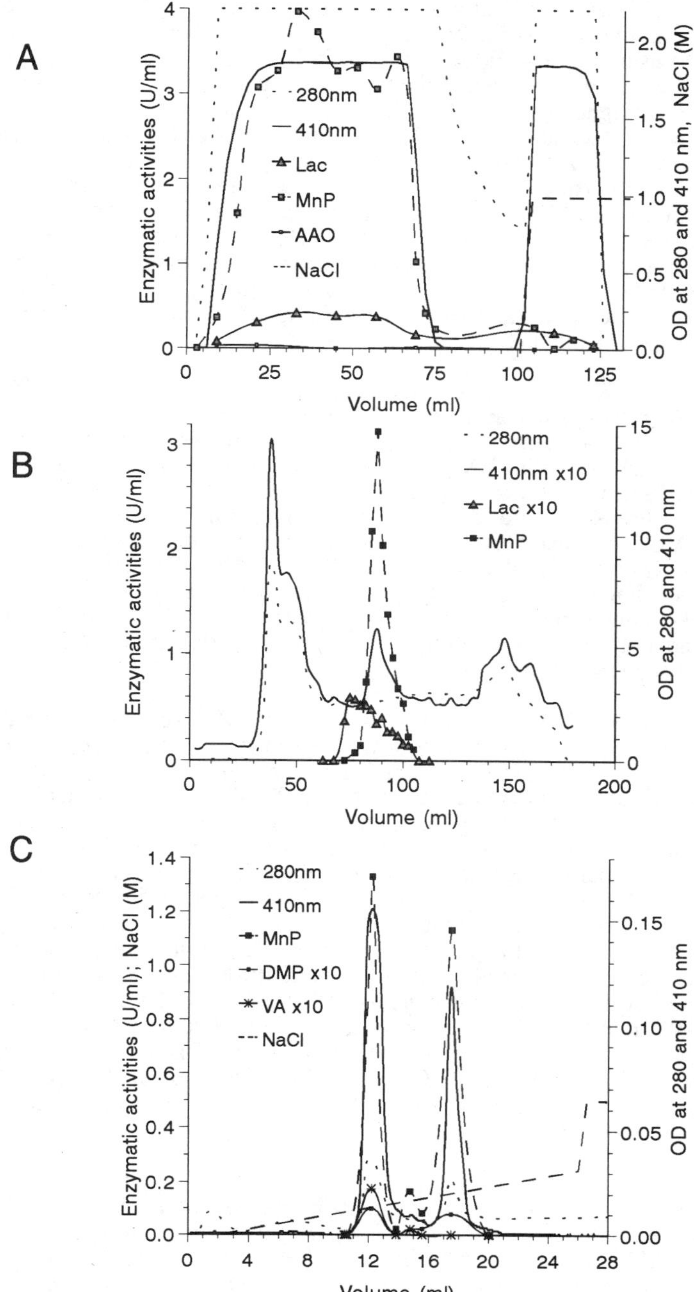

Fig. 1. MnP purification from SSF cultures of *Pleurotus eryngii*, including Q-cartridge (A), Sephacryl S-200 (B) and Mono-Q (C) chromatographies.

liquid cultures and *P. pulmonarius* SSF, but it was asymmetric in the case of *P. eryngii* SSF, suggesting the existence of isoenzymes with different molecular masses. A peak of laccase activity appeared before the MnP peak. In this purification step, most of the protein was eliminated (280 nm profile), as well as some fungal pigment (410 nm) and a large amount of brown-colored compounds from straw degradation absorbing at 280 and 410 nm (SSF purification). MnP purification was completed by Mono-Q chromatography at pH 5, which retained MnP and allowed isoenzyme separation using a slow NaCl gradient. At this purification step, the 410 nm profile was coincident with that of MnP activity, revealing the existence of several MnP isoenzymes. No further hemeproteins were found in the 410 nm profile.

The main difference between purification from liquid and SSF cultures concerned Sephacryl S-200 chromatography. During purification from liquid cultures, the MnP activity represented one of the major peaks at 410 nm. On the contrary, it corresponded with a small peak at 410 nm during enzyme purification from SSF, due to the presence of a large amount of contaminant compounds from straw degradation. In spite of these differences, the Mono-Q chromatography enabled isoenzyme purification in all the cases.

The MnP isoenzyme patterns produced by *P. eryngii* and *P. pulmonarius* in liquid and SSF culture are shown in the four Mono-Q chromatograms presented in Fig. 2 (the 410 nm profile coincided with MnP activity). A single MnP peak with the same elution volume, was obtained from both liquid and SSF cultures of *P. pulmonarius* (called MnP-PPL and MnP-PPS, respectively). However, two main MnP peaks with different elution volumes in each case, were obtained from liquid (MnP-PEL1 and MnP-PEL2) and SSF cultures (MnP-PES1 and MnP-PES3) of *P.*

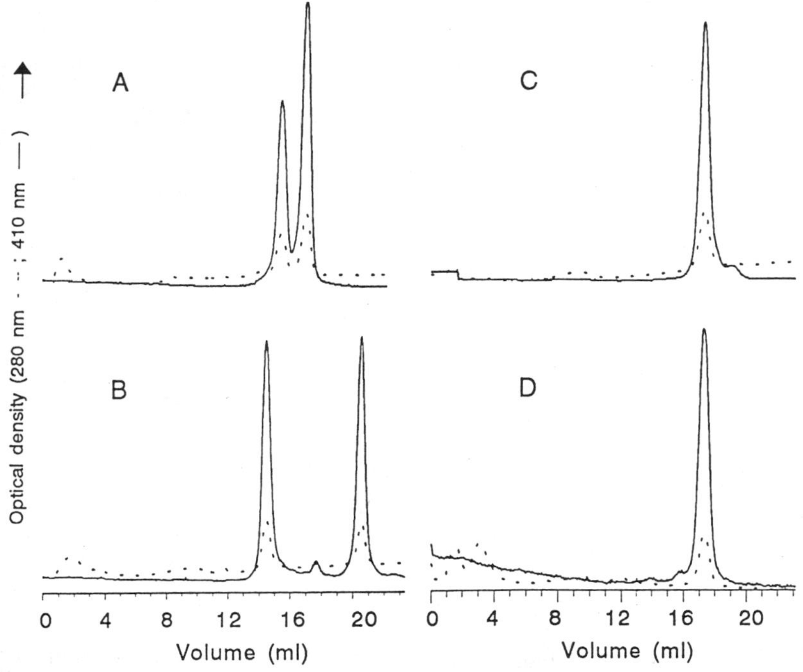

Fig. 2. MnP isoenzymes produced by *Pleurotus eryngii* (A,B) and *P. pulmonarius* (C,D) in liquid medium (A,C) and during straw SSF (B,D) (Mono-Q profiles, 0-0.25 M NaCl gradient from 2 to 32 min at 0.8 ml min^{-1}.)

eryngii. A minor peak (MnP-PES2) was also obtained from SSF cultures of *P. eryngii*. Each of these peaks was collected, concentrated, dialyzed and stored at -80 °C for protein characterization and determination of kinetic constants.

Isoenzyme Characterization. A comparison of the three MnP isoenzymes found in liquid cultures of *P. eryngii* and *P. pulmonarius*, and the four isoenzymes identified from wheat straw after fungal SSF is presented in Table I. Most of them

Table I. Characteristics of the MnP isoenzymes purified from liquid and SSF cultures of *Pleurotus pulmonarius* and *P. eryngii*

	Mono-Q (ml)	*Mr[a] (kDa)*	*pI (pH)*	A_{410}/A_{280}	*N-terminus*
Liquid:					
P. pulmonarius MnP-PPL	18.8	43/41	3.55	4.8	A-CAD-R-T
P. eryngii MnP-PEL1	14.9	43/42	3.70	4.5	A-CAD-R-T
P. eryngii MnP-PEL2	16.5	43/41	3.65	4.5	A-CDD-R-T
SSF:					
P. pulmonarius MnP-PPS	18.8	45/44	3.55	4.9	V-CAT-Q-T
P. eryngii MnP-PES1	14.4	45/43	3.67	4.9	V-CAT-Q-T
P. eryngii MnP-PES2	17.6	45/nd[b]	3.65	4.9	X-PAT-Q-T
P. eryngii MnP-PES3	20.7	42/41	3.80	5.5	V-PAD-N-V

[a]complete and deglycosylated protein; [b] not-determined

showed different elution volumes in Mono-Q chromatography, as demonstrated by co-chromatography of a mixture containing all of them (Fig. 3). The PAGE and IEF of purified isoenzymes showed a single band, revealing that they are homogeneous with respect to molecular mass and charge. Although the two MnP

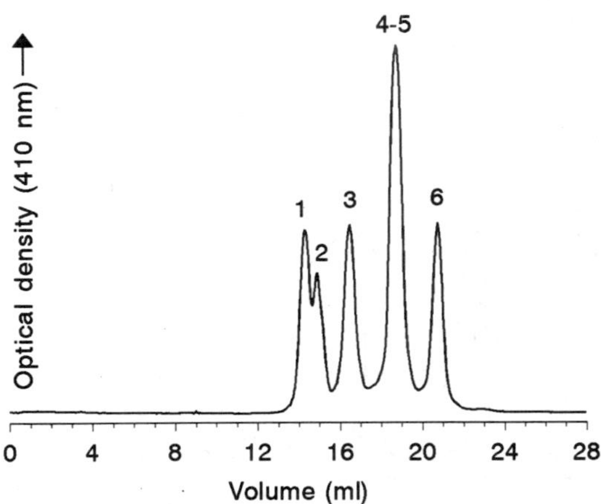

Fig. 3. Co-chromatography of the six major MnP isoenzymes purified from *Pleurotus* species: MnP-PES1 (1), MnP-PEL1 (2), MnP-PEL2 (3), MnP-PPL (4), MnP-PPS (5) and MnP-PES3 (6) (Mono-Q profiles using a 0 to 0.25 M NaCl gradient from 2 to 32 min at 0.8 ml min^{-1}).

isoenzymes from liquid and SSF cultures of *P. pulmonarius* showed the same elution volume and pI (3.55), it was found by SDS-PAGE that they differ in molecular mass (43 and 45 kDa for liquid and SSF isoenzyme, respectively). On the contrary, the two isoenzymes found in liquid cultures of *P. eryngii* showed different pI (3.65 and 3.70) but the same molecular mass (43 kDa). Finally, the two major isoenzymes from *P. eryngii* SSF cultures differ in both pI (3,67 and 3.80) and molecular masses (45 and 42 kDa) (the minor isoenzyme, MnP-PES2, presents a pI of 3.65 and a molecular mass of 45 kDa). Although SDS-PAGE was used to differentiate *Pleurotus* MnP isoenzymes and compare their molecular masses with those reported in the literature, the results from Superose-12 chromatography (39 kDa) and MALDI/TOF-MS (37 kDa) of *P. eryngii* MnP from liquid culture, indicated that an overestimation of MnP molecular-mass is obtained by this method. The six major isoenzymes are glycoproteins, as demonstrated by 2.5-5% decrease of the molecular mass after deglycosylation. The A_{410}/A_{280} ratios of the different isoenzymes were between 4.5 and 5, with the exception of MnP-PES3 which showed a ratio of 5.5. The N-terminal sequences of the isoenzymes produced in liquid cultures were similar (MnP-PEL1: ATDADGRTTANAACCVLF; MnP-PEL2: ATCDDGRTTADAACCILF; and MnP-PPL: ATCADGRTTANAACC VLF). The complete N-terminal sequencing of SSF isoenzymes was hampered by some interferences, but significant differences in the first amino acids were found with respect to those produced in liquid cultures.

Catalytic Properties of the MnP Isoenzymes. The six major MnP isoenzymes purified and characterized from liquid and SSF cultures of *P. eryngii* and *P. pulmonarius* (as well as the minor isoenzyme from *P. eryngii* SSF) presented high peroxidase activity on Mn^{2+} (in the presence of adequate chelators the Mn^{3+} rapidly oxidized phenolic compounds like DMP). Consequently, they were considered as manganese peroxidases (EC 1.11.1.13) although they presented additional Mn-independent activities on phenolic (e.g. DMP) and non-phenolic compounds (e.g. VA) and oxidase activity on NADH.

The extent of Mn-mediated and Mn-independent (verified in the presence of EDTA) peroxidase activities of MnP from *P. eryngii* and *P. pulmonarius* was strongly dependent on the pH and the characteristics and concentration of substrates (according to K_m values presented below). The optimum pH for Mn-independent peroxidase activities on DMP, VA and ABTS was 3 (the activities on the two former substrates was nearly absent at pH 5) while optimum pH for Mn^{2+} peroxidase and Mn-mediated peroxidase (on DMP) activities were 5 and 4.5 respectively (being nearly absent at pH 3). The ratio between Mn-independent and Mn-mediated activities (at 0.1 mM concentration) was very high in the case of ABTS (50% at pH 5) and lower in the case of DMP (15-20% at pH 5). The peroxidase activity on VA was not affected by the presence of Mn^{2+}. Finally, the Mn^{2+}-peroxidase activity also depends on the efficiency of buffers to produce Mn^{3+} chelates. High Mn-mediated peroxidase activity on DMP was found in 100 mM tartrate or 4 mM oxalate (but lower activity in 100 mM succinate). O_2 production was observed as a result of peroxidase activity on Mn^{2+}. This was due to partial oxidation of H_2O_2 by the Mn^{3+} produced by the enzyme. The addition of DMP to the reaction mixture inhibited O_2 release because the Mn^{3+} produced reacted with DMP. No O_2 was produced during Mn-independent oxidation of VA and DMP. Concerning the stoichiometry of the Mn-mediated oxidation of DMP at pH 5 (where Mn-independent activity was very low), it was observed that the activity on DMP (0.1 mM), calculated by the formation of coerulignone, was 44% of the activity on Mn^{2+} (0.1 mM), measured by the formation of Mn^{3+}-tartrate, indicating that four moles of Mn^{3+} were necessary to oxidize 2 moles of DMP producing one mole of coerulignone.

Mn-independent oxidase activity (measured as O_2 consumption for substrate comparison) was observed with 1 mM NADH and NADPH (similar activity), but

Mn^{2+} was necessary for oxidation of 1 mM dithiothreitol and glutathione (near 20% O_2 consumption compared to NADH). The consumption of O_2 was nearly absent with β-mercaptoethanol. NADH oxidation was similar in sodium tartrate, methyl malonate, L-malate and D-malate, but the reaction was much lower in sodium citrate, phosphate or oxalate. In the presence of NADH and Mn^{2+}, H_2O_2-independent oxidation of DMP was obtained at pH 5. The reduction of Nitro Blue Tetrazolium and cytochrome-*c* during NADH oxidation suggests the production of superoxide anion radical, which is probably reduced to H_2O_2 by Mn^{2+}. Superoxide production was supported also by the increased oxidation of phenol red by MnP (pH 3) in the presence of different oxidase substrates and Mn^{2+}. This was due to superoxide reduction by Mn^{2+} producing H_2O_2, used by MnP to oxidize phenol red.

EDTA inhibited all the reactions investigated, except the Mn-independent oxidation of DMP. Sodium metavanadate (1-50 μM) inhibited peroxidase activity on VA, while all other reactions were not affected. Oxalate (over 1 mM) inhibited both Mn-independent peroxidase reactions on DMP and VA.

The kinetic constants of the 7 MnP isoenzymes investigated are presented in Tables II and III. As shown in Table II, the MnP isoenzymes present K_m values for Mn^{2+} in the micromolar range, while those for VA (Mn-independent peroxidase activity) are in the millimolar range (with the exception of MnP-PE3 which did not oxidize VA). The K_m for DMP in the absence of Mn^{2+}, presented intermediate values (attaining 1 mM for MnP-PE3). The K_m values for the H_2O_2 were lower at pH 3 (used for Mn-independent peroxidase reactions) than at pH 5 (used for Mn^{2+} or Mn-mediated peroxidase reactions).

Table II. Substrate affinity of MnP isoenzymes from liquid and SSF cultures of *Pleurotus pulmonarius* and *P. eryngii*: K_m values (μM) for Mn^{2+} peroxidase, Mn-mediated peroxidase activity on DMP, Mn-independent peroxidase activities on DMP and VA, and NADH oxidase activities

	Peroxidase activity								
	on Mn^{2+}		*Mn-mediated*			*Mn-independent*[a]			*Oxidase*
K_m (μM) for:	H_2O_2	Mn^{2+}	H_2O_2	DMP	Mn^{2+}	H_2O_2	DMP	VA	NADH
Liquid culture:									
MnP-PPL	10	24	10	8	12	3	250	3800	60
MnP-PEL1	6	20	6	9	16	5	200	3500	90
MnP-PEL2	9	19	8	9	14	3	300	3000	80
SSF:									
MnP-PPS	9	43	9	9	22	2	250	3800	70
MnP-PES1	9	68	8	8	40	2	200	3500	90
MnP-PES2	nd[b]	120	nd	nd	38	nd	nd	nd	nd
MnP-PES3	10	200	10	10	60	3	1000	0	50

[a] Mn-independent reactions carried out at pH 3 (others at pH 5); [b] not-determined.

A comparison of the maximal peroxidase activities on Mn^{2+}, Mn-mediated peroxidase activity on DMP, Mn-independent peroxidase activities on DMP and VA, and oxidase activity on NADH, of the different *Pleurotus* MnP isoenzymes from liquid and SSF cultures is presented in Table III. The highest activities were obtained for the oxidation of Mn^{2+} in the presence of H_2O_2 (estimated by the formation of Mn^{3+}-tartrate). The Mn-mediated peroxidase activity on DMP was, as an average, 25% of the Mn^{2+}-peroxidase activity measured as Mn^{3}-tartrate. In spite

Table III. A comparison of activities of MnP isoenzymes from liquid and SSF cultures of *Pleurotus pulmonarius* and *P. eryngii*: V_{max} values (U mg^{-1} protein)[a]

	Peroxidase				
	with Mn^{2+}		*without* Mn^{2+}		*Oxidase*
V_{max} (U mg^{-1}) on:	Mn^{2+}	DMP	DMP	VA	NADH
Liquid culture:					
MnP-PPL	115	32	28	17	12
MnP-PEL1	142	32	25	18	10
MnP-PEL2	148	34	24	18	8
SSF:					
MnP-PPS	70	20	9	4	4
MnP-PES1	110	32	8	5	3
MnP-PES2	120	nd[b]	nd	nd	nd
MnP-PES3	114	27	5	0	0

[a]Mn-independent reactions carried out at pH 3 (others at pH 5); [b] not-determined.

of differences in K_m values (Table II), the maximal Mn-mediated and Mn-independent peroxidase activities on DMP were similar for the different isoenzymes. The maximal peroxidase activities on VA and oxidase activities on NADH were lower than the maximal peroxidase activities on DMP. As a rule, the isoenzymes from SSF presented lower V_{max} values for the different substrates.

Discussion

The existence of isoenzymes - i.e. proteins catalyzing the same enzymatic reaction but differing in molecular mass, charge or kinetic constants, and often codified by different genes - is a characteristic of many biological systems. The role of the different isoforms is difficult to establish but they could be involved in metabolic regulation, morphogenesis or environmental adaptation. The existence of isoforms is a common characteristic of many extracellular enzymes from filamentous fungi, where isoenzyme patterns have been used in the past as a characteristic for species identification or strain improvement. For most of the enzymes involved in lignin degradation by white-rot basidiomycetes, different isoforms have been described. This is the case of: i) 6 LiP isoenzymes in *P. chrysosporium* (*22,23*), ii) 4 MnP isoenzymes in N-limited liquid culture of *P. chrysosporium* and a fifth isoenzyme on paper pulp (*24,25*), iii) a variable number of laccase isoenzymes in the different white-rot basidiomycetes (*26*); and iv) two AAO isoenzymes in *P. sajor-caju* and *P. floridanus* (*1*).

Since its description in 1984 (*13*), MnP has been reported in more than 50 species of basidiomycetes, which can be classified (*27*) in different genera of Agaricales (*Agaricus, Armillaria* and *Lentinula*), Cantharellales (*Sparassis*), Ganodermatales (*Ganoderma*), Hericiales (*Hericium*), Hymenochaetales (*Phellinus*), Poriales (*Bjerkandera, Poria* (synonym *Ceriporiopsis*), *Coriolopsis, Daedaleopsis, Dichomitus, Meripilus, Panus, Perenniporia, Phaeolus, Piptoporus, Pleurotus, Polyporus, Rigidoporus* and *Trametes*) and Stereales (*Phanerochaete, Phlebia* (synonym *Merulius*), *Pulcherricium* and *Stereum*) (*22,28-30*). MnP isoenzymes have been purified from at least the 14 fungal species listed in Table IV. The MnP from *P. chrysosporium* (*41-44*), *Phlebia radiata* (*32*) and *T. versicolor* (*37,45*) have

been largely investigated with respect to their catalytic and structural characteristics. The Mn-oxidizing peroxidases of *Pleurotus* species differ from those of the three latter species not only in culture conditions for enzyme production (they are produced in Mn-free glucose-peptone medium) but also in their catalytic properties. Since Mn^{2+} is the best substrate of the seven isoenzymes purified from cultures of *P. pulmonarius* (two isoenzymes, one from liquid and another from SSF cultures) and *P. eryngii* (five isoenzymes, two from liquid and three from SSF cultures), they are considered as manganese peroxidases, although most of them present also Mn-independent enzymatic activities (namely NADH oxidase, DMP peroxidase and VA peroxidase). NADH oxidase activity of MnP isoenzymes has been reported in *P. chrysosporium* (*46*), *Lentinula edodes* (synonym *Lentinus edodes*) (*29,31*) *P. radiata* (*32*) and *P. ostreatus* MnP (*36*) (Table IV). Mn-dependence has been considered as a characteristic of the oxidation of phenolic and related substrates by MnP, although some exceptions have been reported, as discussed below.

The MnP isoenzymes purified from *P. eryngii* and *P. pulmonarius* are hemeproteins with 2.5-5% carbohydrate content. The characteristics of those from liquid cultures have been already reported (*2,16*). Among the 7 isoenzymes studied, only MnP-PES3 (from *P. eryngii* SSF) does not present peroxidase activity on VA, being in this aspect similar to the typical MnP from *P. chrysosporium* and other fungi. However, it differs in the extremely lower NADH oxidase activity (V_{max} < 0.5 U mg^{-1}). Among the *Pleurotus* isoenzymes, MnP-PES3 not only differs in the low NADH oxidase and the absence of VA peroxidase activity but also in other characteristics, showing different N-terminal sequence, the highest pI and A_{410}/A_{280} ratio (close to that of *P. chrysosporium* MnP), and the lowest molecular mass and lowest affinities for Mn^{2+} and DMP (in the absence of Mn^{2+}). Moreover it exhibits an anomalous behavior in Mono-Q chromatography (pH 4.5), since it presented a higher elution volume than the two other *P. eryngii* isoenzymes from SSF but its pI was the highest. A similar situation has been described during purification of MnP and laccase isoenzymes from wheat straw treated with *P. radiata* under SSF conditions (*47*). This behavior could be due to the formation of polyelectrolyte complexes between enzyme and phenolic compounds released from straw degradation, as reported for cellulolytic enzymes (*48*). The nature and stability of these complexes, which should be destroyed during IEF (*47*), should be investigated in detail since they could also affect the catalytic properties of the isoenzymes (this seems to be the case for MnP-PES3). Among the other six isoenzymes produced by *P. eryngii* and *P. pulmonarius*, some differences can be observed between those from liquid and from SSF cultures. These include higher molecular masses of isoenzymes from SSF (45 kDa) compared to those from liquid culture (43 kDa), different N-terminal sequences (A---D-R in liquid, and V---T-Q in SSF isoenzymes), and lower Mn^{2+} affinities, maximal Mn-independent peroxidase activities on DMP and VA, and oxidase activity on NADH of SSF isoenzymes compared with those from liquid. The existence of differences between MnP isoenzymes produced in liquid culture and under SSF conditions has been described also in *P. chrysosporium* (*24*) and *Poria subvermispora* (synonym *Ceriporiopsis subvermispora*) (*38*). Among SSF isoenzymes, MnP-PPS from *P. pulmonarius* is similar to MnP-PES1 from *P. eryngii*, showing comparable N-terminal sequence, molecular mass and catalytic properties but differing in the molecular mass of the deglycosylated protein, the pI and the Mono-Q elution volume. The differences between N-terminal sequences of the MnP isoenzymes (together with differences in pI and other enzyme characteristics) suggest that they are encoded by different genes (five in *P. eryngii* and two in *P. pulmonarius*), as occurs for MnP isoenzymes from other ligninolytic fungi (*49*).

The molecular masses of the MnP isoenzymes from *P. eryngii* and *P. pulmonarius* (estimated by SDS-PAGE) are in the range of MnP from most ligninolytic fungi (42-45 kDa), with the exceptions of *P. radiata* (*32*) and *P.*

Table IV. MnP isoenzymes described from different fungi

		Mr	*pI*	*MIP*	*NADH*	*VA*	*ref.*
Phanerochaete chrysosporium	L	45	4.5	-	+	-	*(25)*
	L	45	4.9	-	+	-	*(25)*
	L	45	4.2				*(25)*
	S	45	4.9				*(24)*
Lentinula edodes	S	45	3.2		+	+	*(31)*
Phlebia radiata	L	49	3.8		+	-	*(32)*
Panus tigrinus	S	43	2.95				*(33)*
	S	43	3.2				*(33)*
Rigidoporus lignosus	S	42	3.5				*(34)*
	S	42	3.7				*(34)*
Phlebia brevispora	L	45	3.3				*(35)*
	L	45	4.2				*(35)*
	L	45	5.3				*(35)*
Pleurotus ostreatus	L	45	3.5		+		*(36)*
	L	45	3.8		+		*(36)*
	L	45	4.3	+	+	-	*(36)*
Trametes versicolor	L	44	2.96				*(37)*
	L	45	2.93				*(37)*
	L	45	3.09				*(37)*
	L	45	3.17				*(37)*
	L	45	3.06				*(37)*
Poria subvermispora	L	53	4.2		+		*(38)*
	L	53	4.3				*(38)*
	L	53	4.4				*(38)*
	L	53	4.6				*(38)*
	S	63	3.2				*(38)*
	S	63	3.3				*(38)*
	S	63	3.4				*(38)*
Agaricus bisporus	S		3.5				*(39)*
Phanerochaete sordida	L	45	3.3				*(40)*
	L	45	4.2				*(40)*
	L	45	5.3				*(40)*
Lentinula edodes	L	59	5.3		-	+	*(29)*
Pleurotus pulmonarius	L	43	3.55	+	+	+	
	S	45	3.55	+	+	+	
Pleurotus eryngii	L	43	3.65	+	+	+	
	L	43	3.7	+	+	+	
	S	45	3.67	+	+	+	
	S	45	3.65	+	+	+	
	S	42	3.8	+	+	-	

Abbreviations: MIP: Mn-independent peroxidase activity on phenols, NADH= oxidase activity on NADH, VA= peroxidase activity on veratryl alcohol, L= liquid culture, S= solid-state fermentation

subvermispora (*38*). Moreover, these *Pleurotus* isoenzymes belong to the group of more acidic MnP, with pI in the range of 3-4 (isoenzymes with higher pI are produced by *P. chrysosporium* and other ligninolytic fungi). Their carbohydrate content is similar to that reported in *T. versicolor* but lower than that found in *P. chrysosporium* MnP. The lower A_{410}/A_{280} ratio compared with *P. chrysosporium* MnP (*50*), except for MnP-PES3, could be related to the tyrosine content of *Pleurotus* MnP, as reported for *P. eryngii* MnP-PEL1 and MnP-PEL2 (*16*). The N-terminal sequences of the MnP isoenzymes from *P. eryngii* and *P. pulmonarius* liquid cultures show the highest identities with those reported for *P. ostreatus* (*51*) and *T. versicolor* (*37*) MnP.

Nevertheless, the most noteworthy characteristic of MnP isoenzymes from *P. eryngii* and *P. pulmonarius*, with the exception of isoenzyme MnP-PES3, concerns the Mn-independent peroxidase activities on phenolic (as DMP) and non-phenolic aromatic compounds (as VA). Among a variety of substrates and dyes, *P. chrysosporium* MnP only exhibits a very low Mn-independent activity on pinacynol (*50,52*). Among the MnP isoenzymes investigated in different fungi, the existence of some Mn-independent activity has been reported only in *P. ostreatus* (ABTS) and *L. edodes* (VA) (*31,36*), but the corresponding activities are not well documented. A more extended Mn-independent activity, first detected in liquid culture MnP (*2,16,53*), is described here for six *Pleurotus* isoenzymes. The results obtained show that their maximal Mn-independent activity on DMP and VA is attained at pH 3 and relatively high substrate concentration since substrate affinities are, respectively, one and two order of magnitude higher than those found for Mn^{2+}. Increased redox-potential at low pH, as suggested for a soybean peroxidase (*54*), could enable oxidation of VA by the six MnP isoenzymes from *Pleurotus*. These isoenzymes share some catalytic properties with LiP, namely Mn-independent peroxidase activity on VA and VA-mediated oxidation of KTBA (α-keto-γ-methylthiobutiric acid) (*55*), suggesting that they could play an important role in lignin degradation by *Pleurotus* species. These results, together with recent findings concerning Mn^{2+} oxidation by LiP (*56*) and the above-mentioned VA oxidation by plant peroxidases (*54*), evidence that catalytic properties of the peroxidases produced by ligninolytic fungi should be carefully re-evaluated in order to establish their participation in lignin biodegradation.

The authors thank J. Varela for N-terminal analyses and C. Rodríguez for her skillful technical assistance. This research has been partially funded by the Spanish Biotechnology Programme and the AIR Project of the European Union "Biological delignification in paper manufacture" (AIR2-CT93-1219).

Literature Cited

1. Martínez, A. T.; Camarero, S.; Guillén, F.; Gutiérrez, A.; Muñoz, C.; Varela, E.; Martínez, M. J.; Barrasa, J. M.; Ruel, K.; Pelayo, M. *FEMS Microbiol. Rev.* **1994**, *13*, 265-274.
2. Camarero, S.; Böckle, B.; Martínez, M. J.; Martínez, A. T. *Appl. Environ. Microbiol.* **1996**, *62*, 1070-1072.
3. Camarero, S.; Martínez, M. J.; Martínez, A. T. In *Advances in Solid State Fermentation;* Roussos, S.; Lonsane, B. K.; Raimbault, M.; Viniegra-Gonzalez, G. Eds.; Kluwer Acad. Publ.: Dordrecht, 1996; pp 335-345.
4. Martínez, M. J.; Muñoz, C.; Guillén, F.; Martínez, A. T. *Appl. Microbiol. Biotechnol.* **1994**, *41*, 500-504.
5. Camarero, S.; Galletti, G. C.; Martínez, A. T. *Appl. Environ. Microbiol.* **1994**, *60*, 4509-4516.

6. Guillén, F.; Evans, C. S. *Appl. Environ. Microbiol.* **1994**, *60*, 2811-2817.
7. Guillén, F.; Martínez, A. T.; Martínez, M. J.; Evans, C. S. *Appl. Microbiol. Biotechnol.* **1994**, *41*, 465-470.
8. Gutiérrez, A.; Caramelo, L.; Prieto, A.; Martínez, M. J.; Martínez, A. T. *Appl. Environ. Microbiol.* **1994**, *60*, 1783-1788.
9. Bourbonnais, R.; Paice, M. G. *Biochem. J.* **1988**, *255*, 445-450.
10. Guillén, F.; Martínez, A. T.; Martínez, M. J. *Appl. Microbiol. Biotechnol.* **1990**, *32*, 465-469.
11. Guillén, F.; Martínez, A. T.; Martínez, M. J. *Eur. J. Biochem.* **1992**, *209*, 603-611.
12. Sannia, G.; Limongi, P.; Cocca, E.; Buonocore, F.; Nitti, G.; Giardina, P. *Biochim. Biophys. Acta Gen. Subj.* **1991**, *1073*, 114-119.
13. Kuwahara, M.; Glenn, J. K.; Morgan, M. A.; Gold, M. H. *FEBS Lett.* **1984**, *169*, 247-250.
14. Kirk, T. K.; Croan, C.; Tien, M.; Murtagh, K. E.; Farrell, R. L. *Enzyme Microb. Technol.* **1986**, *8*, 27-32.
15. Martínez, A. T.; Ruiz, F. J.; Martínez, M. J. *Abs. 5th Intern. Mycol. Congr., Vancouver, 14-21 August* **1994**, p 134.
16. Martínez, M. J.; Ruiz-Dueñas, F. J.; Guillén, F.; Martínez, A. T. *Eur. J. Biochem.* **1996**, *237*, 424-432.
17. Hammel, K. E.; Tardone, P. J.; Moen, M. A.; Price, L. A. *Arch. Biochem. Biophys.* **1989**, *270*, 404-409.
18. Bao, W. L.; Fukushima, Y.; Jensen, K. A.; Moen, M. A.; Hammel, K. E. *FEBS Lett.* **1994**, *354*, 297-300.
19. Guillén, F.; Martínez, M. J.; Martínez, A. T. In *Biotechnology in the Pulp and Paper Industry: Recent Advances in Applied and Fundamental Research*; Messner, K.; Srebotnik, E. Eds.; Facultas-Universitätsverlag: Vienna, 1996; pp 389-392.
20. Kimura, Y.; Asada, Y.; Kuwahara, M. *Appl. Microbiol. Biotechnol.* **1990**, *32*, 436-442.
21. Valmaseda, M.; Martínez, M. J.; Martínez, A. T. *Appl. Microbiol. Biotechnol.* **1991**, *35*, 817-823.
22. Orth, A. B.; Tien, M. In *The Mycota. II. Genetics and Biotechnology*; Kück, U. Ed.; Springer-Verlag: Berlin, 1995; pp 287-302.
23. Tien, M. *CRC Crit. Rev. Microbiol.* **1987**, *15*, 141-168.
24. Datta, A.; Bettermann, A.; Kirk, T. K. *Appl. Environ. Microbiol.* **1991**, *57*, 1453-1460.
25. Pease, E. A.; Tien, M. *J. Bacteriol.* **1992**, *174*, 3532-3540.
26. Thurston, C. F. *Microbiology-UK* **1994**, *140*, 19-26.
27. Hawksworth, D. L.; Kirk, P. M.; Sutton, B. C.; Pleger, D. N. *Dictionary of the Fungi.* CAB international: London, 1995.
28. Peláez, F.; Martínez, M. J.; Martínez, A. T. *Mycol. Res.* **1995**, *99*, 37-42.
29. Buswell, J. A.; Cai, Y. J.; Chang, S. T. *FEMS Microbiol. Lett.* **1995**, *128*, 81-87.
30. Hatakka, A. *FEMS Microbiol. Rev.* **1994**, *13*, 125-135.
31. Forrester, I. T.; Grabski, A. C.; Mishra, C.; Kelley, B. L.; Strickland, W. N.; Leatham, G. F.; Burgess, R. R. *Appl. Microbiol. Biotechnol.* **1990**, *33*, 359-365.
32. Karhunen, E.; Kantelinen, A.; Niku-Paavola, M.-L. *Arch. Biochem. Biophys.* **1990**, *279*, 25-31.
33. Maltseva, O. V.; Niku-Paavola, M.-L.; Leontievsky, A. A.; Myasoedova, N. M.; Golovleva, L. A. *Biotechnol. Appl. Biochem.* **1991**, *13*, 291-302.
34. Galliano, H.; Gas, G.; Seris, J. L.; Boudet, A. M. *Enzyme Microb. Technol.* **1991**, *13*, 478-482.
35. Rüttimann, C.; Schwember, E.; Salas, L.; Cullen, D.; Vicuña, R. *Biotechnol. Appl. Biochem.* **1992**, *16*, 64-76.

36. Bekker, E. G.; Pirtskhalaishvili, D. O.; Élisashvili, V. I.; Sinitsyn, A. P. *Biokhimiya, Engl. Transl.* **1992** (1993), *57*, 863-868.
37. Johansson, T.; Welinder, K. G.; Nyman, P. O. *Arch. Biochem. Biophys.* **1993**, *300*, 57-62.
38. Lobos, S.; Larrain, J.; Salas, L.; Cullen, D.; Vicuña, R. *Microbiology-UK* **1994**, *140*, 2691-2698.
39. Bonnen, A. M.; Anton, L. H.; Orth, A. B. *Appl. Environ. Microbiol.* **1994**, *60*, 960-965.
40. Rüttimann-Johnson, C.; Cullen, D.; Lamar, R. T. *Appl. Environ. Microbiol.* **1994**, *60*, 599-605.
41. Wariishi, H.; Akileswaran, L.; Gold, M. H. *Biochemistry* **1988**, *27*, 5365-5370.
42. Wariishi, H.; Dunford, H. B.; MacDonald, I. D.; Gold, M. H. *J. Biol. Chem.* **1989**, *264*, 3335-3340.
43. Pribnow, D.; Mayfield, M. B.; Nipper, V. J.; Brown, J. A.; Gold, M. H. *J. Biol. Chem.* **1989**, *264*, 5036-5040.
44. Pease, E. A.; Andrawis, A.; Tien, M. *J. Biol. Chem.* **1989**, *264*, 13531-13535.
45. Johansson, T.; Nyman, P. O. *Acta Chem. Scand.* **1987**, *41*, 762-765.
46. Asada, Y.; Miyabe, M.; Kukawa, M.; Kuwahara, M. *Agric. Biol. Chem.* **1986**, *50*, 525-529.
47. Vares, T.; Kalsi, M.; Hatakka, A. *Appl. Environ. Microbiol.* **1995**, *61*, 3515-3520.
48. Blazej, A.; Kosik, M. In *Cellulosics: Pulp, Fibre and Environmental Aspects*; Kennedy, J. F.; Phillips, G. O.; Williams, P. A. Eds.; Ellis Horwood: New York, 1993; pp 397-407.
49. Gold, M. H.; Alic, M. *Microbiol. Rev.* **1993**, *57*, 605-622
50. Glenn, J. K.; Gold, M. H. *Arch. Biochem. Biophys.* **1985**, *242*, 329-341.
51. Asada, Y.; Watanabe, A.; Irie, T.; Nakayama, T.; Kuwahara, M. *Biochim. Biophys. Acta* **1995**, *1251*, 205-209.
52. Glenn, J. K.; Akileswaran, L.; Gold, M. H. *Arch. Biochem. Biophys.* **1986**, *251*, 688-696.
53. Martínez, M. J.; Martínez, A. T. In *Biotechnology in the Pulp and Paper Industry: Recent Advances in Applied and Fundamental Research*; Messner, K.; Srebotnik, E. Eds.; Facultas-Universitätsverlag: Vienna, 1996; pp 417-420. *Abs. 6th Intern. Conf. Biotechnology in the Pulp and Paper Industry, Vienna, 11-15 June* **1995**, p 147.
54. McEldoon, J. P.; Pokora, A. R.; Dordick, J. S. *Enzyme Microb. Technol.* **1995**, *17*, 359-365.
55. Caramelo, L.; Martínez, M. J.; Martínez, A. T. *Abs. IUMS Congr., Jerusalem, 18-23 August* **1996**, *Jerusalem.*
56. Khindaria, A.; Barr, D. P.; Aust, S. D. *Biochemistry* **1995**, *34*, 7773-7779.

Chapter 15

Use of Laccase for Bleaching of Pulps and Treatment of Effluents

M. Luisa F. C. Gonçalves and W. Steiner

"SFB-Biokatalyse" Institute of Biotechnology, Technical University of Graz, Petersgasse 12/I, A–8010 Graz, Austria

Laccases have been shown to facilitate bleaching of kraft pulp in conjunction with organic mediators. Laccases, have also been used for a long time in detoxifying and decolorizing of waters containing phenolic pollutants. Several patented processes have been reported in recent years, and many are currently being developed or improved. This chapter reviews the current literature and discusses recent developments and applications of laccases for bleaching and for effluent treatment in the pulp and paper industry.

Traditionally pulp and paper manufacture produces large quantities of polluted effluents, and releases toxic process chemicals into the environment. Due to rising environmental concerns, stringent statutory regulations have been imposed, and have compelled industry to provide cleaner processes.

The first approach was in waste water treatment, in many cases a combination of chemical, physical and biological processes. This has become less important in recent years, with an increasing tendency of the pulp and paper industry to use less polluting pulping and bleaching techniques. These include reducing or eliminating the use of chlorine or its derivatives in bleaching, by introducing an enzyme stage in the process. Alternative competitive enzyme-based bleaching processes have been developed, that maintain product quality, do not raise running costs and are easy to install.

In the present work we review progress on the use of the enzyme laccase in effluent treatment and pulp bleaching. Laccase is commonly produced by white-rot fungi, organisms capable of mineralizing lignin. Laccases catalyze the removal of one electron from phenolic hydroxyl groups, forming radicals from lignin model compounds that are able to polymerize or depolymerize as they react further. β-1 and

0097–6156/96/0655–0197$15.00/0
© 1996 American Chemical Society

β-O-4 dimers are both degraded by laccase via C-α-C-β, alkyl-aryl cleavages, and C-α oxidation. Aromatic ring cleavage may also be detected after the enzyme's action. Laccase can oxidize non-phenolic substrates in the presence of primary mediators such as HBT (hydroxybenzotriazole), and the enzyme produces manganese chelates which could penetrate the plant cell wall. Laccase is considered by many authors to be capable of degrading lignin together with the enzymes lignin peroxidase and manganese peroxidase *(1)*.

Effluent Treatment with Laccases

Polychlorinated phenols are major components of pulp mill bleach plant effluents and are highly resistant to microbial degradation.

Residual lignin in kraft pulp is highly modified by alkaline condensation reactions during pulping and gives the pulp a characteristic dark brown color. Chlorinated derivatives are formed during bleaching. These are toxic and reported mutagenic, and constitute a waste water treatment problem. Pollution by colored effluents became an important issue in the 1990s, mainly in the USA where some States legislated control of this kind of pollution by kraft pulp industries. One way of avoiding the chlorolignin formation is by implementation of chlorine-free bleaching methods.

Effect of Laccase on Effluent Components. Laccase can be used on both chlorolignin and phenolic compounds. It renders phenolic compounds less toxic via polymerization reactions. Because these enzymes are relatively non-specific, they are capable of cross-coupling pollutant phenols with naturally occurring phenols *(2)*. Furthermore, detoxification may result both from the change in the chemical form of the pollutant phenol and from the physical removal by precipitation or sedimentation of the insoluble complexed products *(3)*. Pre-treating the chlorolignin with laccase also increases the efficiency of precipitation by polyimides *(4)*.

Treatments of Pulp and Paper Mill Effluents Using Laccase. Many of the early methods for waste water treatments did not employ isolated enzymes, but used microorganisms, free or as immobilized biomass. Since fungi were unable to grow or degrade lignin using the effluent compounds only, it had to be supplemented with for example glucose. This is a burden in costs, additional to usually very slow and incomplete treatments.

In 1986, Field patented a method for the "biological treatment of waste waters containing non-degradable phenolic compounds and degradable non-phenolic compounds" (*5*). It consisted of an oxidative treatment to reduce or eliminate toxicity of the phenolic compounds followed by an anaerobic purification. This oxidative pre-treatment could be performed with laccase enzymes and it was claimed to reduce chemical oxygen demand by one thousand fold.

The use of laccase increased efficiency of chlorolignin precipitation was reported first in 1987 (*4*). The following year, the same authors described the removal

of chlorophenols and chlorolignins from bleaching effluents by combined chemical and biological treatments *(6)*. The organic matter from spent bleaching effluents of the chlorination (C), extraction (E), or a mixture of both (C+E) stages, was precipitated as a water insoluble complex with polyethyleneimide. The color, chemical oxygen demand (COD) and "absorbable organic halogens" (AOX) were reduced by 92, 65 and 84% respectively for the chlorination effluent, and by 76, 70 and 73% for the extraction effluent. No significant reduction in biological oxygen demand (BOD5) of treated effluent was detected but fish toxicity was greatly reduced. Enzyme treatment results in co-precipitation of the bulk mono- and dichlorophenols with the liquors of the chlorination and extraction bleaching stages.

In 1988, a review on the use of enzymes for wastewater treatment in the pulp and paper industry *(7),* examined the new possibilities of using enzymes like laccase, peroxidase, and ligninase for this effect.

Forss et al. examined the use of laccase for effluent treatment in 1989 *(8)*. They aerated pulp bleaching waste water in the presence of laccase for one hour at pH 4.8 and subsequently flocculated with aluminium sulfate. High removal efficiencies (80-99%) were obtained for chlorinated phenols, guaiacols, vanilins and catechols. The aeration of debarking wastewater in the presence of laccase reduced the concentration of phenolic compounds by more than 90%.

The ligninolytic system of some white rot fungi was investigated for potential applications in the treatment of bleach plant effluents *(9)*. A medium containing between 10 and 20% of hardwood bleach plant effluent (HWE) from the first extraction stage, with low absorbable organic halide (AOX) content stimulated laccase production in *Phebia radiata*. This effect was even more pronounced when softwood effluent containing about ten times more AOX was used instead (20% v/v effluent, corresponds to 60 mg/L AOX). The effluents were treated with carrier immobilized fungus in 2 L bioreactors and a 10 L Biostat fermentor, resulting in 50 to 60% AOX reduction (about 4 mg/L per day), and disappearance of the dark color. The same researchers later reported the presence of laccase and lignin peroxidase isoenzymes in active form in effluent media *(10)*. The results obtained with treatments using *Phebia radiata* and *Phanerochaete chrysosporium* immobilized on polypropylene carriers and cultivated in 2 L bioreactors were also presented *(11)*.

The use of laccases for wastewater treatment was patented by Call in 1991*(12)*. He claimed that wastewater from delignification and bleaching could be treated with laccases (from *Coriolus versicolor*) in the presence of non aromatic oxidants and reductants and aromatic compounds.

Fungal decolorization of straw soda pulping effluents in 1992, achieved best results with the known laccase producer *Trametes versicolor (13)*. The origin of the effluent color was investigated, being found that the alkali-lignin fraction contributed 77% of the total color. Biomass immobilized in plastic supports removed 80% of color after 4 days treatment of effluent supplemented with 10g/l glucose.

The covalent immobilization of laccase on activated carbon for phenolic effluent treatment was used in 1992 *(14)*, in an attempt to overcome the rapid inactivation the enzyme usually suffers in such a process. The carbon-immobilized laccase removed color from pulp mill bleach plant effluent (extraction stage) at a rate of 115 color ISO units per enzyme unit per hour, and the removal rate increased with

increasing effluent concentration. Although the stability of the laccase was improved by immobilization, its oxidation of the effluent was still very slow, making it inappropriate for color removal from pulping and bleaching effluents at industrial scale and had to be performed on a recirculation mode.

The degradability of chlorine-free bleach effluent lignins by fungi was studied in 1994 *(15)*. Bleachery effluent from a sulfite process pulp mill, which was extracted with alkali and treated with oxygen and hydrogen peroxide was used for the experiments. The effluent lignins were depolymerized and a substantial reduction in the aromatic compound content was caused by *Stagonospora gigaspora*, which produced laccase during the degradation activities.

Novo Nordisk (1995) patented a method for the enhancement of laccase reactions and use of enzyme compositions, among others, for purification of waste water from pulp manufacture *(16)*. The method works by an enzymic polymerization and/or modification of lignin containing material, in which the substrate is oxidized in the presence of laccae and an enhancing agent such as derivates of biphenyl, phenothiazine and phenoxazine.

The first report on paper mill bleach effluent (BPE) decolorization (and synthetic lignin degradation) by marine fungi appeared in 1996*(17)*. An unidentified basidiomycete was capable of 74% and 98% BPE decoloration in 14 days at pH 8.4 and 4.5 respectively. This effect is similar at lower pH to that achieved with *P. chrysosporium*. It was established that marine fungi possessing laccase only, are able to effectively decolorize BPE.

Laccase or laccase-producing fungi based treatments were not always successful. In a study (1984) on the potential of an enzymatic treatment of pulp and paper mill effluent to facilitate decoloration by lime precipitation *(18)*, pretreatment of the pulp and paper mill effluents with laccase prior to lime addition gave very small or no improvement in color removal efficiency, also when the enzymatic treated effluents were spiked with addition of phenolic compounds. An attempt to treat bleach effluent from a sulfite process pulp mill with *T. versicolor* (1994), not only failed to depolymerize or degrade lignin but also increased the amount of chromophores *(15)*. This happened in spite of the activities produced (laccase and manganese peroxidase) being the same as the ones produced by other fungi capable of performing the treatment.

The MYCOPOR and MYCOR reactors *(19)* use immobilized *Phanerochaete chrysosporium* for color removal of pulp mill effluents, and cause also a reduction in chlorolignin and chlorinated organic compounds. This effect has been attributed to the action of manganese peroxidases and ligninases produced by the organism, but the proof that it also produces laccases (*20*) may come to alter this concept.

Bleaching of Pulps with Laccase

The xylanase bleaching effect is well known and has already been implemented in many industrial units, and can reduce up to 25% the need for bleaching chemicals (*21*). Nevertheless, more effective enzymes are needed to enable a total chlorine-free

enzymatic bleaching. Many of the considered lignin degrading enzymes like manganese peroxidase, lignin peroxidase and laccase have been tried and tested.

In the present decade, many processes have been published and patents filed for enzymatic bleaching, mostly for methods employing either manganese peroxidase or laccase. Both enzymes require additional reagents or mediators to direct their oxidative cycles *(22)*.

Like for effluent treatments, some bleaching processes use isolated laccases and others use the enzyme-producing organisms. Even though the fungal treatments are very slow, they are effective.

A patent by Enso-Gutzeit Oy was filed in 1992, entitled "Bleaching of pulp in presence of oxidizing enzyme and transition metal compound" *(23)*. Semi-bleached birchwood pulp was accordingly bleached with a solution of manganese sulfate and laccase, followed by sequential treatment with sodium hydroxide and sulfuric acid, producing bleached pulp with 70% brightness, 4.2% more than pulp bleached without the enzyme step. The method claimed to effectively reduce the quantity of chlorine containing chemicals needed for bleaching of pulp and the content of organic chlorinated compounds in the waste waters.

The effects of fungal and enzymatic treatments on pulp bleachability were studied by Kantelinen (*24*). Improved bleachability of kraft pulps occurred when lignin modifying enzymes from *Phebia radiata* were employed, and the best result was obtained when laccase was used after hemicellulase. This working group also examined the effects of laccase, manganese peroxidase and lignin peroxidase, on the bleachability of oxygen delignified kraft pine pulps (*25*), and concluded that lignin-modifying enzymes do not improve bleachability when acting alone, but increase the brightness one ISO unit further when combined with xylanase treatment, giving a total brightness improvement of 2 to 3.5 ISO units.

The non-chlorine bleaching of kraft pulp was patented in 1994 (*26*), using an enzyme treatment to obtain a brighter pulp with lower lignin content. This process consists on the oxidation of kraft pulp with manganese peroxidase in the presence of laccase or hydrogen peroxide, removal of metal ions at acidic pH and brightening the chelated pulp with alkaline hydrogen peroxide. The enzyme stage lasts from 30 minutes to four hours, at acidic pH and temperatures between 25 and 60°C. A general review (*27*) about the biological bleaching of kraft pulps by white rot fungi and their enzymes and an article on the bleaching of kraft pulps with oxidative enzymes and hydrogen peroxide (*21*) appeared in 1994. Laccase was found to partially delignify soft-wood kraft pulps in a broad range of initial lignin contents, and improved the brightness after subsequent alkaline hydrogen peroxide treatment by up to 10 ISO units.

Call *(28)* registered a first patent in 1990, for a process of enzymatic bleaching of pulps with laccase. The process involved adding oxidizing and reducing agents to an aqueous pulp containing salts, complexing agents, and adjusted to a redox potential in the range 200-500mV, and then adding the enzyme to start the bleaching. At a 1% pulp consistency, the reaction took from fifteen minutes to twelve hours, a remarkable improvement over immobilized biomass and organism employing systems that need incubation times of several days.

Further patent by the same author in 1991 (*12*), was to use laccases in delignification, bleaching and as we already referred waste water treatment. It produced good results with 1% aqueous thermomechanical pulp, and dosages of laccase 100,000 IU, veratryl alcohol 8 μM and hydrogen peroxide 400 μM per g of solids. The reaction time was reduced to 4 hours at pH 5.

The next patent came in 1993 (*29*), describing a process for modifying, break down or bleaching of lignin. In 1994 the same author made two communications, one about the state of the art of enzyme bleaching and disclosure of a breakthrough process *(30)* and another one about the enzymatic bleaching of pulps with the laccase-mediator-system (*31*). The laccase-mediator concept combines the enzyme with a low molecular and environmentally friendly redox mediator, generating strongly oxidizing mediator(s) which specifically degrade lignin leaving the cellulose fibers intact. The method uses pulp consistency between 1 and 20%, and operates optimally at pH 4,5 and 45°C with a retention time of 1 to 4 hours. Pressurized oxygen has to be added to act as co-substrate.

Laccase-Mediator-System-technology was tested successfully at a pilot plant trial (*32*) using pulps of different origins (hard and soft-woods, annual plants, sulfate and sulfite pulps). The performance of this system has been further improved by optimizing the mixture of the system components, protecting the enzyme towards inactivation and enhancing reaction kinetics to obtain a reduction of the mediator quantity, which is the main operating cost of the process. For many pulps, a 50% delignification could be obtained using 5 kg of mediator per ton of pulp. The most efficient mediators found belonged to the N- heterocyclic compounds group. Recent process developments have found a new group of more efficient mediators, whose nature has not yet been disclosed *(33)*.

The Novo Nordisk patent filed in 1995 about enhancement of laccase reactions with aromatic ring-containing compounds, claimed to be a method for bleaching of pulp for paper production, among other applications (*16*), as it was already referred in effluent treatment section.

A study of bleaching unbleached kraft pulp (UKP) from hardwood by white rot fungi was published in 1995 *(34)*, in which a solid state fermentation system produced an increase in pulp brightness by 15 and 30 ISO units after 5 days of treatment, with. *T. versicolor* and *P. chrysosporium*, respectively. The kappa number of the pulp decreased with increasing brightness.

The project "bleaching through enzymes via scaling of technology", is trying to replace the magnesium bisulfite process used in a pulp mill in Portugal by an enzyme system for reducing pollutant discharges *(35)*. The objective is to demonstrate that the existing technologies for pulp bleaching, can be replaced by economically competitive enzymatic based ones. So far the best results observed were a 35% reduction in kappa number of the pulp following alkaline extraction, for a laccase/oxygen/additive system.

Concluding Remarks

Effluent treatments and bleaching processes with laccase have been chronologically listed in Figures 1. and 2., respectively.

Laccases-based methods that have been proposed so far for effluent detoxifying and decolorizing achieve limited results and/or require excessive time to be commercially and industrially competitive.

Bleaching processes using laccase-enhancer (mediator) systems are most promising, although the pilot plant trials so far realized show improvements are still needed before its application at an industrial scale can be successful.

In order for laccases to act as oxidative bleaching enzymes, a slight pressure of oxygen and a mediator (low molecular weight electron carrier) are required. The enzyme operates in an acid pH range, unlike the conventional methods of peroxide brightening and oxygen delignification. For a broader commercial application of laccases, either in bleaching or effluent treatment, an inexpensive source of enzyme and mediator are needed.

Year	Author	Ref.	Result
1986	Field	5	COD reduction by one thousand fold.
1987	Troyanowski et al.	4	Increase in efficiency of Cl-lignin precipitation.
1988	Troyanowski et al.	6	Removal of color, COD and AOX by 92, 65 and 84% respectively; BOD remains constant.
1988	Hakulinen	7	Possibilities to use enzymes for bleaching.
1989	Forss et al.	8	Over 90% decrease in phenolic compounds concentration.
1990	Hatakka et al.	9	AOX reduction, dark color disappears.
1991	Lankinen et al.	10, 11	Effluent treatment in bioreactors; up to 60% AOX reduction, dark color disappearance.
1991	Call and Mücke	12	Patent on laccase waste water treatment.
1992	Terrón et al.	13	Effluent decoloration.
1992	Davis and Burns	14	Color removal at 115 color ISO units per enzyme unit per hour.
1994	Bergbauer and Eggert	15	Degradation of Bleach effluent lignin.
1995	NovoNordisk	16	Patent on laccase waste water treatment.
1996	Raghukumar et al.	17	Up to 98% bleach plant effluent decoloration.

Figure 1. Effluent treatments with laccase.

Year	Author	Ref.	Result
1990	Call	28	Patent for bleaching of pulps
1991	Call	12	Patent for delignification and bleaching of pulps.
1992	Enso-Gutzeit Oy	23	Effective reduction of Cl in bleaching and Cl-compounds in wastewaters.
1993	Kantelinen et al.	24	Improved bleachability of kraft pulp.
1993	Call	29	Patent for modification, break down or bleaching of lignin.
1994	Ranua et al.	25	Brightness improvement with combined enzymes treatment.
1994	Bourbonnais and Paice	26	Patent treatment for brighter pulp with lower lignin content.
1994	Bourbonnais and Paice	21, 27	Delignification of softwood kraft pulp; brightness improvement of up to 10 ISO units.
1994	Call and Mücke	30	Comunication on the disclosure of a break-through process
1994	Call and Mücke	31	Presentation of the Laccase-Mediator-System (LMS)
1995	Call and Mücke	32	LMS pilot plant trial.
1995	NovoNordisk	16	Patent for bleaching of pulp for paper production.
1995	Katagair et al.	34	Solid state fermentation increases pulp brightness up to 30 ISO units in 5 days.
1995	Fischer et al.	35	35% reduction of Kappa with laccaase/oxygen/additive system followed by alkaline extraction.
1996	Call and Mücke	33	Method improvements, more effective mediator (not disclosed).

Figure 2. Bleaching processes with laccase.

The problem of obtaining large quantities of enzyme at a reasonable price seems to have been solved already. Both the patented methods for laccase production by recombinant organisms for use in industrial processes (*36-37*), and the production of laccase from selected basidiomycetes using potato wastes (*38*), offer good solutions. The question now is to overcome the price of mediators and co-substrates, by optimizing even more the process conditions so that they are needed in smaller quantities. Some improvements on enzyme stabilization and on the enhancement of the enzyme kinetics can also be expected for existing processes.

Others than laccases, the lignin peroxidases, manganese peroxidase and cellobiose:quinone oxidoreductase are also extracellular enzymes produced by ligninolytic fungi, and their role in lignin biodegradation is the subject of current research (*39*). If the present trend is maintained, the future will most probably bring new processes where the combined action of several enzymes is used instead of isolated ones, for more environmentally friendly pulping and bleaching processes in the paper industry.

References

1. Young, H. D.; Hah, I. C.; Kang, S. O., *FEMS Microbiology Letters* **1995,** *132(3)*, 183.
2. Haars, A.; Hüttermann, A, *Archives Microbiol.* **1980,** *125*, 233.
3. Bollag, J. M.; Shuttlework, K. L.; Andreson, D. H., *Appl. and Envorinm. Microbiol.* **1988,** *54(12)*, 3086.
4. Troyanowski, A.; Milstein, O.; Majcherczyk, A.; Haars, A.; Huettermann, A., *Colloq. INRA* **1987,** *40(Lignin Enzymic Microb. Degrad.)*, 223.
5. Field, J. A., *Patent* **1986,** *EP 238148.*
6. Troyanowski, A.; Milstein, O.; Majcherczyk, A.; Tautz, D.; Zanker, H.; Huettermann, A., *Water Science Technol.* **1988,** *20,* 161.
7. Hakulinen, R.,. *Water Science Technol.* **1988,** *20 (1),* 251.
8. Forss, K.; Jokinen, K.; Savolainen, M.; Williamson, H., *Pap. Puu* **1989,** *71(10)*, 1108.
9. Hatakka, A. I.; Lankinen, V.P.; Lundell, T.K.; Hietanen, P.; Fabbricius, B. O.; Pellinen, J., *Cellul. Souces Exploit.*; Horwood, London, U.K., 1990, pp. 149.
10. Lankinen, V. P.; Ikeroinen,M. M.; Pellinen, J.; Hatakka, A. I., *Water Science Technol.* **1991,** *24,* 189.
11. Lankinen, V. P.; Valo, M.; Hatakka, A. I, *Biotechnology in Pulp and Paper Industry*; UNI Publishers Co, Ltd., Tokyo, Japan, 1992, pp. 69-74.
12. Call, H. P., *Patent* **1991,** *DE 4137761.*
13. Terrón, M. C.; Calvo, A. M.; Fidalgo, M. L.; Manzanares, P.; Balestero, M.; Martinez, A. T.; Marín, C.; González, A.E., *Biotechnology in pulp and paper industry*; UNI Publishers Co, Ltd., Tokyo, Japan, 1992, pp. 51-56.
14. Davis, S.; Burns, R. G., *Appl. Microbiol. and Biotechnol.* **1992,** *37*, 474.
15. Bergbauer, M.; Eggert, C., *Canadian Journal Microbiol.* **1994,** *40(3)*, 192.
16. Novo Nordisk, *Patent* **1995,** *WO 9501426.*
17. Raghukumar, C.; Chandramohan, D.; Mitchel, Jr. F. C.; Reddy, C. A., *Biotechnology Letters* **1996,** *18(1)*, 105.
18. Campbell, J. C.; Joyce, T. W., *Proeedings. Industrial. Waste Conf* **1984,** *38,* 67.
19. Jaklin-Farcher, S.; Szeker, E.; Stifter, U.; Messner, K., *Biotechnology in Pulp and Paper Industry*; UNI Publishers Co, Ltd., Tokyo, Japan, 1992, pp. 81-85.
20. Sirinivasan, C.; D'Sousa, T. M.; Boominathan, K.; Reddy, C. A., *Applied and Environm. Microbiol.* **1995,** *61(12)*, 4274.
21. Paice, M. G.; Bourbonnnais, R. E.; Reid, I., *Tappi* ***Journal*** **1994,** *78(9),* 1.
22. Paice, M. G.; Bourbonnais, R.; Reid, I. D.;. Archibald, F. S.; Jurasek, L., *Journal of Pulp and Paper Science* **1995,** *21(8),* J280.
23. *Vaheri, M.; Piirainen, O.,(Enso-Gutzeit Oy), Patent* **1992,** *WO 9209741.*
24. Kantelinen, A.; Hortling, B.; Ranua, M.; Viikari, L., *Holzforschung* **1993,** *47(1)*, 29.
25. Ranua, M.; Kantelinen, A.; Nikupaavola, M. L.; Suurnakki, A., *Bioresource Technology* **1994,** *50(1),* 73.

26. Paice, M. G.; Bourbonnnais, R. E., *Patent* **1994,** *CA 2115881.*
27. Paice, M. G.; Reid, I., *FEMS Microbiol. Rev.* **1994,** *13(2-3),* 369.
28. Call, H. P., *Patent* **1990,** *90- 4008893.*
29. Call, H. P., *Patent* **1994,** *WO 94/29501.*
30. Call, H. P.; Mücke, I., *NBC Comunication* **1994,** 1.
31. Call, H. P.; Müecke, I., *Proceedings Pulping conference* **1994,** San Diego, CA, USA.
32. Call, H. P.; Mücke, I.,*Proccedings of the 6th International Conference on Biotechnology in the Pulp and Paper Industry* **1995,** 27.
33. Call, H. P.; Mücke, I., *Symposyum Abstract; Spring 96 Newsletter -Cellulose, Paper and Textile Division of the ACS* **1996**, 147.
34. Katagari, N.; Tsutsumi, Y.; Nishida, T., *Appl. and Envorinm. Microbiol.* **1995,** *61(2)*, 617.
35. Fischer, K.; Furumoto, H.; Brown, I. D., *Proccedings of the 6th International Conference on Biotechnology in the Pulp and Paper Industry* **1995,** 39.
36. Valtion Tecknillinen Tutkimuskeskus, *Patent* **1992,** WO 92021046.
37. Novo Nordisk, *Patent* **1995,** *WO 9507988.*
38. Trojanowski, J.; Mayer, F.; Huttermann, A., *Starch-STARKE* **1995,** *47(3),* 116.
39. Reid, I.; *Canadian Jounal of Botany* **1995**, *73(S1E),* S1011.

Pulp Bleaching

Chapter 16

Reduction of Active Chlorine Charges in Bleaching of Xylanase-Pretreated Sulfite Pulp

L. P. Christov[1,2] and B. A. Prior[2]

[1]Sappi Management Services, P.O. Box 3252, Springs 1560, South Africa
[2]Department of Microbiology and Biochemistry, University of the Orange Free State, P.O. Box 339, Bloemfontein 9300, South Africa

Biobleaching of acid bisulfite pulp with *Aureobasidium pullulans* xylanases and reduced amounts of chlorine dioxide (up to 50% as active chlorine) can produce dissolving pulp with improved properties such as alkali solubility (S_{10} and S_{18}) and brightness over the conventionally ECF-bleached control. The bleachability of the xylanase-pretreated pulp was evaluated in a subsequent treatment with oxygen, hydrogen peroxide or three ECF-bleaching sequences. A brightness gain of 1.7, 1.1 and 1.6 points was achieved by the xylanase pretreatment respectively in the X-O$D_1E_oD_2$P, X-O$D_1E_oD_2$H and X-O$D_1E_oD_2$ bleaching combinations. The brightness of the X-O$D_1E_oD_2$P-bleached pulp (94.2%) was approximately 2 and 3 units higher than that of the X-O$D_1E_oD_2$H and X-O$D_1E_oD_2$ pulp, respectively. The S_{10}, S_{18} and S_{10}-S_{18} values of the X-O$D_1E_oD_2$P dissolving pulp were reduced relative to the control by up to 22, 13 and 31%, respectively. In all instances the viscosities of the X-O$D_1E_oD_2$H pulp were lower than those of the X-O$D_1E_oD_2$ and X-O$D_1E_oD_2$P dissolving pulps.

The most important application of dissolving-grade pulps is in viscose rayon production. However, high purity cellulose-based products such as cellulose esters (acetates and nitrates), cellulose ethers (carboxymethyl cellulose), and graft and cross-linked cellulose derivatives are also of considerable industrial interest (*1*).

Compared to other types of pulp dissolving pulp contains less lignin and hemicellulose (xylan and glucomannan) as a result of extensive pulping and bleaching. While lignin is almost completely removed (<0.05% of pulp) after bleaching of both prehydrolysis-kraft and sulfite pulps (*2*), part of the hemicellulose (normally less than 10% of pulp) still remains in pulp (*3*). Excessive amounts of hemicellulose in dissolving pulp can cause serious problems in the viscose processes. For instance, the high hemicellulose content can affect the swelling of the pulp in the mercerization; the viscose filterability and drainage of sodium hydroxide during steeping; the completion of the reactions during xanthation and the strength properties of the viscose end product (*4*). Haze, false viscosity, discolorization and poor filterability of cellulose acetates and nitrates can be introduced by the hemicellulose contaminants in cellulose. Therefore a more complete removal of

0097–6156/96/0655–0208$15.00/0
© 1996 American Chemical Society

residual hemicelluloses from dissolving pulp would facilitate its processing and would improve the quality of the final product.

The composition and structure of the wood hemicelluloses are more complicated than that of the cellulose. In hardwoods, the major hemicellulose component is the *O*-acetyl-4-*O*-methyl-glucuronoxylan (15-30% of the dry wood weight) whereas glucomannan constitutes only 2-5% of wood. Softwoods contain about 20% *O*-acetyl-galactoglucomannan and 5-10% arabino-4-*O*-methylglucuronoxylan (*5*).

Bleaching of Pulps

The bleaching of kraft and sulfite pulps for dissolving pulp manufacture can be regarded as a purification process aimed at final delignification, and an increase in the α-cellulose content and brightness. It involves the use of a number of bleaching chemicals which with regard to environmental aspects can be classified into non-chlorine containing reagents such as oxygen, ozone, hydrogen peroxide or chlorine-containing reagents such as elemental chlorine, chlorine dioxide and sodium or calcium hypochlorite (*6*). Bleaching of pulp with the latter group causes the formation of chlorolignins as well as low molecular weight aromatic and aliphatic compounds, many of which, especially the chlorinated phenols, catechols and guaiacols are toxic to the receiving waters (*7*).

In order to reduce the negative environmental impact of bleaching, new pulping and bleaching techniques are being developed. The aim is to minimize or eliminate the use of chlorine-containing agents thereby implying extended cooking time or non-chlorine bleaching. An alternative approach is to use biological systems such as enzymes and microorganisms which can selectively remove hemicelluloses and/or lignin from pulp without affecting the cellulose.

Biobleaching of Kraft Pulps

Most of the research effort has been focused on the biobleaching of kraft pulps. For instance, xylanases have been shown to improve pulp properties such as breaking length, fibrillation, porosity and beatability (*8*), viscosity (*9*), brightness (*10*). Pretreatment with xylanases was reported to enhance kraft pulp bleachability and reduce the amount of total active chlorine needed in bleaching by approximately 15% (*11*).

According to the proposed mechanism of biobleaching of kraft pulps with xylanases (*12*) the enzyme can hydrolyze that fraction of xylan that has been reprecipitated on the fiber surface during the kraft pulping. This enables the bleaching chemicals to attain an improved accessibility to residual lignin resulting in enhanced lignin degradation and diffusion from pulp. In addition, the size of the lignin-carbohydrate complexes will be reduced by the enzyme action which would then improve their diffusion in the alkali extraction step (*13*).

Only about 20% of xylan in birch kraft pulps could be removed even when using very high loadings of *Trichoderma reesei* xylanases (*14*). Possible reasons for this could be the modifications in the structure and rearrangements and aggregations of the xylan chains during its readsorption on the cellulose fibers (*15*).

Biobleaching of Sulfite Pulps

Little attention has been paid to enzymatic prebleaching of sulfite pulps for dissolving pulp production in the past (*16*). It was found that the enzymatic hydrolysis of xylan in dissolving pulp is limited by the poor accessibility of the substrate (*16, 17*). When unbleached sulfite pulp of eucalyptus wood was treated

with *Aureobasidium pullulans* xylanases in non-optimized conditions, the pentosan content and kappa number decreased by up to 35 and 24%, respectively, if very high enzyme charges and prolonged incubation times were applied (*18*). Using triple-repeated consecutive xylanase-oxygen treatments the pentosan content of sulfite pulp was reduced two-fold, the kappa number decreased by 60%, brightness was enhanced by 18 points, and the α-cellulose content was enriched by 3 points (*19*).

Significant amounts of hemicellulose (xylans and glucomannans) are dissolved in the sulfite pulping process. The bonds between the pentose units (arabinose and xylose) are hydrolyzed much more rapidly than the glycopyranosidic ones (*20*). On the other hand the glucuronic acid-xylose and xylose-acetic acid linkages are relatively more resistant to the acid hydrolysis conditions (*21*) and little cellulose is lost in the sulfite pulping (*22*). Therefore the degradation products of the hemicellulose acid hydrolysis would appear in the cooking liquor in the following approximate order:

arabinose> galactose> xylose> mannose> glucose> acetic acid> glucuronic acid

In the above order glucose would be the hydrolysis product originating mostly from the glucomannan rather than cellulose polymer. Also, the glucomannans are more resistant to acid dissociation than xylans. Hence, in sulfite cooking of birch only 45% of the original xylan remains in pulp after 20 min cooking and its original DP of 200 is reduced to less than 100 (*23*).

During acid sulfite pulping, redeposition of xylans onto the fiber surface has not been observed (*20*). The possible reasons for this would be that the harsh cooking conditions and presence of acid-resistant residual acetyl and 4-*O*-methylglucuronic acid groups function as a barrier against the adsorption and intercrystallization of xylan onto the cellulose micromolecules.

Therefore the remaining xylan in sulfite pulp is less accessible and the effects produced on hemicellulose by xylanases can be regarded as a result of changes in the pulp fibers rather than on the fiber surface only (*19*). Hence, the xylan localized in the primary and mainly secondary cell walls as well as the xylan fraction involved in lignin-carbohydrate formations may be enzymatically attacked (Christov, L.P. *et al. Holzforschung*, in press.). Major factors influencing the degree of enzymatic removal of xylan from sulfite pulp would be the size, i.e. penetration capabilities of xylanases and their substrate specificity on one hand, and the accessibility, physical and chemical state of the xylan substrate, on the other (*18*).

Here we report on the effects of the xylanase pretreatment of sulfite pulp with the *A. pullulans* enzyme preparation on the savings of active chlorine achieved during bleaching of dissolving pulp.

Materials and Methods

Enzyme Production and Preparation. *Aureobasidium pullulans* NRRL Y-2311-1 was a kind gift from C. P. Kurtzman (Northern Regional Research Center, Peoria, IL, USA). The maintenance of the stock culture and inocula preparation were described previously (*18*). The enzymes were produced in a medium containing 1% xylose, 0.25% $(NH_4)_2SO_4$, 0.1% yeast extract and 0.5% KH_2PO_4 (pH 5) for 2 days. The supernatant of the culture broth was concentrated by ultrafiltration (10 000 molecular mass cutoff) at 4°C, subsequently dialized against deionized water, preserved with sodium azide and stored at -20°C (*17*).

Enzyme Assays. Xylanase activity was determined by measuring the reducing sugars released by the enzymes from 1% oat spelts xylan according to the Somogyi (*24*)-Nelson (*25*) method. Cellulase activity was determined in a similar manner using Avicel cellulose or carboxymethyl cellulose.

Enzymatic Pretreatment of Pulp. The conditions for the application of the enzyme preparation of *A. pullulans* on pulp are given in Table I. Prior to use and after each treatment step, pulp was thoroughly washed with distilled water until a neutral pH of the wash waters was achieved. The buffer used to adjust pH was sodium acetate (*26*). The pulp-enzyme slurry was mixed every 15 min. The enzyme reaction was terminated by boiling for 10 min and pulp was filtered and washed.

Chemical Bleaching of Pulp. The bleaching of sulfite pulp after the enzyme pretreatment was carried out in sequence $OD_1E_oD_2$, $OD_1E_oD_2H$ or $OD_1E_oD_2P$ under the conditions shown in Table I. The controls were bleached in the same way but omitting the enzyme pretreatment step.

Analyses of Pulp Properties. Unbleached acid bisulfite pulp produced from *Eucalyptus grandis* wood was obtained from Sappi Saiccor (Pty.) Ltd., South Africa. Pulp properties were tested according to the standard methods of the Technical Association of the Pulp and Paper Industry (TAPPI, Atlanta, GA, USA): alkali solubility S_{10} and S_{18} at 25°C (TAPPI T235 cm-85); kappa number (TAPPI T236 cm-85); brightness (TAPPI T452 om-87); and viscosity (TAPPI T230 om-89). Each experiment was analysed in duplicate.

Results and Discussion

Assessment of the *A. pullulans* Xylanases on Pulp. Evaluation of the bleachability of the xylanase-pretreated sulfite pulp was made by subsequent treatment with oxygen or hydrogen peroxide (Table II). In both cases there was a brightness gain which can be ascribed due to the enzyme prebleaching. This brightness increase was more pronounced when pulp was bleached with peroxide (3.6 points) than oxygen (1.4 points). This is not surprising since hydrogen peroxide is a better bleaching agent than oxygen. It also explains the difference of nearly 10% in lowering the kappa number using peroxide as compared with oxygen. The additional delignification effect produced by the xylanase was about 5% in the case of peroxide bleaching and 3% in the case of oxygen. Alkali solubilities S_{10} and S_{18} were also improved especially in the oxygen bleaching due to the higher load of NaOH on pulp: 1.8% (O) against 0.7% (P) (see Table I). The hemicellulose content of the control of unbleached pulp (measured as alkali solubility S_{18}) was reduced by approximately 6% and 10% when pulp was bleached with peroxide alone (P) or in conjunction with xylanases (XP), respectively (Table II). Therefore the difference of 4% would represent the enzyme contribution to the degradation and removal of hemicellulose from pulp. Similarly, S_{10} was reduced additionally by 3% when pulp was first pretreated with xylanases and then bleached with peroxide. On the other hand viscosity of sulfite pulp was not significantly affected by the xylanase pretratments as compared to the controls of oxygen- and peroxide-bleached pulps (Table II).

Biobleaching of Sulfite Pulp with *A. pullulans* Xylanases in Sequence X-$OD_1E_oD_2$. Table III summarizes the results obtained from the biobleaching experiment performed in sequence X-$OD_1E_oD_2$ with reduced charges of chlorine dioxide (as active chlorine) at D_1 and D_2 bleaching stages. The reduction of active chlorine was from 1.1% to 0% at D_1 and from 0.6% to 0% at D_2 respectively. Pulp properties such as alkali solubilty, brightness and viscosity were analysed following the decrease in the amount of chlorine dioxide used for bleaching. No correlation between the latter, on one hand, and the alkali solubility or viscosity, on the other, was found. As expected, brightness was substantially influenced by the change in the active chlorine doses on pulp. A bleach-boosting effect of 1.6 brightness points over

Table I. Conditions for biobleaching of sulfite pulp with *A. pullulans* xylanase (X) in sequence X-$OD_1E_oD_2$, X-$OD_1E_oD_2H$ or X-$OD_1E_oD_2P$

Stage	Consistency (%)	Temperature (°C)	Time (min)	pH_{final}	Charge on pulp
X	9	55	180	4.7	15 IU/g pulp
O	11	100	60	9.5	1.8% NaOH + 0.8% O_2
D_1	10	65	40	2.4	1.1% act.Cl
E_o	11	100	140	10.5	2.8% NaOH + 0.8% O_2
D_2	11	65	180	3.8	0.6% act.Cl
H	12	55	120	9.5	0.5% act.Cl
P	12	70	180	9	1.5% H_2O_2 + 0.7% NaOH

Table II. Properties of sulfite pulp pretreated with *A. pullulans* xylanase (X) and subsequently bleached with oxygen (O) or hydrogen peroxide (P)[a]

Treatment	S_{10} (%)	S_{18} (%)	Kappa number	Brightness (%, ISO)	Viscosity (cP)
Unbleached pulp	11.2	7.1	6.7	63.3	27.1
P	10.7	6.7	3.4	76.7	24.4
XP	10.4	6.4	3.1	80.3	24.4
O	9.6	5.4	4.0	68.2	27.2
XO	9.4	5.4	3.8	69.6	27.6

[a] Each value represents the mean of duplicate independent determinations; see Table I for treatment conditions

Table III. Properties of xylanase-pretreated dissolving pulp bleached in sequence X-$OD_1E_oD_2$ with different charges of active chlorine at D_1 and D_2 stages[a]

D_1 (%)	D_2 (%)	Reduction (%)	S_{10} (%)	S_{18} (%)	Brightness (%, ISO)	Viscosity (cP)
1.1	0.6	0	7.8	4.1	91.4	29.2
1.1	0.3	18	7.7	4.0	91.3	25.2
1.1	0.0	35	7.7	4.0	81.8	23.1
0.825	0.6	16	7.6	4.1	91.3	34.0
0.825	0.3	34	7.9	4.1	90.9	26.1
0.825	0.0	51	7.6	4.1	83.2	30.8
0.55	0.6	32	7.8	4.3	90.9	26.7
0.55	0.3	50	7.8	4.2	89.3	27.6
0.55	0.0	68	7.8	4.2	83.9	26.5
0.225	0.6	51	7.5	4.1	85.0	23.7
0.225	0.3	69	7.4	4.1	82.6	26.8
0.225	0.0	87	7.5	4.1	75.5	27.9
0.0	0.6	65	7.8	4.3	79.3	25.8
0.0	0.3	82	7.7	4.2	73.3	28.9
0.0	0.0	100	7.6	4.2	72.9	34.0
1.1[b]	0.6	0	9.1	4.6	89.8	27.1

[a] Each value represents the mean of duplicate independent determinations; see Table I for treatment conditions

[b] Control: $OD_1E_oD_2$

the control was observed with the xylanase-pretreated pulp sample bleached under the same conditions. However, the maximum brightness of 91.4% achieved with the biobleaching of sulfite pulp still does not fulfill the level of 93% required in the dissolving pulp manufacture. The alkali solubility S_{10} of the xylanase-prebleached samples was reduced by 13-19% over the control whereas S_{18} decreased by 7-13%, respectively (Table III). The S_{10}-S_{18} value which indicates the amount of degraded cellulose was also reduced from 4.5%, for the control, to 3.8% or less, for the biobleached dissolving pulp. This would suggest that a portion of the long-chain xylan fraction measured as S_{10} was enzymatically removed from pulp (Christov, L.P. *et al. Holzforschung*, in press.).

Biobleaching of Sulfite Pulp with *A. pullulans* Xylanases in Sequence X-$OD_1E_oD_2P$. In order to improve the brightness of dissolving pulp, hydrogen peroxide was introduced as a final bleaching step and sulfite pulp was biobleached with *A. pullulans* xylanases in sequence X-$OD_1E_oD_2P$ (Table IV). Maximum brightness of 94.2% was obtained when the xylanase-pretreated pulp was bleached without reduction in the active chlorine charges. The brightness gain over the control was 1.7 units. Similar increase (1.6 units) in the brightness was achieved with the X-$OD_1E_oD_2$-bleached pulp (Table III). Also, as observed with the X-$OD_1E_oD_2$ pulp, the decrease in the dose of chlorine dioxide on pulp had no impact on the alkali solubility and viscosity values (Table IV). However, it was possible to reduce the total active chlorine charge two-fold (by 51%) while still maintaining a brightness level of 93%. The dose of chlorine dioxide at D_1 stage could be reduced by up to 80% (from 1.1 to 0.225%) whereas at D_2 stage it was lowered by up 50% (from 0.6 to 0.3%). Senior *et al.* (*27*) reported a 33% reduction in chlorine dioxide in the first D-stage during bleaching of hardwood kraft pulp in sequence (CD)ED_1ED_2 following xylanase treatment. Using the same bleaching combination, chlorine dioxide savings of 40% in the (CD)-stage were obtained in bleaching of xylanase-pretreated softwood kraft pulp (*28*). In a bleach kraft mill trial experiment the first chlorination stage was completely replaced with an enzymatic pretreatment stage thereby a high brightness pulp (88%) was produced at comparable chemical dosages (*29*). In our biobleaching experiments on hardwood sulfite pulp a complete elimination of D_1 and/or D_2 was unattainable since the brightness dropped below the 93% level by up to 8 units (Table IV). Apparently, the distribution of the active chlorine charges between D_1 and D_2 during bleaching is of importance to obtain a maximum brightness. On the other hand, the second chlorine dioxide stage D_2 could be skipped (63% reduction of the total active chlorine) after a combined fungus-xylanase pretreatment of sulfite pulp due to the strong biobleaching effect produced by the fungus *Ceriporiopsis subvermispora* (Christov, L.P. *et al. Holzforschung*, in press.).

As a result of the xylanase pretreatment, the alkali solubilities S_{10} and S_{18} of the biobleached dissolving pulp decreased respectively by 9-27% and 4-15%. However, these variations were restricted in the samples having a brightness of 93% or higher: 12-22% for S_{10} and 6-13% for S_{18}, respectively. Similarly, the (S_{10}-S_{18}) values of these samples were appreciably reduced by 16-31% relative to the control (Table IV).

The viscosity of the biobleached dissolving pulp produced using the same active chlorine charges at D_1 (1.1%) and D_2 (0.6%) as the control increased by 2.3 cP over the control (from 30.9 to 33.2 cP). A similar increase in viscosity was observed with the X-$OD_1E_oD_2$-bleached pulp as compared to the control (Table III). Viscosity increase due to enzyme pretreatment of pulp was reported previously as well (*9, 29*). Apparently the removal of some hemicellulose from pulp, in particular xylan, which has a relatively low DP compared to the cellulose, did cause the viscosity increase. On the other hand the significant fluctuation in the viscosity

Table IV. Properties of xylanase-pretreated dissolving pulp bleached in sequence X-$OD_1E_oD_2P$ with different charges of active chlorine at D_1 and D_2 stages[a]

D_1 (%)	D_2 (%)	Reduction (%)	S_{10} (%)	S_{18} (%)	Brightness (%, ISO)	Viscosity (cP)
1.1	0.6	0	7.8	4.4	94.2	33.2
1.1	0.3	18	7.9	4.5	94.0	31.8
1.1	0.0	35	7.6	4.2	90.9	33.6
0.825	0.6	16	7.3	4.2	93.5	31.6
0.825	0.3	34	7.7	4.1	92.5	26.6
0.825	0.0	51	7.4	4.1	90.9	34.5
0.55	0.6	32	8.2	4.5	94.1	28.4
0.55	0.3	50	8.5	4.3	91.0	21.1
0.55	0.0	68	7.8	4.3	90.9	33.7
0.225	0.6	51	7.9	4.5	93.0	32.3
0.225	0.3	69	8.3	4.5	90.9	25.9
0.225	0.0	87	6.8	4.1	87.7	37.2
0.0	0.6	65	8.2	4.4	91.2	28.1
0.0	0.3	82	8.2	4.6	89.1	29.5
0.0	0.0	100	7.4	4.1	84.8	42.3
1.1[b]	0.6	0	9.3	4.8	92.5	30.9

[a] Each value represents the mean of duplicate independent determinations; see Table I for treatment conditions

[b] Control: $OD_1E_oD_2P$

values of pulp produced with reduced amounts of chlorine dioxide, a phenomenon observed with the X-O$D_1$$E_o$$D_2$-bleached dissolving pulp as well (Table III), cannot be easily explained. It can be suggested, however, that the viscosity has been influenced by the particular combination of the active chlorine charges applied on pulp at the D_1 and D_2 stages since the other bleaching conditions were identical in all instances (see Table I).

Comparison of Different Biobleaching Sequences. Table V illustrates the major differences in the properties of dissolving pulps bleached in the sequences O$D_1$$E_o$$D_2$, O$D_1$$E_o$$D_2$P and O$D_1$$E_o$$D_2$H with or without enzyme pretreatment with *A. pullulans* xylanases. No reductions of the active chlorine loadings on pulp have been used in the preparation of these pulps. Notably the bleaching combinations X-O$D_1$$E_o$$D_2$P and O$D_1$$E_o$$D_2$ ending with hydrogen peroxide proved to be the most efficient in terms of S_{10}, S_{18} and brightness. The brightness of the X-O$D_1$$E_o$$D_2$P-bleached dissolving pulp (94.2%) was by approximately 2 and 3 units higher than that of the X-O$D_1$$E_o$$D_2$H and X-O$D_1$$E_o$$D_2$ pulp, respectively. The brightness gain over the controls, due to the biobleaching with xylanases, was the highest also with the X-O$D_1$$E_o$$D_2$P pulp (1.7 units) as compared to the other two pulps: 1.6 units for the X-O$D_1$$E_o$$D_2$ and 1.1 units for the X-O$D_1$$E_o$$D_2$H pulp. For comparison, a brightness increase of 2 units was reported in bleaching of hardwood kraft pulp pretreated with the *A. pullulans* enzymes (*30*). Another advantage of using the hydrogen peroxide as the last bleaching stage would be an improved brightness stability (*30*).

The introduction of P or H as a final bleaching step in the bleaching sequence O$D_1$$E_o$$D_2$ did increase slightly S_{10} and S_{18}. This was more obvious in the case of sodium hypochlorite where there was a significant drop in the viscosity as well. In addition to its bleaching function, the hypochlorite is used also to control the DP of the final dissolving pulp since conversion to cellulose derivatives and some resultant properties are dependent upon the molecular chain length of the cellulose (*31*). This is especially needed if the dissolving pulp is used for viscose rayon manufacture. If a higher viscosity of dissolving pulp is desired (higher than 25 cP), as it is required for example in the cellulose acetate production, then the bleaching sequence O$D_1$$E_o$$D_2$P would be more appropriate. In both bleaching combinations, however, the xylanase pretreatment would improve the quality of the final dissolving pulp.

Conclusions

Chemical savings of up to 51% (as active chlorine) could be achieved when sulfite pulp was biobleached with *A. pullulans* xylanases in sequence X-O$D_1$$E_o$$D_2$P. The dose of chlorine dioxide at D_1 stage could be reduced by up to 80% whereas at D_2 stage it was lowered by up 50%. However, a complete elimination of D_1 and/or D_2 was unattainable. The dissolving pulp produced had improved brightness and alkali solubilities S_{10} and S_{18} over the enzymatically untreated control. The reduction of the chlorine dioxide loadings at D_1 and D_2 stages of bleaching had no apparent impact on the viscosity and S_{10}/S_{18} values of dissolving pulp. However, the particular combination of active chlorine charges used in the D_1 and D_2 bleaching stages may play an important role in production of dissolving pulp with definitive pulp properties such as brightness and viscosity. A comparison of the bleaching sequences O$D_1$$E_o$$D_2$P and O$D_1$$E_o$$D_2$H applied in dissolving pulp production indicates that the first one is superior to the second one in terms of brightness and alkali solubility (S_{10} and S_{18}). Moreover, a replacement of the hypochlorite with peroxide as a final bleaching step would result in a further reduction of the AOX-levels of the bleach plant effluent.

Table V. Biobleaching of sulfite pulp with *A. pullulans* xylanase (X)[a]

Treatment sequence	S_{10} (%)	S_{18} (%)	Brightness (%, ISO)	Viscosity (cP)
$OD_1E_oD_2$	9.1	4.6	89.8	27.1
X-$OD_1E_oD_2$	7.8	4.1	91.4	29.2
$OD_1E_oD_2P$	9.3	4.8	92.5	30.9
X-$OD_1E_oD_2P$	7.8	4.4	94.2	33.2
$OD_1E_oD_2H$	9.5	5.2	90.9	18.0
X-$OD_1E_oD_2H$	8.8	4.9	92.0	19.8

[a] Each value represents the mean of duplicate independent determinations; see Table I for treatment conditions

Acknowledgments

This work was sponsored by Sappi Saiccor (Pty.) Ltd. We thank Ron Braunstein (Sappi Management Services) for testing the samples for brightness, Clint Yunnie (Sappi Saiccor) for analysing the viscosity, Dalene van den Berg and Helen Borchers for excellent technical assistance. We thank Sappi Management Services for giving permission to publish the work.

Literature Cited

1. McGinnis, G.D.; Shafizadeh, F. In *Pulp and Paper Chemistry and Chemical Technology*; Casey, J.P., Ed.; John Willey & Sons: New York, NY, 1980, Vol. 1; pp 1-38.
2. Sjöström, E. *Wood Chemistry: Fundamentals and Applications*; Academic Press: New York, NY, 1981; pp 146-157.
3. Rydholm, S.A. *Pulping Processes*; John Wiley & Sons: Ney York, NY, 1965; pp 1172-1178.
4. Hinck, J.F.; Casebier, R.L.; Hamilton, J.K. In *Pulp and Paper Manufacture*; Kocurek, M.J.; Ingruber, O.V.; Wong, A., Eds.; Sulfite Science & Technology; Technical Section CPPA: Montreal, QC, 1985, Vol. 4; pp 213-243.
5. Timell, T.E. *Wood Sci Technol.* **1967**, *1*, 45.
6. Fengel, D.; Wegener, G. *Wood: Chemistry, Ultrastructure, Reactions*; Walter de Gruyter: Berlin & New York, 1984; pp 463-466.
7. Suntio, L.R.; Shiu, W.Y.; Mackay, D. *Chemosphere* **1988**, *17*, 1249.
8. Noe, P.; Chevalier, J.; Mora, F.; Comtat, J. *J. Wood Chem. Technol.* **1986**, *6*, 167.
9. Paice, M.G.; Bernier, R.; Jurasek, L. *Biotechnol. Bioeng.* **1988**, *32*, 235.
10. Shoham, Y.; Schwartz, Z.; Khasin, A.; Gat, O.; Zosim, Z.; Rosenberg, E. *Biodegradation* **1992**, *3*, 207.
11. Viikari, L.; Kantelinen, A.; Sundquist, J; Linko, M. *FEMS Microbiol. Rev.* **1994**, *13*, 335.
12. Kantelinen, A.; Hortling, B.; Sundquist, J.; Linko, M.; Viikari, L. *Holzforschung* **1993**, *47*, 318.
13. Paice, M.G.; Gurnagul, N.; Page, D.H.; Jurasek, L. *Enzyme Microb. Technol.* **1992**, *14*, 272.

14. Kantelinen, A.; Rantanen, T.; Buchert, J.; Viikari, J. *J. Biotechnol.* **1993**, *28*, 219.
15. Puls, J.; Poutanen, K.; Lin, J.J. In *Biotechnology in the Pulp and Paper Manufacture*; Kirk, K.T.; Chang, H.-M.; Butterworth-Heinemann: Boston, MA, 1990; pp 183-190.
16. Paice, M.G.; Jurasek, L. *J. Wood Chem. Technol.* **1984**, *4*, 187.
17. Christov, L.P.; Prior, B.A. *Biotechnol. Lett.* **1993**, *15*, 1269.
18. Christov, L.P.; Prior, B.A. *Appl. Microbiol. Biotechnol.* **1994**, *42*, 492.
19. Christov, L.P.; Prior, B.A. *Enzyme Microb. Technol.* **1996**, *18*, 244.
20. Annergren, G.A.; Rydholm, S.A. *Sven. Papperstidn.* **1959**, *62*, 737.
21. Bryce, J.R.G. In *Pulp and Paper Chemistry and Chemical Technology*; Casey, J.P., Ed.; John Willey & Sons: New York, NY, 1980, Vol. 1; pp 291-376.
22. Sjöström, E. *Wood Chemistry: Fundamentals and Applications*; Academic Press: New York, NY, 1981; pp 116-120.
23. Meier, H. *Sven. Papperstidn.* **1962**, *65*, 589.
24. Somogyi, M. *J. Biol. Chem.* **1952**, *195*, 19.
25. Nelson, N. *J. Biol. Chem.* **1944**, *153*, 375.
26. Christov, L.P.; Prior, B.A. *Biotechnol. Lett.* **1995**, *17*, 821.
27. Senior, D.J.; Hamilton, J.; Bernier, R.L.; du Manoir, J.R. *Tappi J.* **1992**, *75*, 125.
28. Tolan, J.S.; Canovas, R.V. *Pulp Paper Can.* **1992**, *93*, 39.
29. Turner, J.C; Skerker, P.S.; Burns, B.J.; Howard, J.C.; Alonso, M.A.; Andres, J.L. *Tappi J.* **1992**, *75*, 83.
30. Yang, J.L.; Lou, G.; Eriksson, K.-L. *Tappi J.* **1992**, *75*, 95.
31. Larsen, L.E.; Partridge, H. deV. In *The Bleaching of Pulp*; Singh, R.P., Ed.; Tappi Press: Atlanta, GA, 1979; pp 101-112.

Chapter 17

Use of Xylanase in the Totally Chlorine-Free Bleaching of Eucalyptus Kraft Pulp

M. B. Roncero, T. Vidal, A. L. Torres, and J. F. Colom

Especialidad Papelera y Gráfica, Escuela Técnica Superior de Ingenieros Industriales de Terrassa, Universidad Politécnica de Cataluña, Colón, 11 E–08222 Terrassa, Spain

Environmental pressures are forcing the pulp and paper industry to develop new technologies that reduce or eliminate the presence of various contaminants in bleaching plant effluents. Oxygen delignification techniques, replacement of elemental chlorine with chlorine dioxide, ozone, hydrogen peroxide and new bleaching agents as well as the use of xylanase enzymes for biobleaching, reduce or eliminate the production of chlorinated organic substances. This paper compares the sequence XOZP with OZP in the bleaching of *Eucalyptus globulus* kraft pulps. It has been studied the influence of enzymatic treatment on the consumption of bleaching agents: ozone and hydrogen peroxide. Chemical, physical, optical and refining properties of pulps, as well as COD and colour of effluent are studied as well. The xylanase treatment is positive and it is possible to manufacture fully bleached pulps at high brightness and viscosity without using chlorine compounds at a low ozone and hydrogen peroxide consumption.

Reducing the environmental effect of mill effluents has been a major concern for the pulp and paper industry in recent years. Chlorine-free bleaching represents an alternative for eliminating undesirable chlorinated compounds. Enzymatic prebleaching of pulp opens new possibilities for non-polluting rearrangement of bleaching processes. Especially the use of xylanases for improving the bleachability of kraft pulp has become important *(1 - 5)*.

Xylanase enzymes are liquid solutions of protein which hydrolize a portion of the xylan in the pulp. There has been recent evidence of direct brightening but in general it is considered that enzymes decrease the subsequent amount of oxidative chemicals required to bleach the pulp *(3, 6 -7)*.

The incorporation of xylanase enzymes into a pulp mill bleaching sequence is simple and economically feasible *(8)*. The effectiveness of xylanases for totally chlorine-free bleaching is currently being studied using peroxide and ozone. This paper

0097–6156/96/0655–0219$15.00/0
© 1996 American Chemical Society

describes laboratory results of a enzyme treatment for preventing dioxin formation during the bleaching of pulp. XOZP and OZP sequences in the bleaching of *Eucalyptus globulus* kraft pulps are compared. Chemical, physical, optical and refining properties of pulps, as well as COD and colour of effluent are also studied. The results are compared with the ECF (XODPD and ODPD) sequences *(9 - 18)*.

Materials and Methods

Pulp. An industrial *Eucalyptus globulus* kraft pulp of ENCE mill (Pontevedra, Spain) with a kappa number of 13.6, brightness of 31.7 % ISO and viscosity of 1,090 cm^3/g was used. The bleaching sequences have been performed in the laboratories of the E.T.S.I.I. of Terrassa (Spain) and they were: OAZ$_{0.4}$RP, OAZ$_{0.55}$RP, XOAZ$_{0.4}$RP (TCF sequences) and OD$_1$PD$_2$, XOD$_1$PD$_2$ (ECF sequences). In order to evaluate the physical properties, the fully bleached pulps were refined with a PFI mill and evaluated according to standard methods.

Terminology of Bleaching Stages. The terminology of bleaching stages used in this paper is as follows: X : Xylanase pretreatment (Pulzyme HC); O : Oxygen delignification; A : Treatment with oxalic acid without final wash; Z : Ozone bleaching stage; R : Post-treatment with sodium borohydride; P : Hydrogen peroxide bleaching stage; D : Chlorine dioxide bleaching stage.

Conditions in Bleaching Stages. Three TCF sequences (OAZ$_{0.4}$RP, OAZ$_{0.55}$RP, XOAZ$_{0.4}$RP) have been carried out. The OAZ$_{0.4}$RP and XOAZ$_{0.4}$RP sequences were to study the effect of xylanase pretreatment, when a same charge is used in the ozone bleaching stage. The OAZ$_{0.55}$RP has been performed to study the influence of enzymatic treatment on the consumption of bleaching agents (ozone) at the same final brightness. Conditions in bleaching stages are shown in Table I. Treatment with oxalic acid without final wash before Z-stage is carried out because this additive prevents viscosity loss during ozone bleaching, and increases brightness values, in comparison with blank tests *(19)*. During ozone bleaching some carbonyl groups are created on the cellulose chains. These groups are responsible for the rapid depolymerization of the cellulose chains in a subsequent peroxide stage (alkaline stage). It is necessary to reduce these groups and to make the cellulose quite stable during a P stage. Sodium borohydride is used *(19-22)*.

Results and Discussion

Effect of Xylanase Pretreatment on Pulp Properties. Results of brightness, viscosity and kappa number after oxygen delignification stage for enzymatic treated and untreated pulps are presented in Table II. After O-stage, XO-pulp has higher brightness and viscosity , and lower kappa number than O-pulp. Addition of a xylanase pretreatment (Table III) to an OAZ$_{0.4}$RP bleaching sequence reduced the final kappa

Table I. Conditions in bleaching stages

STAGE	X	O	A	Z	R	P
Consistency, % o.d.p.	10	10	10	40	10	10
Temperature, °C	60	110	amb.	amb.	amb.	85
Reaction time, min.	120	60	10	-	60	180
Initial pH	7 - 8	-	-	-	-	10.5
Enzyme, EXU/kg	500	-	-	-	-	-
NaOH, % o.d.p.	-	1.5	-	-	-	1.8
$MgSO_4 \cdot 7H_2O$, % o.d.p.	-	0.5	-	-	-	0.2
O_2 pressure, kg/cm^2	-	6	-	-	-	-
Oxalic acid, % o.d.p.	-	-	2	-	-	-
Ozone, % o.d.p.	-	-	-	0.4 or 0.55	-	-
In. Conc. O_3, mg/l	-	-	-	35	-	-
Flow O_3/O_2, l/h	-	-	-	150	-	-
$NaBH_4$, % o.d.p.	-	-	-	-	0.5	-
Na_2CO_3, % o.d.p.	-	-	-	-	1	-
H_2O_2, % o.d.p.	-	-	-	-	-	2
Acid wash with CO_2	no	no	no	no	no	yes

Table II. Effect of xylanase pretreatment

	XO	O
Kappa number	7.1	8.4
Brightness, % ISO	53.9	52.1
Viscosity, cm^3/g	956	912

Table III. Final results of pulp properties of studied TCF sequences

Sequence	$OAZ_{0.4}RP$*	$OAZ_{0.55}RP$**	$XOAZ_{0.4}RP$*
Kappa number	1.3	0.6	0.7
Brightness, % ISO	87.3	90.3	90
Viscosity, cm^3/g	813	793	807
Residual H_2O_2, %	0.6	0.6	0.99

*: Z-Stage with 0.4% ozone; **: Z-Stage with 0.55% ozone

Table IV. Comparison physical properties of ECF and TCF sequences for 3,000 and 4,500 PFI revolutions

3,000 PFI revolutions

Sequence	$OAZ_{0.4}RP$	$OAZ_{0.55}RP$	$XOAZ_{0.4}RP$	ODPD	XODPD
°SR	25	27	25	25	23
WRV (%)	155	155	148	153	148
Power consump., W·h	38	38	38	42	40
Bulk, cm^3/g	1.65	1.63	1.58	1.56	1.56
Permeability B., ml/min	1,685	1,428	1,845	2,391	2,779
Tensile Index, N·m/g	60.3	62	55.9	72.7	61.7
Folding Endurance	3.14	3.28	3.01	3.09	2.95
Tear Index, $mN{\cdot}m^2/g$	10.7	11.1	12.5	11.8	13.1
Burst Index, kN/g	4.14	4.3	4.05	4.43	3.83

4,500 PFI revolutions

Sequence	$OAZ_{0.4}RP$	$OAZ_{0.55}RP$	$XOAZ_{0.4}RP$	ODPD	XODPD
°SR	33	36	32	33	30
WRV (%)	166	160	160	158	154
Power consump., W·h	56	56	57	62	61
Bulk, cm^3/g	1.6	1.55	1.55	1.50	1.51
Permeability B., ml/min	565	595	960	1,232	1,682
Tensile Index, N·m/g	65.4	73.2	67.4	82.3	68.9
Folding Endurance	3.56	3.52	3.39	3.18	3.15
Tear Index, $mN{\cdot}m^2/g$	-	11.6	12.2	11.0	12.6
Burst Index, kN/g	4.46	4.99	4.47	4.69	4.39

Table V. Comparison of pulp properties (ECF)

Sequence	XODPD	ODPD
Kappa number	0.84	0.67
Brightness (% ISO)	88.3	87.7
Viscosity (cm^3/g)	923	868

by almost half (from 1.3 to 0.7) and icreased brightness by almost three points (87.3 to 90). In a side-by-side comparison, it was necessary to increase the Z-stage bleaching from 0.4% to 0.55% in order to obtain an equivalent kappa and brightness. Even so, the xylanase-treated pulp showed higher viscosity. Diminutions of kappa number and viscosity, and increase of brightness follow the same pattern in these three TCF pulps. $OAZ_{0.4}RP$ and $XOAZ_{0.4}RP$ have the same viscosity (about 800 cm^3/g), but brightness is three points higher with enzymatic treatment (X) and the kappa number is lower in $XOAZ_{0.4}RP$. At the same reagents consumption, xylanase treated pulp has higher brightness than when xylanase is not applied.

OAZ0.55RP (untreated pulp) and XOAZ0.4RP (enzymatic treated pulp) have the same viscosity and brightness, but a higher ozone charge is required in untreated pulp (0.55 % O_3) than in treated pulp (0.4 % O_3). XOAZ0.4RP has a lower hydrogen peroxide consumption, so it is posible to reduce H_2O_2 charge in this sequence, with no impact on the final results. So, at the same final brightness, the enzymatic treated pulp can save around 27 % in ozone charge. Alternatively savings are possible in hydrogen peroxide charge.

We can conclude that the introduction of xylanase has positive effects and makes it possible to produce high bleached pulp and to reduce chemicals consumption without using molecular chlorine.

Effect of Xylanase Pretreatment in Physical and Refining Properties. Power consumption is the same in the three pulps (Table IV), but °SR and WRV are higher for untreated pulps (OAZ0.4RP, OAZ0.55RP) than for the treated pulp (XOAZ0.4RP) due to the presence of more hemicelluloses (Figures 1a and 1b). Thus, refining is easier with untreated pulp. Results of bulk property do not show any possible effect of enzymatic pretreatment about this characteristic. For treated pulp the permeability and tear properties (Figures 1c and 1d) are higher than in the case of untreated pulp, because the papers of treated pulp have a more open structure, due to the lesser content of hemicelluloses, and as a consequence less fiber bonding. At the same reagents consumption, tensile and burst properties make no difference when xylanase is applied. But at same brightness, these properties are higher than when enzymatic treatment is not applied. In burst index there is a slight difference, no very important. Results of folding endurance are similar in three pulps, but results are slightly lower in xylanase treated pulp. Perhaps, the main differences in refining and physical properties are related to the quantity of hemicelluloses in pulp fibers.

Comparison of Results with ECF Sequences. Results of TCF (OAZ0.4RP, OAZ0.55RP, XOAZ0.4RP) sequences are compared with ECF (XODPD, ODPD) sequences. Tables III and V present the results of TCF and ECF bleached pulps. The same conclusions are valid in TCF and ECF sequences. At the same reagent consumption levels for ECF sequences, higher brightness and viscosity are achieved with enzymatic treatment. The use of xylanase in *Eucalyptus globulus* kraft pulp bleaching improves bleachability in subsequent stages. Physical and refining properties of TCF and ECF pulps for 3,000 and 4,500 PFI revolutions, are shown in Table V. Because the power consumption is higher in ECF pulps, refining is easier with TCF, for about the same °SR and WRV. The same conclusions about physical properties for

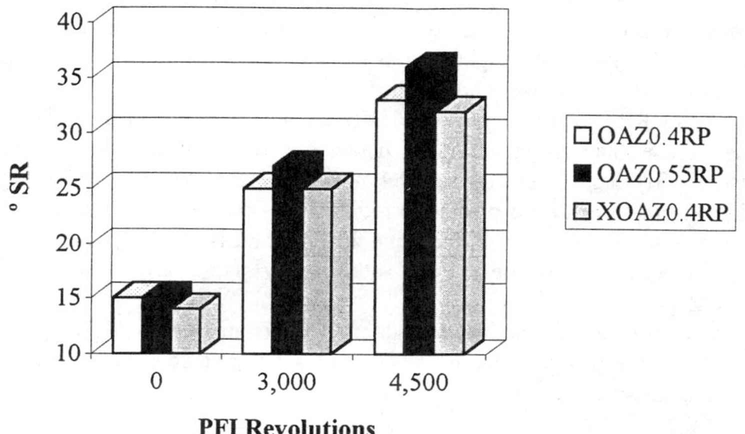

Figure 1a. °SR vs. PFI revolutions of TCF sequences.

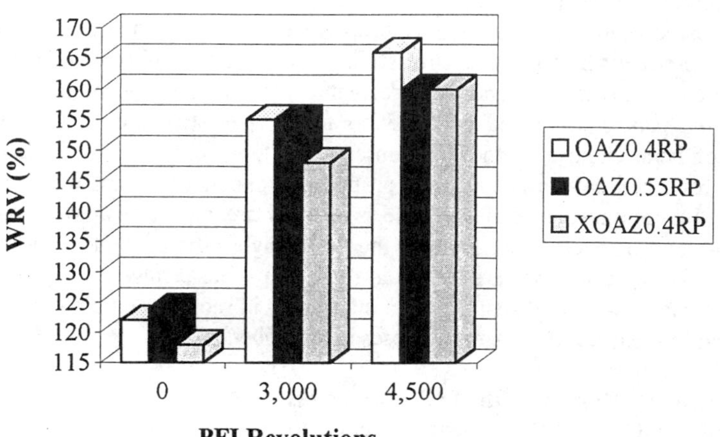

Figure 1b. WRV vs. PFI revolutions of TCF sequences.

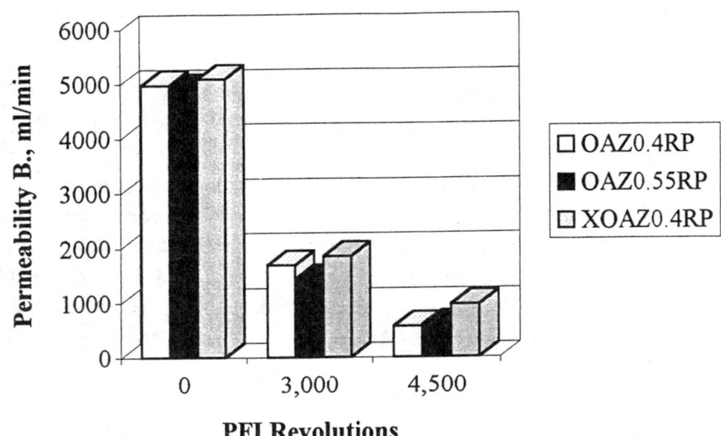

Figure 1c. Permeability vs. PFI revolutions of TCF sequences.

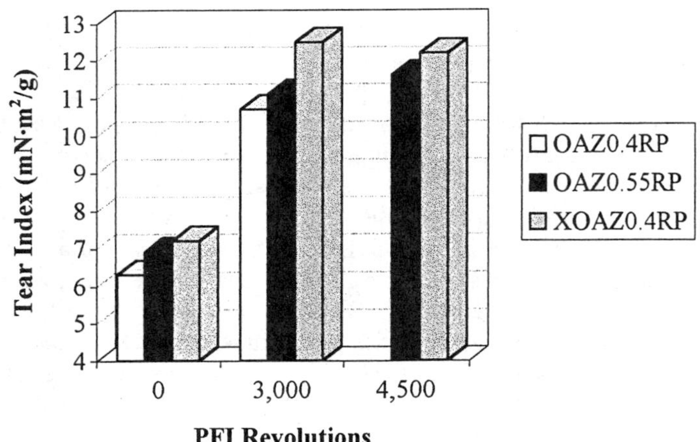

Figure 1d. Tear Index vs. PFI revolutions of TCF sequences.

TCF sequences hold true for ECF sequences. Permeability and tear properties are higher for treated pulp (XODPD) than untreated pulp (ODPD), while for tensile and burst properties the effect is opposite.

We conclude that the physical properties are similar and very high for the TCF and ECF sequences studied, so we can obtain high bleached TCF pulps comparable to ECF pulps without resorting to chlorine compounds.

Effect of Enzyme on Effluents. The effluents from ECF and TCF bleaching were characterized by determining color and COD. In the case of COD, the effluent from enzymatic treated pulp had approximately twice the total COD value that untreated pulp. The enzymatic treatment accounted for 50% of total COD in the TCF sequence and 51% in the ECF sequence. The higher COD occurs because xylanase hydrolyzes a fraction of the hemicellulose xylan in the pulps. This releases carbohydrates and some lignin. In the ECF sequences, the X and O stages contributed to the highest COD values: XO-stage represents a 85% of total COD in treated pulp and O-stage represents a 70% of total COD in untreated pulp. The D (chlorine dioxide) and P (hydrogen peroxide) contibuted to the lowest total COD value. In the TCF sequences, the X, O and R stages contributed to the highest COD values, whereas the P stage presents a small percentage of the COD value. In the case of color, the effluent from enzymatical treatment stage contributed to about 28% in ECF and 32% in TCF sequences. These values of color can be explained by a probable bad washing of the brown stock. The X, O, R and Z stages contributed to the highest color values in effluents. Otherwise, P-stage presents the lowest color value, because of the presence of hydrogen peroxide residual in the effluent, which works as a bleaching agent.

Conclusions

Xylanase pretreatment (X) of a TCF bleached pulp following a sequence XOAZRP is effective and makes it possible to produce high bleached pulp without using chlorine compounds. Xylanase improves brightness and viscosity and lowers the final kappa number when the same chemical charges are applied. When using xylanase it is possible to reach a predetermined brightness with a significant saving of chemicals. In the case studied the evaluated savings are 27% in ozone charge and 28% in hydrogen peroxide charge for a final brightness of 90% ISO.

In TCF pulps, physical properties are slightly different for the treated and untreated pulps. At the same brightness it is possible to reduce chemicals consumption, retaining the strength properties. Refining is easier with untreated pulp due to the presence of more hemicelluloses.

The effects of xylanases on physical and optical properties are similar for TCF and ECF sequences.

The effluent from enzymatic treated pulp had approximately twice the total COD value that untreated pulp, because xylanase treatment is used to hydrolyze a certain fraction of the hemicellulose xylan in the pulps. This results in the release of carbohydrates and some lignin linked to these carbohydrates, which contributes to higher COD in effluents.

The effluent from enzymatic treated pulp contributed to about 28% in ECF and 32% in TCF sequences, of the total color value.

References

1. Gliese, T.; Kleemann, S. *Wochenblatt für Papierfabrikation* **1995,** 526-532.
2. Yang, J. L.; Lou, G; Eriksson, K. L. *Tappi Journal* **1992,** *75*, 95-101.
3. Suurnäkki, A.; Kantelinen, A.; Bichert, J.; Viikari, L. *Tappi Journal* **1994,** *77*, 111-116.
4. David, J. *Pulp and Paper* **1992,** *66*, 111-114.
5. Siles, J. F.; Vidal, T.; Colom, J. F. *VII Conf. Téc. Latin. de Celulosa y Papel.*, Caracas, Venezuela, **1994**.
6. Tolan, J. S.; Olson, D., Dines, R.E. *Pulp & Paper Canada* **1995,** *96*, 107-110.
7. Bajpai, P.; Bhardwaj, N. K.; Maheshwari, S.; Bajpai, P. K. *Appita* **1993,** *46*, 274-276.
8. Tolan, J. S. *Journal of Pulp and Paper Science* **1995,** *21*, J132-J137.1.
9. Siles, J., Vidal, T., Colom, J. F., Torres, A. L. *International Symposium on Wood and Pulping Chemistry*, Helsinki **1995,** *II*, 391-396.
10. Yang, J. L.; Sacon, V. M.; Edward, S.; Eriksson, K-E L. *Tappi Journal* **1993,** *76*, 91-96.
11. Turner, J. C.; Skerker, P. S.; Burns, B. J.; Howard, J. C.; Alonso, M. A.; Andrés, J. L. *Tappi Journal* **1992,** *75*, 83-89.
12. Nelson, S. L.; Wong, K. K. Y., Saddler, J. N., Beatson, R. P. *Pulp & Paper Canada* **1995,** *7*, 42-45.
13. Tolan, J. S.; Foody, B. E. *Journal of Pulp and Paper Science* **1995,** *21*, J191-J196.
14. Allison, R. W.; Clark, T. A.; Wratall, S. H. *Appita* **1993,** *46*, 269-272.
15. Allison, R. W.; Clark, T. A.; Wratall, S. H. *Appita* **1993,** *46*, 349-353.
16. Paice, M. G.; Bourbonnais, R.; Reid, I. D.; Archibald, F. S.; Jurasek, L. *Journal of Pulp and Paper Science* **1995,** *21*, J280-J284.
17. Cates, D. H.; Eggert C.; Yang, J. L.; Eriksson, K.-E. L. *Tappi Journal* **1995,** *78*, 93-98.
18. Siles, F. J., Torres, A.L., Colom, J.F., Vidal, T. *Afinidad* **1996,** *LIII*, 93-102.
19. Roncero, B., Vidal, T., Colom, J.F. *International Pulp Bleaching Conference*, Washington, **1996**
20. Colodette, J.L., Singh, U.P., Ghosh, A.K., and Singh, R.P. *Pulp and Paper* **1993,** *75*.
21. Chirat, C., Viardin, M.-T., and Lachenal, D. *Paperi Ja Puu - Paper and Timber* **1994,** *76*, 417.
22. Lindholm, C.-A. *Paperi Ja Puu - Paper and Timber* **1992**, *74*,. 224.

Chapter 18

Biological Bleaching of Kraft Pulp with Lignin-Degrading Enzymes

R. Kondo[1], K. Harazono[1], K. Tsuchikawa[2], and K. Sakai[1]

[1]Department of Forest Products, Kyushu University, Hakozaki, Higashi-ku, Fukuoka 812–81, Japan
[2]Tokushu Paper Manufacturing Company Limited, Shizuoka 411, Japan

In vitro bleaching of an unbleached hardwood kraft pulp was performed with manganese peroxidase from fungus *Phanerochaete sordida* strain YK-624. When the kraft pulp was treated with a partly purified MnP in the presence of $MnSO_4$, Tween 80 and sodium malonate with continuous H_2O_2 addition at 37℃ for 24 h, pulp brightness increased by about 15 points and the kappa number decreased by about 6 points in comparison with untreated pulp. When the pulp was treated without the additon of $MnSO_4$, the pulp brightness increased by about 10 points in the presence of 2 mM oxalate, a good manganese chelator and reducing reagent, while the brightness did not significantly increase in the presence of 50 mM malonate. To establish a totally chlorine-free bleaching process, oxygen-bleached kraft pulp (OKP) was treated with four-stage bleaching process consisting of sequential MnP treatment, alkaline extraction, MnP treatment and hydrogen peroxide treatment stage. Full bleached kraft pulp (brightness 91%, yield 97%) could be obtained from OKP by the combination of enzyme treatment and hydrogen peroxide bleaching. To improve in vitro bleaching of the kraft pulp with MnP, the characterization of various MnPs from some white-rot fungi was examined. MnP from *Ganoderma* sp. YK-505 was superior to MnPs from *P. sordida* YK-624 and *P. chrysosporium* in stabilities against high temperature and high concentration of H_2O_2. MnP from *Ganoderma* sp. YK-505 differed in pH-activity profile from other MnPs. These results suggest that MnP from *Ganoderma* sp. YK-505 has different structure from those other fungi, and may be useful for biotechnological applications and studies of the relationship between structure and function.

The kraft process, at present the most common commercial chemical delignification method, produces a dark pulp because of the color of residual modified lignin residues. These residues are normally bleached or removed in multistage bleaching procedures using a combination of chlorination and alkaline-extraction stepes. The effluent from

0097–6156/96/0655–0228$15.00/0
© 1996 American Chemical Society

such bleaching processes is of growing environmental concern because it contains numerous chlorinated organic substances, including mutagenic chlorinated phenols and dioxins (*1,2*). Today, general concern about the environmental impact of chlorine bleaching effluents has led to a trend toward elementary chlorine-free or totally chlorine-free bleaching methods. Considerable interest has been focussed on the white-rot basidiomycete fungi, since they are the only group of organisms known to be capable of preferential degradation of native lignins and complete degradation of wood.

In vivo bleaching of kraft pulp with white-rot fungi

Work in our laboratory has demonstrated that the white-rot fungus, IZU-154 can decrease both residual lignin color (brightness) and concentration (kappa number) in hardwood kraft pulp and softwood kraft pulp (*3,4*). Thus, fungal treatment yields pulps with substantially increased brightness (biobleaching). Introducing the IZU-154 treatment into the kraft pulp bleaching process made it possible to bleach unbleached kraft pulp without any chlorine-based chemicals, and the pulp bleached by this process had satisfactory optical and strength properties (*5*). To date, *Phanerochaete sordida*(*6*), *Coriolus (Trametes) versicolor* (*7-9*) and unidentified isolate(*10*) have been shown to brighten kraft pulp remarkably. However, the process is rather slow compared with chemical bleaching (days instead of hours), and the attack of the cellulose by the fungus cannot be avoided completely. To overcome these drawbacks of the fungal process, the enzymolozy of fungal delignification should be determined.

Lignin-degrading enzymes contributed to biobleaching

Paice et al. and we have reported that manganese peroxidase (MnP) and laccase activities but not lignin peroxidase (LiP) activity are detected during biobleaching of unbleached hardwood kraft pulp by white-rot fungi with bleaching ability(*8,11*). Furthermore, we developed a cultivation system in which a membrane filter was used to prevent direct contact between hyphae and kraft pulp while allowing extracellular enzymes to attack the kraft pulp. By using this system we found that the level of secreted MnP activity in the filterable components was substantial during in vitro bleaching with *P. sordida* YK-624 (*12*). Some reports have also shown that MnP plays an important role in the bleaching of the pulp by white rot fungi (*13-15*). MnP first discovered in *P. chrysosporium* (*16,17*) is known to be secreted by many lignin-

degrading fungi, including *T. versicolor* (*15*), *P. sordida* (*18*), and others (*19,20*). MnP is a heme-containing enzyme and oxidizes Mn^{2+} to Mn^{3+} which chelated with an organic acid, in turn, oxidizes phenolic substrates, including lignin model compounds (*21,22*), dehydropolymerizate (*23*) and high-molecular-weight chlorolignin (*24*).

In vitro bleaching of kraft pulp with MnP

There have been several reports of using in vitro enzyme treatments to bleach hardwood kraft pulp. Arbeloa et al. showed that treatment of unbleached kraft pulp with LiPs facilitated subsequent chemical bleaching (*25*). Bourbonnais et al. demonstrated that unbleached kraft pulp was delignified with isolated laccase from *T. versicolor* in the presence of 2,2-azinobis-(3-ethylbenzthiazoline-6-sulphonate) and that methanol was released (*26*). It has also been reported that some delignification of hardwood kraft pulp by MnP was observed, but the extensive brightening observed with the fungus was not achieved with MnP (*15*). Direct utilization of LiP and laccase of *Phlebia radiata* in kraft pulp bleaching was not successful (*27*).

We performed in vitro bleaching of the kraft pulp with MnP from the fungus *P. sordida* YK-624, which was isolated from decayed wood obtained from a forest and exhibited remarkable bleaching ability with the pulp (*11,12*). When the kraft pulp was treated with a partly purified MnP in the presence of $MnSO_4$, Tween 80 and sodium malonate with continuous H_2O_2 addition at 37°C for 24 h, pulp brightness increased by about 15 points and the kappa number decreased by about 6 points in comparison with untreated pulp. When the pulp was treated with MnP without Tween 80, the brightness increased by only about 4 points, to 35.5%. Therefore, the surfactant is also an important factor on bleaching of pulp with MnP. The effect of various surfactant on the brightening of the pulp with MnP is shown in Fig. 1 (*28*). It was observed that the use of Tween 80 or the combination with an unsaturated fatty acid and Tween 20 improved the brightness of the pulp after bleaching with MnP compared with the use of Tween 20 only. Recently, Bao et al. indicated that nonphenolic lignin was oxidized by a lipid peroxidation system that consisted of MnP, Mn^{2+} and unsaturated fatty acid esters present in Tween 80 (*29*). The detergent action of Tween 80 and/or the unsaturated fatty acids it contains may be responsible for the enhancement of MnP's bleaching effect.

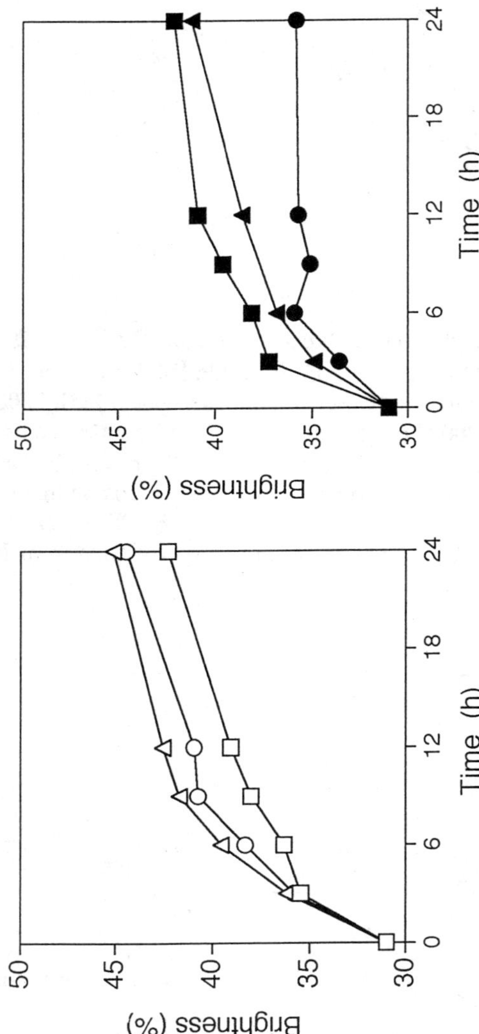

Figure 1. Effect of surfactant on bleaching of UKP with MnP. UKP was suspended at a consistency of 1% in 50 mM malonate buffer (pH 4.5) containing 100 U of MnP, 0.1 mM $MnSO_4$ and 0.05% surfactant, and 10 mM aqueous H_2O_2 was added at a rate of 3 ml/h at 30℃ for 24 h. Symbols: □, Tween 20; △, Tween 20 + linolenic acid; ○, Tween 80; ■, linolenic acid; ▲, CHAPS; ●, without surfatant. (Reproduced with permission from Ref. 28. Copyright 1996. Japan Tappi.)

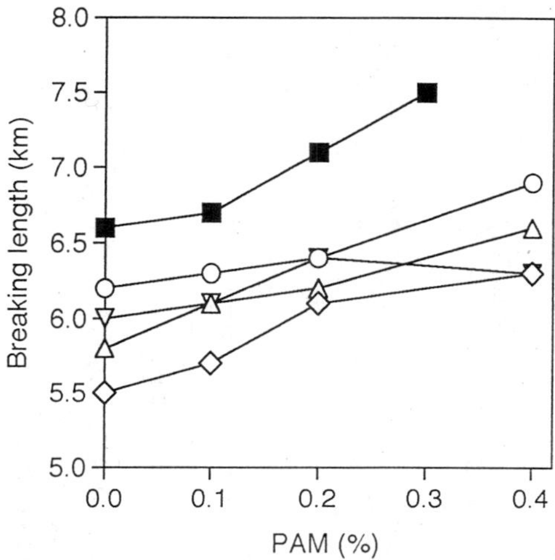

Figure 2. Effect of PAM addition on breaking length. Symbols: ■, O-MEMP; ○, O-MEM; ◇, O-CED; ▽, O-P; △, OKP. [MnP treatment (M): 20 g of oxygen bleached kraft pulp (OKP) (pulp consistency, 2%), 2000 U of MnP, 0.1 mM $MnSO_4$, 200 U of glucose oxidase and 2.5 mM glucose at 30℃ and 150 rpm for 24 h. Alkaline extraction (E): 2.5% aqueous NaOH at a pulp consistency of 10% for 1 h at room temperature. Peroxide treatment (P): 4% aqueous H_2O_2 at a pulp consistency of 15%. CED treatment: 2.1%Cl_2, 1.5%NaOH, 0.5%ClO_2.] (Reproduced with permission from Ref. 30. Copyright 1996. Japan Tappi.)

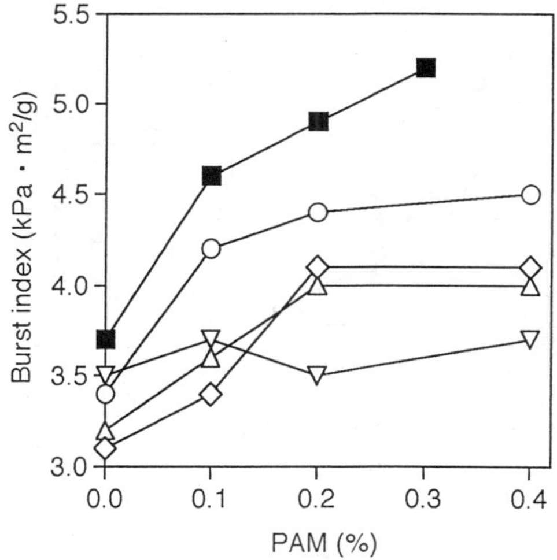

Figure 3. Effect of PAM addition on burst index. Symbols are same as Figure 2.

TCF bleaching process with introduction of enzyme treatment of MnP

To establish a totally chlorine-free bleaching process, oxygen-bleached kraft pulp (OKP) was treated with four-stage bleaching process consisting of sequential MnP treatment, alkaline extraction, MnP treatment and hydrogen peroxide treatment stage. Full bleached kraft pulp (brightness 91%, yield 97%) could be obtained from OKP by the combination of enzyme treatment and hydrogen peroxide bleaching. The results suggest that it should be possible to establish a chlorine-free bleaching process. The physical properties of this in vitro bleached pulp showed higher values in breaking length and burst index than those of a pulp bleached with chlorine-based chemicals. Moreover, the addition of polyacrylamide to the in vitro bleached pulp resulted the great improvement on strength properties as shown in Fig. 2-4 (*30*).

In vitro bleaching of kraft pulp with MnP without addition of $MnSO_4$

There are some differences between in vivo bleaching with fungi and in vitro bleaching with MnP with respect to treatment conditions. Major differences between the two bleaching systems were the use of additional $MnSO_4$ and high concentrations of malonate as a Mn^{3+}-chelator during in vitro bleaching with MnP. Since the solid-state fermentation system used in our work were added only water and mycelia to the pulp (*3-5,11*), the manganese in the pulp must have been used by the MnP catalytic system. We reported that unbleached hardwood kraft pulp was not brightened by *P. sordida* YK-624 when metal ions had previously been removed from the pulp by acid treatment but that the ability of the fungus to bleach this pulp was restored when Mn salts were added back to the pulp (*6*). Chemical species of Mn present in the pulp is unclear. Insoluble MnO_2 accumulates in wood decayed by some white rot fungi (*31*). Mn^{3+} oxidized by MnP is transformed to Mn^{4+} (MnO_2) and Mn^{2+} due to the hydrolysis and disproportionation (*32*).

Organic acids such as lactate, malonate and oxalate, which chelate Mn^{3+} generated by MnP, stabilizing Mn^{3+} in aqueous solution, play important roles in the MnP system. The white rot fungi, including *P. chrysosporium* and *T. versicolor,* secrete organic acids which function as Mn^{3+} chelators (*33-35*). It was reported that millimolar concentrations of oxalate are produced by various fungi such as *P. chrysosporium* and *T. versicolor* (*34,36-38*). The physiological concentrations of

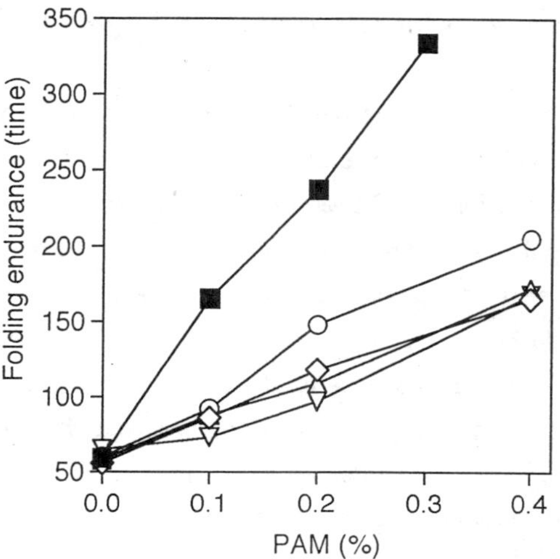

Figure 4. Effect of PAM addition on folding endurance. Symbols are same as Figure 2.

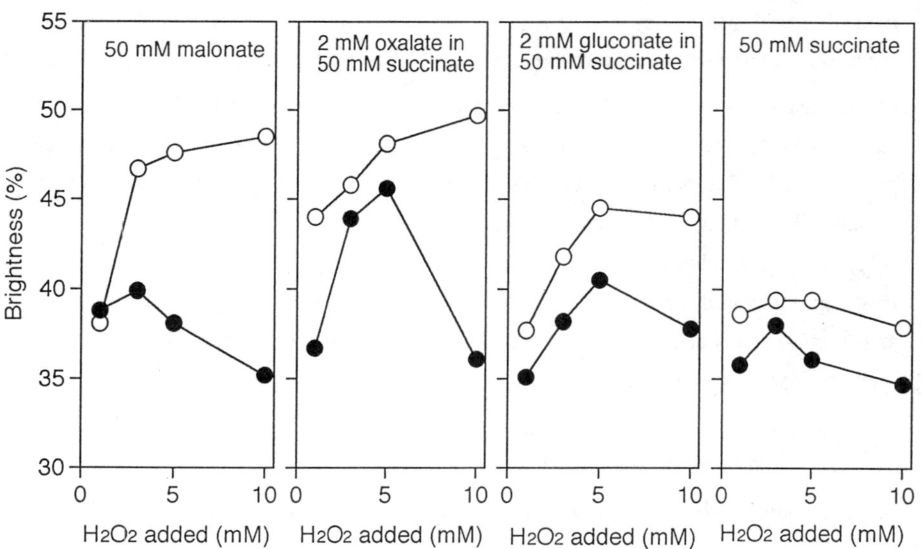

Figure 5. Effect of the addition of $MnSO_4$ on the brightness of UKP treated with MnP. UKP was suspended at a consistency of 1% in 50 mM malonate buffer (pH 4.5) containing 100 U of MnP, 0.1 mM $MnSO_4$ and 0.05% surfactant, and various concentrations of aqueous H_2O_2 were added at a rate of 3 ml/h at 30℃ for 24 h. Symbols: ●, without $MnSO_4$; ○, 0.1 mM $MnSO_4$ added. (Reproduced with permission from Ref. 40. Copyright 1996. American Society for Microbiology.)

oxalate stimulate optimal MnP activity (*33-35,37*). Oxalate is also known to have reductive activity(*39*). Therefore, we tried to bleach kraft pulp with MnP without addition of $MnSO_4$ by using oxalate with a good Mn^{3+}-chelating and reductive ability. The effect of the addition of $MnSO_4$ on bleaching of kraft pulp with MnP in the presence of various organic acids was examined by changing the concentration of aqueous H_2O_2 added (Fig. 5) (*40*). In 50 mM malonate buffer, an increase in brightness was observed with the addition of $MnSO_4$ but not without additional $MnSO_4$. When the pulp was treated with the continuous addition of 3 and 5 mM aqueous H_2O_2 in the presence of 2 mM oxalate, the brightness increased by about 13 points without the addition of $MnSO_4$. This indicates that MnP bleaching of kraft pulp can be performed without the addition of $MnSO_4$ if we use oxalate, which can also reduce insoluble MnO_2.

Oxygen-bleached hardwood kraft pulp was bleached with MnP with addition of $MnSO_4$ for the establishment of a totally chlorine-free bleaching process as previously mentioned. The addition of hydrogen peroxide bleaching was needed after treatment with MnP. The metals in pulp should be controlled by using chelating agents, e. g., EDTA, prior to hydrogen peroxide bleaching. Since manganese content in the pulp decreases by bleaching with MnP without the addition of $MnSO_4$, the use of oxalate during bleaching with MnP will reduce the amount of EDTA or allow omission of the EDTA treatment step during the subsequent hydrogen peroxide bleaching stage.

Characterization of MnPs from various white-rot fungi for the improvement of enzyme bleaching

Many ligninolytic fungi produce MnP in various cultures such as synthetic liquid and solid-state media (*19,20*). For the ubiquity of MnP among various fungi, a MnP isozyme which has a different function from well-known MnP from *P. chrysosporium* (*41-43*) may be present. Enzymological and kinetic studies of the isozyme could be useful for understanding catalytic characteristics and improving stability of peroxidases. In previous study, higher selectively ligninolytic fungi was screened and isolated from decayed wood samples, and biological bleaching of hardwood kraft pulp with those fungi was carried out (*11*). The result showed the relationship between pulp brightness increases and MnP activities was observed during the bleaching with almost fungi which remarkably brightened hardwood kraft pulp. On the other hand, since *Ganoderma* sp. YK-505 deviated from the relationship,

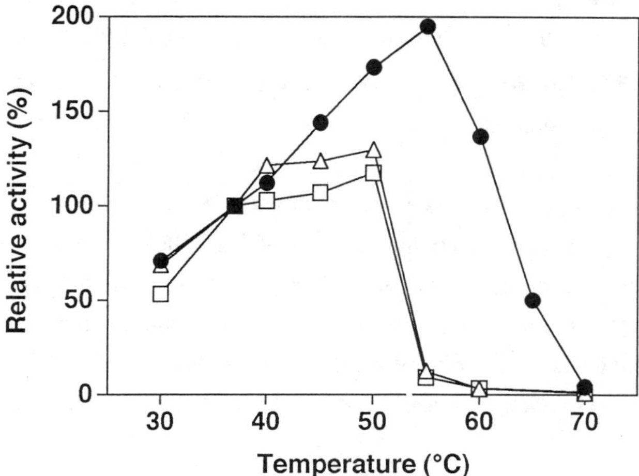

Figure 6. Effect of temperature on the activity of MnP from *Ganoderma* sp. YK-505 (●), *P. sordida* YK-624 (△) and *P. chrysosporium* (□).Reaction mixtures contained 1 mM 2, 6-dimethoxyphenol, 1 mM $MnSO_4$, 0.2 mM H_2O_2 and 50 mM malonate (pH 4.5). Control activity(100%) was obtained at 37℃.

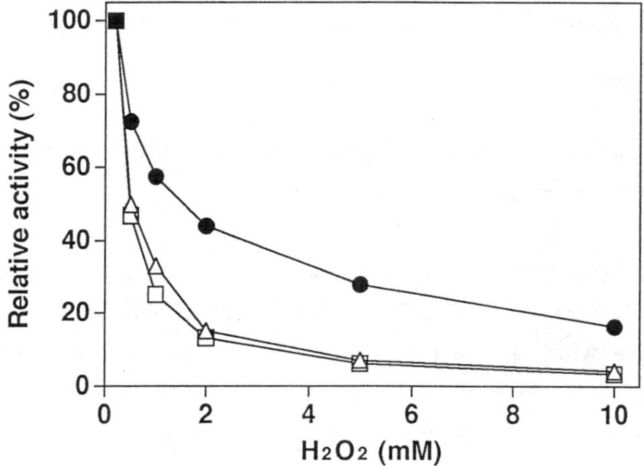

Figure 7. Effect of H_2O_2 concentration on the activity of MnP from *Ganoderma* sp. YK-505 (●), *P. sordida* YK-624 (△) and *P. chrysosporium* (□).Reaction mixtures contained 1 mM 2, 6-dimethoxyphenol, 1 mM $MnSO_4$ and 50 mM malonate (pH 4.5). Control activity(100%) was obtained with 0.2 mM H_2O_2.

we expected MnP from the fungus might have a different property from those of other white rot fungi. Therefore, properties of partially purified MnPs from liquid cultures of three white rot fungi, *Ganoderma* sp. YK-505, *P. sordida* YK-624 and *P. chrysosporium* were compared.

Thermostability of MnPs from three fungi was determined. Enzyme reaction was done at various temperatures for 3 minutes after incubation at each temperature for 10 minutes (Fig. 6). The temperature profile of MnP from *P. sordida* YK-624 was similar to that of MnP from *P. chrysosporium*. High activities were exhibited in the temperature range 40-50℃. The activities of MnPs from both fungi were not significantly detected at temperatures more than 50℃. In the case of MnP from YK-505, higher the temperature was, higher the activity was observed. Optimum activity was exhibited at 55℃. About 50% activity was also observed at 65℃. MnP from each fungus was incubated at 30, 40 and 50℃, and then the remaining activity was measured using standard assay at 37℃. All MnPs were stable at 30℃ for 24 h. When MnPs were incubated at 40℃, MnPs from *P. sordida* YK-624 and *P. chrysosporium* decreased in 40-60% activity but MnP from YK-505 maintained the activity. When MnPs were incubated at 50℃, activities of MnPs from *P. sordida* YK-624 and *P. chrysosporium* were not detected immediately for 30 minutes. On the other hand, a decrease rate of activity of MnP from YK-505 was slowly and it was capable to detect MnP activity after incubation for 12 h.

Continuous addition of aqueous H_2O_2 to pulp suspension is essential to bleaching of the kraft pulp with MnP. MnP activities during bleaching of the pulp decreased as a function of time by continuous addition of aqueous H_2O_2, although MnP activity did not decrease when pulp suspension was stirred without addition of H_2O_2. MnP isozyme tolerant to high concentrations of H_2O_2 may be useful for improving the bleaching of kraft pulp with MnP. The influence of H_2O_2 concentration on MnP activity is shown in Fig.7. Although higher concentrations of H_2O_2 decreased the enzyme activity of all MnPs, MnP from YK-505 was superior to MnPs from other fungi against high concentration of H_2O_2. Using guaiacol as a substrate, MnPs of *P. sordida* YK-624 and *P. chrysosporium* showed an optimal pH of 4.0. MnP activities decreased as the pH was raised. On the other hand, MnP from YK-505 had different pH dependency from other MnPs, and showed optimal pHs between 4.5 and 6.5 (Fig. 8).

Use of MnP from *Ganoderma* sp. YK-505, which has stabilities against high temperature and high concentration of H_2O_2, may improve the bleaching of kraft pulp with MnP. It would be interested in studying the relationship of structure/function of MnP from the fungus YK-505.

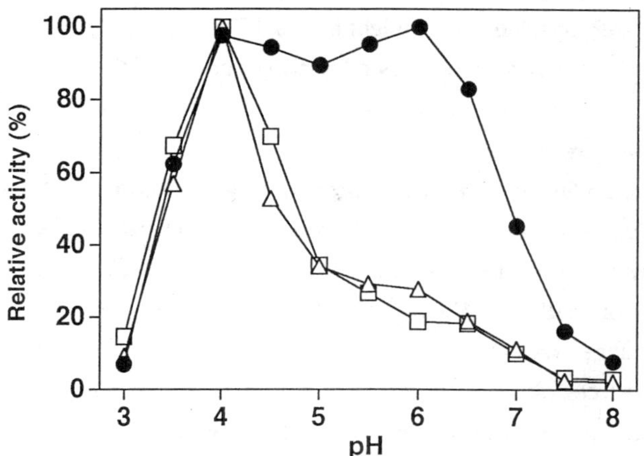

Figure 8. Effect of pH on the activity of MnP from *Ganoderma* sp. YK-505 (●), *P. sordida* YK-624 (△) and *P. chrysosporium* (□).Reaction mixtures contained 3 mM guaiacol, 1 mM $MnSO_4$ and 50 mM malonate (pH 4.5). Control activity(100%) was obtained at pH 4.0.

Literature cited

1. Campin, D. N., S. J. Buckland, D. J. Hannah, and J. A. Taucher. **1991**. *Wat. Sci. Tech.* **24**, 65-74.

2. Rappe, C., S. Swanson, B. Glas, K. P. Kringstad, F. D. Sousa, L. Johansson, and Z. Abe. **1989**, *Pulp Paper Can.* **90**, T273-278.

3. Fujita, K. , R. Kondo, K. Sakai, Y. Kashino, T. Nishida, and Y. Takahara. **1991**. *TAPPI (Tech. Assoc. Pulp Pap. Ind.) J.* **74(11)**, 123-127.

4. Fujita, K. , R. Kondo, K. Sakai, Y. Kashino, T. Nishida, and Y. Takahara. 1993. *TAPPI (Tech. Assoc. Pulp Pap. Ind.) J.* **76(1)**, 81-84.

5. Murata, S., R. Kondo, K. Sakai, Y. Kashino, T. Nishida, and Y. Takahara. **1992**. *TAPPI (Tech. Assoc. Pulp Pap. Ind.) J.* **75(12)**, 91-94.

6. Hirai, H., R. Kondo, and K. Sakai. **1995**. *Mokuzai Gakkaishi* **41**, 69-75.

7. Addleman, K., and F. Archibald. **1993**. *Appl. Environ. Microbiol.* **59**, 266-273.

8. Paice, M. G., L. Jurasek, C. Ho., R. Bourbonnais, and F. Archibald. **1989**. *TAPPI (Tech. Assoc. Pulp Pap. Ind.) J.* **72(5)**, 217-221.

9. Reid, I. D., M. G. Paice, H. Christopher, and L. Jurasek. **1990**. *TAPPI (Tech. Assoc. Pulp Pap. Ind.) J.* **73(8)**, 149-153.

10. Iimori, T., Kaneko, R., Yoshikawa, H., Machida, M., Yoshioka, H., and Murakami, K. **1994.** *Mokuzai Gakkaishi,* **40**, 733-737.

11. Hirai, H., R. Kondo, and K. Sakai. **1994.** *Mokuzai Gakkaishi* **40,** 980-986.

12. Kondo, R., K. Kurashiki, and K. Sakai. **1994.** *Appl. Environ. Microbiol.* **60,** 921-926.

13. Johansson, T., and P. O. Nyman. **1987.** *Acta Chemi. Scan.* **B41,** 762-765.

14. Katagiri, N., Y. Tsutsumi, and T. Nishida. **1995.** *Appl. Environ. Microbiol.* **61,** 617-622.

15. Paice, M. G., I. D. Reid, R. Bourbonnais, F. S. Archibald and L. Jurasek. **1993.** *Appl. Environ. Microbiol.* **59,** 260-265.

16. Kuwahara, M., J. K. Glenn, M. A. Morgan, and M. H. Gold. **1984.** *FEMS Lett.* **169,** 247-250.

17. Paszczynski, A., V. B. Huynh, and R. Crawford. **1985.** *FEMS Microbiol. Lett.* **29,** 37-41.

18. Rüttimann-Johnson, C., D. Cullen, and R. T. Lamar. **1994.** *Appl. Environ. Microbiol.* **60,** 599-605.

19. Orth, A., D. J. Royse, and M. Tien. **1993.** *Appl. Environ. Microbiol.* **59,** 4017-4023.

20. Peláez, F., M. J. Martínez, and A. T. Martínez. **1995.** *Mycol. Res.* **99,** 37-42.

21. Tuor, U., H. Wariishi, H. E. Schoemaker, and M. H. Gold. **1992.** *Biochemistry* **31,** 4986-4995.

22. Wariishi, H., K, Valli, and M. H. Gold. **1989.** *Biochemistry* **28,** 6017-6023.

23. Wariishi, H., K. Valli, and M. H. Gold. **1991.** *Biochem. Biophys. Res. Commun.* **176,** 269-276.

24. Lackner, R., E. Srebotnik, and K. Messner. **1991.** *Biochem. Biophys. Res. Commun.* **178,** 1092-1098.

25. Arbeloa, M., J. Leclerc, G. Goma, and J. C. Pommier. **1992.** *TAPPI (Tech. Assoc. Pulp Pap. Ind.) J.* **75**(3), 215-221.

26. Bourbonnais, R., and M. G. Paice. **1992.** *Appl. Microbiol. Biotechnol.* **36,** 823-827.

27. Kantelinen, A., B. Hortling, M. Ranua, and L. Viikari. **1993.** *Holzforschung* **47,** 29-35.

28. Harazono, K., R. Kondo, and K. Sakai. **1996.** *Japan Tappi* **50**, 697-706.

29. Bao, W., Y. Fukushima, K. A. Jensen Jr., M. A. Moen, and K. H. Hammel. **1994.** *FEMS Lett.* **354,** 297-300.

30. Tsuchikawa, K., R. Kondo, and K. Sakai. **1996.** *Japan Tappi* **50**, 899-905.

31. Blanchette, R. A. **1984.** *Phytopathology* **74,** 725-730.

32. Glenn, J. K., L. Akileswaran, and M. H. Gold. **1986.** *Arch. Biochem. Biophys.* **251(2),** 688-696.

33. Kishi, K., H. Wariishi, L. Marquez, H. B. Dunford, and M. H. Gold. **1994.** *Biochemistry* **33,** 8694-8701.

34. Kuan, I, and M. Tien. **1993.** *Proc. Natl. Acad. Sci. USA* **90,** 1242-1246.

35. Roy, B. P., and F. S. Archibald. **1993.** *Appl. Environ. Microbiol.* **59,** 1855-1863.

36. Dutton, M. V., C. S. Evans, P. T. Atkey, and D. A. Wood. **1993.** *Appl. Microbiol. Biotechnol.* **39,** 5-10.

37. Maxwell, D. P. , and D. F. Bateman. **1968.** *Phytopathology* **58,** 1635-1642.

38. Shimada, M., D. Ma, Y. Akamatsu, and T. Hattori. **1994.** *FEMS Microbiol. Rev.* **13,** 285-296.

39. Taube, H. **1948.** *J. Am. Chem. Soc.* **70,** 3928-3935.

40. Harazono, K., R. Kondo, and K. Sakai. **1996.** *Appl. Environ. Microbiol.* **62,** 913-917.

41. Aitken, M. D., and R. L. Irvine. **1990.** *Arch. Biochem. Biophys.* **276,** 405-414.

42. Paszczynski, A., V. Huynh, and R. Crawford. **1986.** *Arch. Biochem. Biophys.* **244,** 750-765.

43. Wariishi, H., H. B. Dunford, I. D. MacDonald, and M. H. Gold. **1989.** *J. Biol. chem.* **264,** 3335-3340.

Chapter 19

Assessment of Board-Making Parameters of Pulps Prepared by Soda Cooking, Fungal Delignification, or Enzymatic Treatment of Wheat-Straw

M. E. Guadalix[1], G. Almendros[1], M. J. Blanco[2], S. Camarero[3], M. J. Martinez[3], A. T. Martinez[3], T. Vares[4], A. Hatakka[4], M. Pelayo[5], and E. Gago[5]

[1]Centro de Ciencias Medioambientales, Consejo Superior de Investigaciones Cientificas, Serrano 115 dpdo, E–28006 Madrid, Spain
[2]Instituto de Energias Renovables, Centro de Investigaciones Energéticas, Mediaoambientales y Technológicas, Avenida Complutense 22, E–28040 Madrid, Spain
[3]Centro de Investigaciones Biológicas, Consejo Superior de Investigaciones Cientificas, Velázquez 144, E–28006 Madrid, Spain
[4]Department of Applied Chemistry and Microbiology, University of Helsinki, P.O. Box 56, FIN–00014, Finland
[5]Sociedad Anónima Industrias Colulosa Aragonesa, San Juan de la Peña, E–50002 Zaragoza, Spain

The comparative assessment of boardmaking parameters has been carried out with soda and biological refined pulps prepared from wheat straw. When compared with non-treated straw, biological pretreatments (solid-state fermentation with *Pleurotus eryngii* and *Phlebia radiata*, or enzyme treatments) save energy during mechanical refining. Nevertheless, except for mild cooking conditions (0.1 *M* NaOH, 40 °C) energy consumption is higher in biomechanical than in semichemical pulps. The factorial assessment of factors related to the quality of soda pulps, revealed similar effects of NaOH concentration (C) and temperature (T) which were much higher than the effect of reaction time (t). The most significant interaction was C×T and, in some cases, C×t. Simple and multiple linear regression models suggested that the intensity of the 1720 cm^{-1} infrared band may be used in discriminating between chemical and biological pulps. Standard soda pulps still contain about 25% wt alkali-soluble material, suggesting that additional washing at room temperature could substantially improve the delignification performance. The extent of the exothermic peak *ca.* 360 °C (thermogravimetric analyses) paralleled pulp quality whereas that at about 325 °C (breakdown of loosely attached O-containing groups) decreased with the intensity of soda cooking. The extent of the high temperature peaks (up to 500 °C) seems to be related to lignin, which had a reduced value as an indicator of pulp quality.

0097–6156/96/0655–0241$15.00/0
© 1996 American Chemical Society

Despite development of industrial technologies to transform straw into paper and paperboard pulps, the optimization of such processes as regards to paper quality, energy consumption and release of lignin-containing coloured effluents has not been achieved so far with the current technology.

Soda pulping of wheat straw takes advantage of the high solubility of straw lignins in alkaline solutions (*1,2*) and it is a widely used method to individualize straw fibres. By controlling cooking factors and the refining degree, pulps can be obtained for the manufacture of different types of paper and paperboards.

In fact, the presence of lignin-hemicellulose complexes in grasses, involving easily removable cinnamyl esters and a large proportion of phenolic groups in lignin (*3—5*), favour its chemical and biological removal. Due to the above reasons, the biological transformation of a lignocellulosic biomass into value added, partially delignified substrates may lead to future technologies in the pulp and paper industries, with the development of inexpensive, environmentally-friendly processes in which the degrading abilities of ligninolytic fungi may be especially successful in the case of a herbaceous biomass. Amongst the lignocellulose degrading organisms, the ligninolytic fungi provide a series of industrial possibilities both for liquid and solid cultures and for the production of specific enzyme assemblages (mainly ligninolytic peroxidases, aryl-alcohol oxidases, laccases, esterases and xylanases) (*6,7*). This contributes to the more or less complete separation of fibres and the effective depolymerization of lignin in the decaying substrate. The biological treatment can be carried out by solid-state fermentation (SSF) of the straw after fungal inoculation or by the more rapid alternative treatment of the straw with enzyme mixtures obtained in fermentors.

The precise evaluation of the boardmaking characteristics of the different types of pulps, obtained either by chemical or biological processes, cannot be determined straight away by using only a few standard quality parameters. In general, combinations of physical and chemical characteristics are required to define the suitability of pulps for good quality paper. In particular, the possibility of forecasting paperboard properties from routine laboratory tests is of particular interest in the research on biological pretreatments, where the improvement of pulping properties is not necessarily related to the neat concentration of lignin in the straw (*8*). For example, cultivation of lignin-degrading white-rot fungi on grasses, e.g. reed canary grass (*Phalaris arundinacea*), prior to soda-anthraquinone pulping, has recently been studied in order to check the potential of fungi to improve the quality of material for pulping, and the effect of fungal treatment on the properties of printing paper (*9, 11*). Fungal treatment increased the yield, and decreased the fine fraction and the kappa number after cooking. Handsheet properties were better than in controls without fungal treatment (mixtures contained grass pulp, softwood pulp and talc). However, these laboratory tests are time-consuming and very expensive, and thus, there is an urgent need for simple laboratory tests to predict the properties of final paper.

The present study describes the characteristics of semichemical pulps obtained under a variety of industrial conditions in order to compare their characteristics with those from biologically pretreated straw as well as to evaluate quality parameters reflecting the changes induced by different pulping procedures.

Experimental

Chemical Pulping. Wheat straw (*Triticum aestivum*) collected in Zaragoza (Spain) was chopped to 4 cm and used in a laboratory-scale preparation of chemical and biological pulps. In the case of chemical pulps, and in order to obtain samples representative of different industrial conditions, the levels of the pulping factors (i.e., soda concentration, temperature and reaction time) were arranged according to a 2^3 factorial design. The extent of the main effects and interactions were determined in a series of independent boardmaking tests and laboratory parameters. A total of 8 pulps were prepared in replication by using 5 dm^3 laboratory flasks heated in an autoclave at temperatures of between 40 and 100 °C, cooking times were between 1 and 4 h, and NaOH concentrations between 0.1 and 1 *M* (1:15 wt:v straw suspensions). After cooking, the black liquors were removed by filtering at reduced pressure through a 250 μm screen and the pulps were washed with 80 dm^3 water per kg of straw. The characteristics of the pulping procedures and sample labelling are indicated in Table I.

Solid-state Fermentation. For the treatment of straw with the fungi *Phlebia radiata* and *Pleurotus eryngii*, the straw was moistened with water (300%), inoculated and the cultivations were carried out as described elsewhere (*12*), except that cultivation was scaled up to 1 kg of straw in a 12 dm^3 (working volume) bioreactor, with a constant flow of humidified air from the bottom of the bioreactor. In order to compare the results obtained with the biological treatments, autoclaved straw (at 120 °C for 15 min) was used as a control.

Enzyme Treatments. Two types of enzyme pretreatments were carried out. For the first treatment, two enzyme crude extracts (obtained from *Pleurotus eryngii* grown in glucose-tartrate and glucose-peptone media) were mixed,filtered and sterilized to 0.45 μm, and 700 cm^3 were added to each bottle (containing 125 g of straw) and incubated for 48 h at 28 °C in a rotary shaker at 10 rpm. The estimation of enzyme activities in the crude extract showed 220 U of laccase kg^{-1}, 850 U of aryl-alcohol oxidase kg^{-1}, 2300 U of manganese peroxidase kg^{-1}. The cofactors used were 0.05 m*M* Mn^{2+} and 10 m*M* anisyl alcohol. The second enzymatic pretreatment was carried out with commercial xylanase: a solution of Novo Pulpzyme® HB in Sorensen II citrate buffer (0.05 *M,* pH 7) was prepared, and a total of 1.4 dm^3 of this solution was added to each bottle (125 g of straw) containing 112.5 U (0.9 U g^{-1} of straw) and incubated at 45 °C for 24 h in an orbital shaker at 200 rpm.

Assessment of Pulp Quality. Specific boardmaking parameters were determined at the laboratories of SAICA (Zaragoza, Spain) according to the Tappi standard methods (*13*). These parameters corresponded to specific energy consumption (T-423), the short compression test index (iSCT, linear strength required for bending the paperboard, T-826), the Concora medium test index (iCMT, the strength required to crush the ridges of corrugated paperboard, T-826), and the Gurley porosity (permeation time of 100 cm^3 of air at a fixed pressure, T-460, T-536). The tests were carried out with handsheets of 20 cm of diameter and a grammage of 140 g m^{-2}. In order to give comparable data of energy consumption in samples of different beating degree, energies were intrapolated to 60 °SR in refining curves.

Standard Laboratory Analyses. In order to prevent the interference of soluble minerals (mainly bicarbonates and other salts derived from the NaOH or the culture media) in most analytical techniques used (e.g., thermogravimetry, IR spectroscopy and organic carbon analysis), the material to be analyzed was dialyzed, freeze-dried and milled to pass a 500 μm screen.

The elementary composition (%C, %H and %N) was determined with a Carlo Erba EA1108 microanalyzer and calculated in terms of ash-free percentages. The %O was considered the difference to 100 and the atomic H/C and O/C ratios were obtained (*14*). The ash was determined by calcination for 6 hours at 600°C. A water repellency measurement of homogenized materials, the water drop penetration time (WDPT), was also taken (*15*).

Straw pulp fractions were determined gravimetrically after sequential treatments of 2 g of sample. The lipid fraction was extracted in a Soxhlet with benzene-ethanol 2:1 (v:v), filtered, dehydrated with anhydrous Na_2SO_4 and weighed. The water-soluble extract was removed from the above residue in the same extractor (*13*). Then, the alkali soluble fraction was obtained by 3 successive extractions with 0.5 *M* NaOH at room temperature for 4 h. Residual lignin was determined in the NaOH-insoluble fraction by Saeman's hydrolysis (*13*). After ash determination, holocellulose was calculated by difference.

Differential thermogravimetry (DTG) was evaluated as a technique providing information about the relative amounts and the degree of transformation of pulp constituents. It is assumed that the successive exothermic effects reflect the thermal removal of functional groups and the breakdown of progressively more stable linkages in the lignocellulosic matrix. From 10 mg of sample the DTG curves were obtained with a Perkin-Elmer TGS-2 thermobalance, programmed for a constant heating rate of 20°C min^{-1} in the atmosphere.

The infrared (IR) spectra were acquired at between 4000–400 cm^{-1} with a Bruker FTIR spectrophotometer (pellets from 3 ± 0.05 mg sample and 200 mg KBr). A total of 100 spectra were accumulated, and the final spectra were subjected to baseline subtraction (valleys *ca.* 1900 and 750 cm^{-1}). For quantitative comparisons, and in order to minimize the effect of sample concentration, the peak intensities in the original spectra were divided by the absorption value corresponding to the alkyl stretching band at 2920 cm^{-1}.

Results and Discussion

Boardmaking Tests. Table II shows the large energy consumption for refining biological pulps compared with soda pulps. Among soda pulps, the two exceptions (CP1 and CP5) correspond to the straw subjected to the lowest concentration and temperature. This suggests that biological treatments may lead to results comparable to soda pulping when it is intended to obtain pulps with low SR number. In the other cases, chemical pulps tend to be better as suggested by boardmaking than biological pulps: the greatest values for both the iCMT and the iSCT were obtained in harshest chemical conditions. Gurley porosity was comparatively low in biological pulps. Energy saving with the different treatments is evident in the refining plots (Fig. 1), showing clear differences in terms of the factors controlled during soda pulping. The much lower slope of the refining curves for biological compared with chemical pulps indicates higher energy consumption to achieve a greater SR number.

Table I. Wheat straw pulps prepared by soda cooking and fungal or enzymatic treatments

Ref.	Treatment
AS	Control: autoclaved wheat straw
CP1	Soda cooking: 0.1 *M* NaOH, 40 °C, 1h
CP2	Soda cooking: 1 *M* NaOH, 40 °C, 1h
CP3	Soda cooking: 0.1 *M* NaOH, 100 °C, 1h
CP4	Soda cooking: 1 *M* NaOH, 100 °C, 1h
CP5	Soda cooking: 0.1 *M* NaOH, 40 °C, 4h
CP6	Soda cooking: 1 *M* NaOH, 40 °C, 4h
CP7	Soda cooking: 0.1 *M* NaOH, 100 °C, 4h
CP8	Soda cooking: 1 *M* NaOH, 100 °C, 4h
EP1	Autoclaving + enzymatic pulping: enzyme crude (*Pleurotus eryngii*)
EP2	Autoclaving + enzymatic pulping: Pulpzyme® (commercial xylanase)
FP1	Autoclaving + fungal treatment: *P. eryngii*, 18-20 days SSF
FP2	Autoclaving + fungal treatment: *Phlebia radiata*, 18-20 days SSF

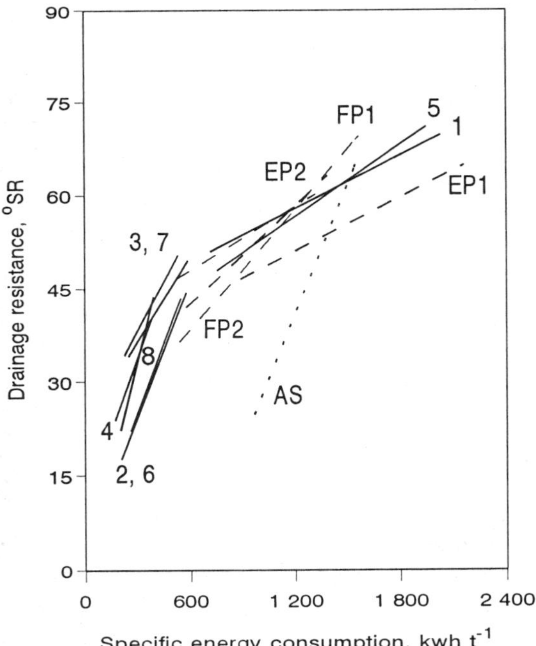

Fig. 1. Refining curves (specific energy *vs.* refining degree) of wheat straw pulps obtained by soda (1–8) and biological pretreatments (EP: enzymatic, FP: fungal). Sample labels refer to Table I.

Table II. Energy required for refining pulps, and properties of the handsheets prepared from the semichemical and biomechanical pulps [a]

Sample	SE [b]	iCMT [c]	iSCT [d]	Gurley [e]
AS	1409	0.03	9.3	2.9
CP1	1342	0.57	16.9	4.0
CP2	714	2.70	46.7	580.3
CP3	609	2.35	48.5	560.6
CP4	490	3.31	50.5	379.3
CP5	1234	0.71	16.4	3.6
CP6	715	2.96	53.2	925.0
CP7	689	2.40	41.6	256.0
CP8	458	3.44	56.7	601.0
EP1	1760	1.04	19.1	10.5
EP2	1184	0.05	10.8	1.6
FP1	1242	0.03	9.9	3.4
FP2	1219	0.04	10.2	6.5

[a] Sample labels refer to Table I; [b] Specific energy, kwh t^{-1}, for refining degree equivalent to 60 °SR.; [c] Concora Medium Test Index, N m^2 g^{-1}; [d] Short Compression Test Index, N m g^{-1}; [e] Gurley porosity, s 100 cm^{-3}.

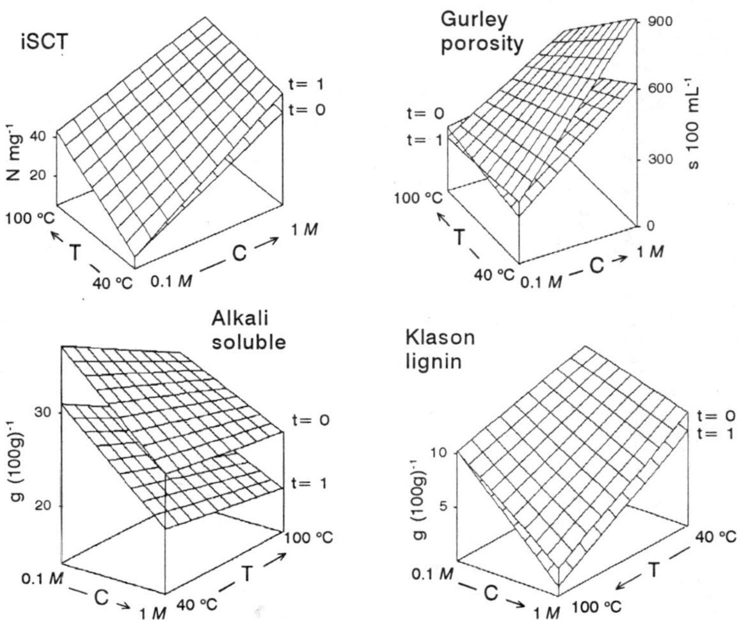

Fig. 2. Response surfaces illustrating the effects of temperature (T), concentration (C) and time (t) in four parameters determined in soda pulps.

Pulp Chemical Characteristics. Table III and Fig. 2 show laboratory parameters of the pulps, and Table V summarizes the results of the factorial design with regard to the effect of the three independent parameters controlled in soda pulping: in general, concentration (C) and temperature (T) had the greatest influence, whereas that of the time (t) was comparatively small. In most cases, the effect of C is enhanced significantly by T, as suggested by the relatively frequent C×T interaction.

The concentration of straw lipids decreased significantly by the soda treatments (as expected from soap formation). Fungal treatments also caused some lipid degradation whereas no large changes were observed with the enzymatic pretreatments. On the other hand, the amount of water-soluble products is not very representative in the case of chemical pulps, since the values largely depend on the concentration of residual NaOH and carbonates after washing the pulps, which enhanced the extraction yields with water. Table V shows that, as in the case of lipids, soda concentration is the only independent parameter with a significant bearing on the amount of water soluble material in the pulps.

The alkali-soluble fraction (including fractions such as alkali-lignin+cinnamics+some hemicellulose) is present in substantial concentrations in straw (*ca.* 42%). A noteworthy result is that, after soda cooking and washing the pulps with water, on avg. *ca.* 25% of alkali-soluble material remains in all the conditions examined. This fact suggests very incomplete removal of residual black liquors in the standard washing of the pulps, but it can also be due to additional solvolysis of the pulps in the successive treatments with NaOH. Table V shows that the concentration of alkali-soluble products in chemical pulps is influenced significantly by the 3 independent parameters examined. The time-dependent removal of the extractive fractions agrees with the above-indicated partial effect of soda cooking, which is efficient even at low concentrations and temperature, but enhanced with repeated treatments.

Concerning the values for the residual carbohydrate referred to as holocellulose (ie, the hydrolyzable, alkali-insoluble fraction determined in the residue of the above exhaustive NaOH extraction), the results suggest that enzymatic treatments lead to some degradation of carbohydrates, presumably due to the selective removal of hemicellulose, whereas the fungal SSF led to some increase in the total holocellulose. In spite of the fact that pulping treatments should exert a delignifying effect, the amount of residual lignin determined by Saeman's hydrolysis increased in most of the pulps with regard to the control. This is clearly connected with the fact that this lignin fraction was determined after NaOH extraction. Tables III and V suggest that heavy cooking (1 *M* and 100 °C) is required for a significant decrease in this residual lignin.

Soda cooking led to a decrease in the ash content of the straw, as expected from the transformation of straw opal into soluble silicates. The biological treatments, on the contrary, cause a relative increase in ash, concomitant with weight loss due to organic matter mineralization. Of the independent parameters studied, concentration and temperature were the most relevant with regard to the solubilization of minerals in chemical pulping (Table V).

The WDPT often provides interesting information connected with other quality parameters (15). In this case, its value decreased after strong soda pulping (leading to a more hydrophilic material), but showed some increase after fungal treatment which could correspond to an accumulation of fungal lipids and chitin).

Both H/C and O/C ratios increased with soda cooking more than with biological

Table III. General analytical characteristics of straw pulps obtained by soda and biological treatments

Sample	LIP [a]	WS [b]	ALK [c]	CEL [d]	LIG [e]	Ash	WDPT [f]	C [g]	H [g]	O [g]	H/C [h]	O/C [h]
AS	6.1	7.1	41.9	36.9	7.9	3.0	2.0	47.3	6.4	46.2	1.62	0.73
CP1	2.6	7.0	33.7	47.0	9.7	1.9	8.6	46.0	6.5	47.3	1.71	0.77
CP2	3.4	12.9	28.8	43.2	11.7	1.9	2.8	46.4	6.6	46.8	1.71	0.76
CP3	2.9	8.9	28.3	49.2	10.6	1.5	10.8	45.2	6.4	48.2	1.71	0.80
CP4	3.6	10.2	24.2	55.1	6.9	1.0	0.6	44.8	6.4	48.5	1.72	0.81
CP5	3.2	7.3	31.0	48.5	10.0	1.4	2.0	45.9	6.5	47.4	1.71	0.77
CP6	3.0	11.4	22.7	52.2	10.7	1.7	1.2	45.3	6.6	47.8	1.76	0.79
CP7	1.9	7.2	25.8	53.8	11.3	1.9	8.5	45.2	6.6	48.0	1.76	0.80
CP8	2.9	11.1	25.3	56.5	4.3	1.1	0.4	45.6	6.2	48.0	1.64	0.79
EP1	6.5	13.0	47.0	21.4	12.0	3.3	2.8	46.7	5.9	47.2	1.52	0.76
EP2	5.1	12.3	45.0	28.8	8.8	3.0	1.4	46.4	6.2	47.1	1.62	0.76
FP1	2.8	9.0	39.1	42.5	6.6	3.5	3.7	46.3	6.3	47.3	1.63	0.76
FP2	4.0	5.0	41.3	43.2	6.5	4.4	10.8	47.0	6.3	46.7	1.62	0.74

[a] Lipids, g $100g^{-1}$: benzene-ethanol 2:1 (v:v) extract, Soxhlet extraction, 8 h; [b] Water-soluble, g $100g^{-1}$: Soxhlet extraction, 8 h.; [c] Alkali-soluble, g $100g^{-1}$: 0.5 *M* NaOH; [d] Holocellulose after alkaline extraction, g $100g^{-1}$; [e] Residual lignin, g $100g^{-1}$: determined in the pulp (by 72% w H_2SO_4 treatment) after NaOH extraction; [f] Water drop penetration time, s; [g] Ash-free percentages, g $100g^{-1}$; [h] atomic ratios. Sample labels refer to Table I.

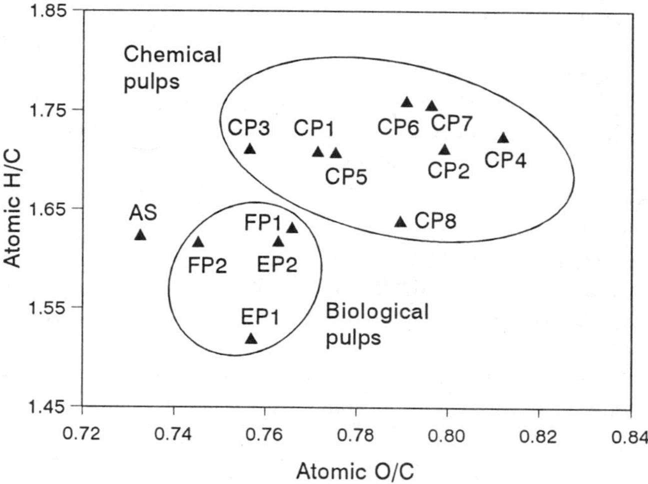

Fig. 3. Van Krevelen diagram obtained from the elementary composition of straw chemical pulps (CP) and biological pulps prepared after fungal pretreatment (FP) or enzymatic pretreatment (EP), in comparison with the original, autoclaved wheat straw (AS). Sample labels refer to Table I.

pretreatment. The van Krevelen (*14*) diagram (Fig. 3) suggests some clusters of samples as well as the more o less successful removal of lignin fractions, indicated by the linear distance between straw and pulp points, the latter shifting towards plot regions characteristic of the stoichiometry of carbohydrates.

Spectroscopic and Thermogravimetric Characterization of Straw Pulps. The FTIR spectra of wheat straw and derived pulps show characteristic lignin and carbohydrate bands. The most prominent peaks were essentially the same in the different samples, but showed some quantitative differences. Fig. 4 shows bands for lignin aromatic units (1510 cm^{-1}), alkyl bending vibrations 1460 cm^{-1}, carbonyl groups conjugated and unconjugated to aromatic rings (1660 and 1720 cm^{-1}, respectively) and typical aromatic ring vibrations (1510, 1600 cm^{-1}). Carbohydrate greatly contributes to the bands in the region *ca.* 1130 cm^{-1} (C–O stretching).

The most striking differences between pulps correspond to the practical lack of the unconjugated carbonyl stretching band (1720 cm^{-1}) in semichemical pulps after soda cooking, which corresponds to the removal of ester-forming structures (mainly cinnamic acids and hemicellulose). Table V indicates that such decrease in the carbonyl band of the soda pulps is mainly concentration- and temperature-dependent. On the other hand, the intensities of this band in the biological pulps suggest that the fungi used did not produce cinnamyl esterases.

The intensity of the 1510 cm^{-1} band may provide an indirect measurement of the amount of substituted aromatic units (*16*). The peak intensity (Table IV) suggests the concentration-dependent efficiency of soda cooking in the removal of lignin to be much greater than in biological pulps.

Fig. 5 shows typical DTG curves of crude wheat straw and some of the pulps obtained. The curves show a small loss of weight with a maximum *ca.* 75 °C, which could reflect the removal of hygroscopic water, followed by a prominent exothermic effect with a maximum at *ca.* 360 °C (effect II), which is attributed to the thermal degradation of carbohydrates. Most of the ill-defined, lower temperature shoulder *ca.* 325 °C (effect I) may correspond to processes, such as the release of external O-containing groups, ester breakdown, further thermal decarboxylation and loss of a fraction of tightly-bonded constitutional water (*17*). This is supported by the heavy decrease of this effect in terms of the intensity of chemical pulping. Discriminating values between pulp types were small in the entire peak areas I+II (Table IV), which suggest that this ill-defined transition between the effects corresponds to very independent thermal reactions. In the same way, the improved discriminating value of the I/II area ratio with regard to the individual areas inversely reflects the efficiency of pulping (i.e., the success of the selective disruption of different types of linkages between constitutional straw macromolecules).

The effects at greater temperatures show a much lower discriminating value. They probably reflect thermal disruption of lignin—which also showed a low value as an indicator of pulping performance—and some pseudomelanoidins formed in previous heating stages. Concerning the effect of the independent cooking parameters, Table V shows that the extent of the different thermal effects is governed mainly by temperature during soda cooking.

Table IV. Characteristics of straw and pulps obtained by soda and biological treatments: Band intensity in the infrared spectra, and peak area measurements from the differential thermogravimetric curves

Sample	1720[a]	1510[a]	0[b]	I[c]	II[d]	III[e]	IV[f]	I+II	III+IV	I/II
AS	0.9	0.7	3.8	21.9	52.0	10.1	11.7	73.9	62.1	0.4
CP1	0.5	0.7	3.3	22.8	51.5	7.6	14.7	74.4	59.2	0.4
CP2	0.4	0.6	4.3	17.0	57.1	4.5	17.1	74.1	61.5	0.3
CP3	0.4	0.6	3.8	13.2	61.4	3.4	18.2	74.6	64.8	0.2
CP4	0.3	0.6	3.6	11.1	64.7	3.6	17.0	75.8	68.3	0.2
CP5	0.6	0.7	4.1	20.2	53.9	4.8	17.0	74.1	58.7	0.4
CP6	0.4	0.6	4.0	18.2	57.5	3.6	16.6	75.8	61.2	0.3
CP7	0.4	0.6	4.0	14.5	60.2	3.2	18.0	74.7	63.4	0.2
CP8	0.3	0.5	3.8	12.3	64.6	3.6	15.6	77.0	68.3	0.2
EP1	0.9	0.7	3.4	24.4	50.4	6.2	15.5	74.8	56.6	0.5
EP2	0.9	0.8	3.3	23.0	51.2	8.8	13.7	74.2	59.9	0.4
FP1	0.9	0.7	4.5	22.3	50.3	6.0	17.0	72.6	56.3	0.4
FP2	0.9	0.7	4.3	22.8	57.3	2.2	13.3	80.1	59.6	0.4

[a] Optical density of IR bands (cm^{-1}) relative to 2920 cm^{-1} alkyl stretching band; [b] Relative weight losses corresponding to the successive thermal effects in the DTG curve, (Fig. 4). Temperature ranges in °C. Percentages of the total curve area; [b] Weight loss in the 40–148 °C range; [c] in the 148–340 °C range; [d] in the 340–370 °C range; [e] in the 370–440 °C range; [f] in the 440–580 °C range.

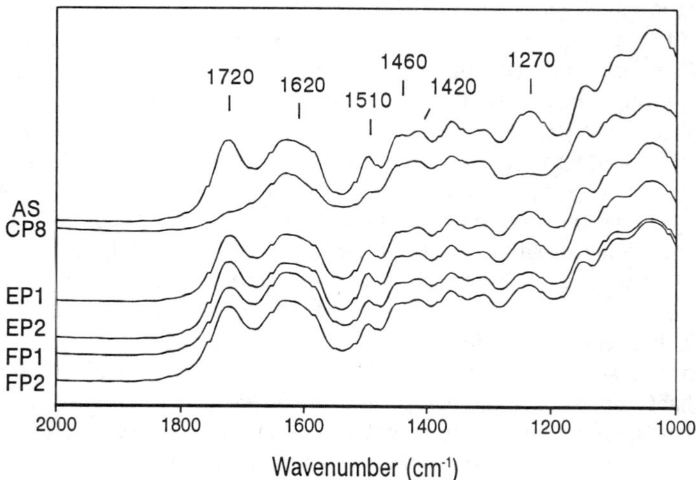

Fig. 4. Comparison between infrared spectra of chemical pulps (CP) and biological pulps (EP: enzymatic treatment, FP: fungal pretreatments) from autoclaved wheat straw (AS). Sample labels refer to Table I.

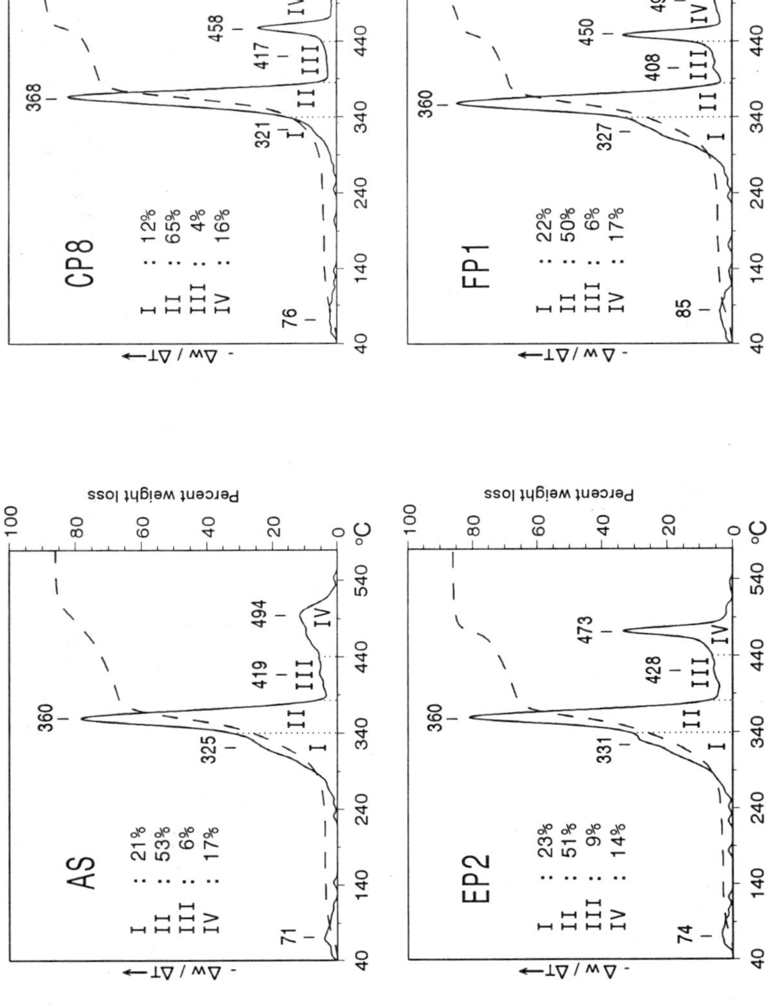

Fig. 5. Thermogravimetric (dashed line) and differential thermogravimetric (continuous line) curves of wheat straw pulps obtained by soda treatments (CP8), enzymatic pretreatment (EP2) and fungal pretreatment (FP1) of wheat straw (AS). The weight loss percentages during the main thermal effects are indicated by Roman numbers. Sample labels refer to Table I.

Table V. Main effects and interactions between factors influencing some properties of wheat straw in chemical pulping

	Average	C	T	C×T	t	C×t	T×t
Specific energy, kwh t^{-1}	781.3	-374.2	-439.7	199.2	ns	ns	ns
iCMT, $Nm^2\ g^{-1}$	2.3	1.6	1.1	-0.6	0.1	0.1	-0.1
iSCT, Nm g^{-1}	41.3	20.9	16.0	-12.4	ns	5.0	ns
Gurley porosity, s $(100\ cm^3)^{-1}$	413.7	415.4	71.0	-333.5	65.3	217.9	-106.8
Ash, g $(100\ g)^{-1}$	1.6	-0.3	-0.4	-0.4	ns	ns	0.3
Lipids, g $(100\ g)^{-1}$	2.9	0.6	ns	ns	ns	ns	ns
Water solubility, g $(100\ g)^{-1}$	9.5	3.7	ns	ns	ns	ns	ns
Alkali solubility, g $(100\ g)^{-1}$	27.5	-4.4	-3.1	2.2	-2.5	ns	ns
Holocellulose, g $(100\ g)^{-1}$	50.7	ns	5.9	ns	4.1	ns	ns
Residual lignin, g $(100\ g)^{-1}$	9.4	-2.0	-2.2	-3.4	ns	-1.2	ns
WDPT, s	4.4	-6.2	1.4	-2.9	-2.7	1.8	1.4
Atomic H/C	1.7	ns	ns	ns	ns	ns	ns
Atomic O/C	0.8	ns	ns	ns	ns	ns	ns
1720 cm^{-1} IR band intensity	0.4	-0.1	-0.1	ns	ns	-0.1	ns
1510 cm^{-1} IR band intensity	0.6	-0.0	-0.1	ns	ns	-0.1	ns
DTG I	16.2	-3.0	-6.8	ns	ns	ns	ns
DTG II	58.9	4.2	7.7	ns	ns	ns	ns
DTG III	4.3	-0.9	-1.7	1.3	-0.9	ns	0.9
DTG IV	16.8	ns	0.8	-1.4	ns	-1.0	-0.8
DTG I+II	75.0	1.2	0.9	0.5	0.6	0.8	ns
DTG III+IV	63.2	3.3	6.2	0.9	ns	ns	ns
DTG (I+II)/(III+IV) ratio	1.2	-0.1	-0.1	ns	0.0	ns	ns
DTG I/II ratio	0.3	-0.1	-0.2	ns	ns	ns	ns

Ranges for the independent variables: C (concentration) = 0.1 *M* NaOH (-); 1.0 *M* NaOH (+); T (temperature) = 40 °C (-); 100 °C (+); t (time) = 1 h (-); 4 h (+). Variable labels refer to Tables I–IV; ns: non-significant ($P > 0.05$).

Table VI. Significant ($P < 0.01$) linear correlation indices [a] between characteristics determined in chemical and biological pulps

r	Variables [b]
0.9480	Alkali-soluble ‖ IR 1720 cm^{-1} intensity
0.9391	SEC ‖ I/II (DTG effects)
0.9323	SEC ‖ I (DTG effect)
0.9088	IR 1510 cm^{-1} intensity ‖ I (DTG effect)
-0.9076	Alkali-soluble ‖ holocellulose
0.8995	IR 1510 cm^{-1} intensity ‖ I/II (DTG effects)
-0.8837	iSCT ‖ I/II (DTG effects)
-0.8834	iSCT ‖ I (DTG effect)
-0.8804	iCMT ‖ I (DTG effect)
-0.8786	iSCT ‖ IR 1720 cm^{-1} intensity
-0.8772	iCMT ‖ I/II (DTG effects)
-0.8745	iCMT ‖ IR 1720 cm^{-1} intensity
0.8717	SEC ‖ Alkali-soluble
-0.8638	iSCT ‖ Alkali-soluble
0.8565	Alkali-soluble ‖ IR 1510 cm^{-1} intensity
0.8531	IR 1720 cm^{-1} intensity ‖ I (DTG effect)
-0.8528	Alkali-soluble ‖ H/C atomic ratio
-0.8524	iCMT ‖ Alkali-soluble
0.8523	Alkali-soluble ‖ I/II (DTG effects)
-0.8489	iSCT ‖ IR 1510 cm^{-1} intensity
0.8483	IR 1510 cm^{-1} intensity ‖ I/II (DTG effects)
0.8442	Alkali-soluble ‖ I (DTG effect)
-0.8439	Alkali-soluble ‖ H/C atomic ratio
0.8432	Alkali-soluble ‖ Ash
0.8418	iSCT ‖ II (DTG effect)
-0.8397	IR 1510 cm^{-1} intensity ‖ II (DTG effect)
0.8272	Ash ‖ IR 1510 cm^{-1} intensity
0.8246	SEC ‖ IR 1720 cm^{-1} intensity
-0.8166	O/C ‖ I (DTG effect)
-0.8119	IR 1510 cm^{-1} intensity ‖ III+IV (DTG effects)
-0.8029	holocellulose ‖ IR 1510 cm^{-1} intensity
-0.8023	SEC ‖ holocellulose
0.7988	iCMT ‖ O/C atomic ratio
-0.7975	Alkali-soluble ‖ O/C atomic ratio
0.7968	SEC ‖ IR 1510 cm^{-1} intensity
0.7966	iSCT ‖ O/C atomic ratio
0.7963	holocellulose ‖ H/C atomic ratio
-0.7958	O/C atomic ratio ‖ I/II (DTG effects)
-0.7897	holocellulose ‖ I/II (DTG effects)
-0.7842	Gurley porosity ‖ Alkali-soluble
-0.7799	O/C atomic ratio ‖ IR 1510 cm^{-1} intensity
-0.7737	iCMT ‖ Ash
-0.7683	IR 1510 cm^{-1} intensity ‖ II (DTG effect)
-0.7663	holocellulose ‖ I (DTG effect)
-0.7614	iSCT ‖ Ash
0.7581	Lipids ‖ Alkali-soluble
-0.7553	SEC ‖ O/C atomic ratio

[a] Supervised selection from the correlation matrix, discarding redundant information and spurious correlations ($P < 0.01$ indices due to outliers in the plots or due to bimodal distribution of the variables (independent clusters for chemical and biological pulps); [b] SEC: Specific energy consumption; labels of the other variables refer to Tables II–IV.

Table VII. Coefficients of multiple regression [a] functions between boardmaking and laboratory parameters determined in straw pulps

Variable [b]	SE	iCMT	iSCT	Gurley
Lipids	-36.59	ns	0.57	0.96
Water soluble	-71.49	ns	1.11	2.62
Alkali soluble	-217.35	-1.18	ns	ns
Holocellulose	-264.85	-1.25	1.61	3.64
Residual lignin	-63.41	-0.15	0.27	ns
Ash	-1.71	0.58	ns	ns
H/C	0.89	0.18	ns	ns
O/C	-0.25	0.07	ns	-0.11
IR1510	-3.01	-0.54	-0.41	-1.20
IR1720	3.42	-0.43	0.56	3.72
I	3.00	0.28	ns	1.42
II	2.38	0.54	ns	ns

[a] Backwards multiple regression. Stepwise variable selection; ns= non-significant coefficients ($P>0.05$). Numerical values of the variables have been normalized beforehand; [b] Variable labels refer to Tables II–IV.

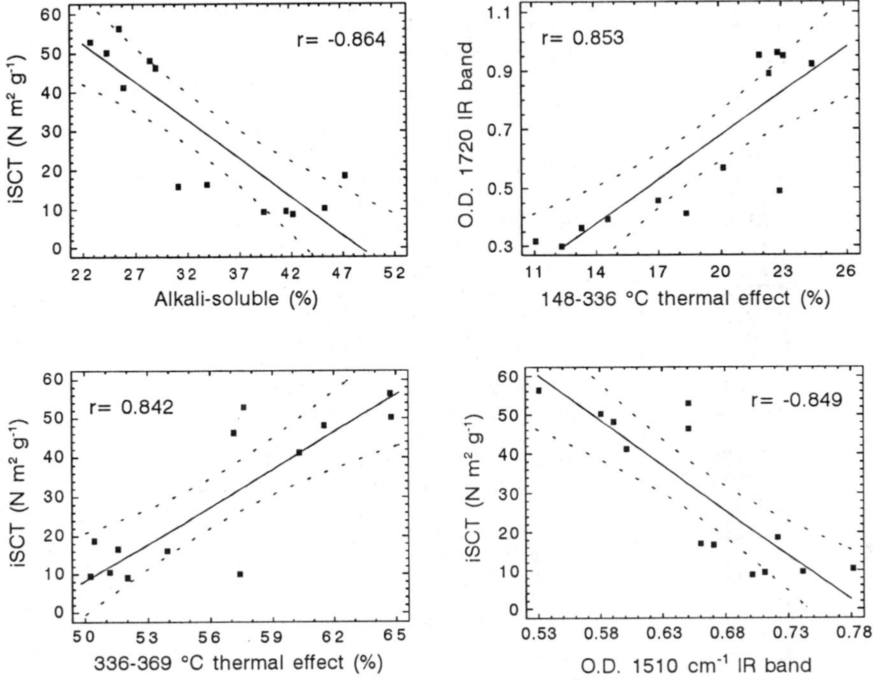

Fig. 6. Correlation plots between pulp quality parameters. Dotted lines indicate 95% confidence limits. Variable names refer to Tables II–IV.

Relationships Between Boardmaking Tests and Laboratory Parameters. Table VI and Fig. 6 show that most laboratory parameters correlate with boardmaking tests to determine pulp quality. This occurs with the total content in alkali-soluble fractions and holocellulose, as well as with parameters provided by IR spectrography and DTG analysis (mainly the band intensity at 1510 and 1720 cm^{-1} and the extent of the thermal effects I and II). The positive correlation between the alkali-soluble fractions and the thermal effect I supports the above-indicated fact, including the split-off of ester-bonded straw fractions. Additional correlations suggest that the H/C and O/C atomic ratios are connected with the carbohydrate enrichment produced by the pulping processes.

The forecasting possibilities of the laboratory parameters can be greatly improved by using multiple regression models (Table VII), assuming that the boardmaking tests (mainly based on the mechanical properties of handsheets prepared from the pulps) are more accurately expressed as linear combinations of a set of laboratory parameters which individually make a significant, but partial contribution to the properties of a complex physico-chemical system, such as the final paperboard. After computing multiple correlation functions by using backwards automatic rejection of the variables contributing with the lowest significant weight, Table VII shows the sets of laboratory parameters that are related to pulp quality as defined by boardmaking industrial standards. The different weights of such dependent factors suggest the importance of pulp extractive fractions (mainly solubility in alkali and water), much more relevant than residual lignin, as well as the above indicated bearing of the IR and DTG parameters, to all appearances connected with the disruption of linkages between straw macromolecules.

Acknowledgements

The authors wish to thank Mr E Barbero (Instituto de Química Orgánica General, CSIC) for the elementary analysis and Novo Nordisk (Spain) for providing the xylanase (Pulpzyme HB). This research was funded by the EU under grant AIR2-CT93-1219.

Literature Cited

1. Lapierre, C.; D. Jouin; B. Monties. *Physico-Chemical Characterization of Plant Residues for Industrial and Feed Use.* Chesson, A. and Ørskov, E.R. (Eds), **1989**, 118. Elsevier Appl. Sci., Amsterdam.
2. Xiao-an, L.; L. Zhong-Zheng; T. Die-Sheng. *Cellulose Chem. Technol.*, **1989**, *23*, 559.
3. Jeffries, T.W. *Biodegradation,* **1990**, *1*, 163.
4. Lam, T.B.T.; K. Iiyama; B.A. Stone. *Phytochemistry,* **1990**, *29*, 429.
5. Lam, T.B.T.; K. Iiyama; B.A. Stone. *Phytochemistry,* **1992**, *31*, 1179.
6. Hatakka, A. *FEMS Microbiol. Rev.*, **1994**, *13*, 125.
7. Martínez, A.T.; S. Camarero; F. Guillén; A. Gutiérrez; C. Muñoz; E. Varela; M.J. Martínez; J.M. Barrasa; K. Ruel; M. Pelayo. *FEMS Microbiol. Rev.*, **1994**, *13*, 265.

8. Martínez, A.T.; J.M. Barrasa; G. Almendros; A.E. González. *Advances on Biological Treatments of Ligno-cellulosic Materials.* Coughlan, M.P. and Amaral Collaço, M.T. (Eds), **1990**, 129. Elsevier. London.
9. Hatakka, A.; A. Mettälä; L. Paavilainen. *Proc. of The 8th International Symposium on Wood and Pulping Chemistry*, June 6-9, **1995**, Helsinki, Finland, Vol. I, pp. 399.
10. Hatakka, A.; A. Mettälä; T. Härkönen; L. Paavilainen. *The Sixth International Conference on Biotechnology in the Pulp and Paper Industry: Advances in Applied and Fundamental Research*, **1996**, (in press).
11. Vares, T.; M. Kalsi; A. Hatakka. *Appl. Environ. Microbiol*, **1995**, *61*, 3515.
12. Valmaseda, H.; G. Almendros; A.T. Martínez. *Biomass Bioenergy*, **1991**, *1*, 261
13. TAPPI. *Test methods 1992-1993.* Tappi, **1993**. Atlanta.
14. van Krevelen, D.W. *Fuel,* **1950**, *29*, 269.
15. Savage, S.M.; J. Osborn; J. Letey; C. Heaton. *Soil Sci. Soc. Amer. Proc.*, **1972**, *36*, 674.
16. Marton, J.; H. Sparks. *Tappi,* **1967**, *50*, 363.
17. Blanco, M.J.; G. Almendros. *J. Agr. Food. Chem.*, **1994**, *42*, 2454.

Other Applications

Chapter 20

Purification and Characterization of Laccase from a Newly Isolated Wood-Decaying Fungus

M. Luisa F. C. Gonçalves and W. Steiner

"SFB-Biokatalyse" Institute of Biotechnology, Technical University of Graz, Petersgasse 12/I, A–8010 Graz, Austria

Laccases are widespread in nature, being found in many plants and fungal species. Enzymatic pulp bleaching by the action of laccases can be enhanced by mediators. The search for organisms producing extremely stable enzymes that are compatible with conditions necessary in such processes, led to the isolation of a *Polyporus sp.* strain growing on an apple tree during winter. In addition to high laccase, traces of mannanase and xylanase activities were also detected. The laccases produced are not denatured by SDS, and were isolated by electrophoresis in the presence of this detergent. The main laccase was further purified and partially characterized with respect to pI (3.5), molecular mass (45 kDa) and optimum pH (4.5) and temperature (25°C).

Laccases (benzenediol:oxygen oxidoreductases, EC 1.10.3.2) are either mono- or multimeric copper-containing enzymes which use molecular oxygen as a terminal electron acceptor. Laccase catalyses the removal of a hydrogen atom from the hydroxyl group of *orto* and *para* mono- and poly-phenolic substrates and from aromatic amines, to form free radicals, capable of undergoing further de-polymerization or polymerization, demethylation or quinone formation. This rather broad substrate specificity of laccases may be additionally expanded by addition of redox mediators in their reaction mixtures *(1)*. Laccase mediator systems oxidize non-phenolic compounds that are otherwise not attacked, and are able to degrade lignin in kraft pulps *(2)*.

Laccases have widespread applications, ranging from effluent decoloration and detoxification to pulp bleaching, removal of phenolics from wines and dye transfer blocking functions in detergents and washing powders, many of which have been

0097–6156/96/0655–0258$15.00/0
© 1996 American Chemical Society

patented. Fungal laccases are of interest in synthetic chemistry and medicine, oxidizing steroid hormones and transforming antibiotics and alkaloids *(3)*. The importance of these enzymes for biotechnology, results also from their considerable retention of activity in organic solvents with applications in organic synthesis, and their participation in biosensors and immunoenzyme assays *(4-7)*.

Laccases in nature can be of plant, fungal, insect and bacterial origin. The first literature reference to the enzyme was made by Yoshida in 1883 *(8)*, by reporting a "diastase-like" activity requiring air, for the polymerization of *Rhus vernicifera* extracts. The biological functions of laccases have not been completely understood, since they are involved in many distinct processes such as lignification in plants, and sporulation, rhizomorph formation, lignin degradation and pigment formation in fungi *(9-14)*. Laccase mediates for example the synthesis of melanin in *Cryptococcus neoformans*, melanin production being the major virulence factor for this organism that causes life-threatening infections in about 10% of AIDS patients *(15)*.

At various stages of development in fungi laccase activity has been observed intra- and/or extracellularly, sugesting that synthesis of the enzyme is developmentally regulated *(16)*. In recent years, the extracellular fungal laccases of *Trametes (Coriolus, Polyporus) versiclor (17)*, *Polyporus sanguineus (18)* and *P. anceps (3)*, *Pleurotus ostreatus (19)*, *Agaricus bisporus (20)*, *Arctium lappa (21)*, *Botrytis cinerea (1)*, *Cerrena unicolor (22)*, *Curvularia sp. (23)*, *Ceriporiopsis subvermispora (2)*, *Cryphonectria parasitica (24)*, *Cryptococcus neoformans (15)*, *Heterobasidion annosun (Fomes annosus) (25)*, *Lentinus edodes (26)*, *L. tigrinus* IBR 101 *(27)*, *Neurospora crassa (28)*, *Pycnoporus coccineus (29)*, *Phanerochaete chrysosporium (30)*, *Pholiota mutabilis*, *Podospora anserina* and *Rhizoctonia paticola (31)*, *Trichoderma sp.(32)*, *Aspergillus nidulans (33)*, and *Schizophyllum commune (34)* have been isolated, purified using various methods and partially characterized. Many of the studied fungi produced more than one laccase form. In some basidiomycetes, the enzyme production could be induced by addition of small amounts of distinct protein synthesis inhibitors like xylidine, gallic acid or *p*-dianisine to culture medium *(35)*. In most cases the constitutive and induced forms of laccase are distinguishable.

General characteristics of the fungal laccases studied are a molecular mass ranging from 46.5 (*P. chrysosporium*) to 80 (*P. anserina)* kDa, the most frequently molecular mass value found being 66 kDa (five of the above mentioned fungi). The optimum pH for activity was generally situated in the acidic region, between 3 and 5 depending on the substrate, with the exception of *R. praticola*, that had an enzyme with pH optimum in the neutral region. The isoelectric points vary from 2.9 (*P. ostreatus*) to 4.8 (*C. subvermispora*). The optimum temperature range was usually between 50 and 60°C. Some genes for laccases from wood-rotting fungi have been sequenced totally or partially, and vectors for cloning of laccase genes have been constructed and expressed for the commercial production of the enzyme (*36-37*). Recently (*38*) laccases were also identified in enzyme systems from marine fungi, alone or in conjunction with xylanases and mannanases.

In our work, screening for laccase-producing fungi isolated and identified a wild-type strain of a white-rot fungus with high levels of laccase production. The objective was to search for a laccase producer whose enzymes are stable and

compatible with the conditions necessary in industrial processes, namely in enzymatic pulp bleaching and effluent treatments.

Materials and Methods

Organisms. Samples of wood, living trees and soil from Austrian forests were screened for laccase producing fungi. *Polyporus sp.*, from an apple tree, was isolated in our laboratory and identified by Central Bureau Voor Schimmelcultures, Baarn, Netherlands. The fungus was deposited in our culture collection under *Polyporus* H7 and fungal mycelium was used for weekly transfes to new Saboraud plates.

Culture Conditions. Organisms being screened were inoculated onto Petri dishes containing potato dextrose or Saboraud's agar and 0.02% of a laccase substrate. The coloration effect produced by the different fungi was always compared with the one produced by *Pleurotus ostreatus*, a known laccase producer. For laccase production a modification of the bran-based media described elsewhere (*2*) was used. The medium contained (per liter) wheat bran flakes (45g) and yeast extract (15g). The sterilized medium had a pH of 6.5, unless buffered with citrate-phosphate to pH 4.5. The effect of addition of 2,5-xylidine to medium after 4 days of incubation, to give a final concentration of 0.4 mM, monitoring the laccase activity on the following 6 days, was examined as described in results. Fermentations were incubated at 25°C on a rotary shaker (150rpm) in 300ml Erlenmeyer flasks containing 100ml of media.

Enzyme Purification and Characterization. Cultures were harvested after the peak of laccase activity was achieved, filtered through a sieve to separate the bran flakes and centrifuged (10 000 rpm, 20 minutes) to remove the mycelia. The supernatant was frozen, defrosted and filtered, for removal of precipitated polysaccharide. The next step was a concentration of at least six-fold in a 30 kDa Ultrafiltration Filtron unit. Preparative SDS-PAGE on 12% gels was then performed by the method of Laemmli (*39*). The samples were treated with 1% SDS and incubated at 37°C for 15 minutes. Laccase was recovered from gels by elution or by using GenElute Agarose spin columns (Sigma). The laccase so obtained gave a single band in SDS-PAGE.

The laccase pH optimum and pH stability were assayed as described in the results. Optimum pH was determined using syringaldazine as a substrate with 50 mM citrate buffer in the range 3 to 6 an with phosphate buffer of the same molarity in the range 4.5 to 6.9.

Molecular mass was estimated by SDS-PAGE (with NOVEX *Mark12*™ wide - range molecular weight standards). The isolelectric point value was determined by analytical isoelectric focusing PAGE, with a pH gradient of 3 to 9 using markers for pH (from Bio-Rad Laboratories) in this range.

The effect of temperature on activity was determined by incubating samples at temperatures between 15 and 90°C for 15 minutes and measuring immediately the remaining activity using the standard procedure. The concentration of dissolved oxygen was kept constant, since its solubility is also a function of the temperature.

The effect of temperature on enzyme activity is then determined only by the balance between the effects of temperature on rates of reaction and inactivation of the enzyme.

Enzyme Assay. Quantitatively, laccase activity was assayed colorimetrically with syringaldazine as substrate as reported elsewhere (*40*), but using 50 mM citrate buffer (pH 4,5) at 37°C The activity was calculated in the international units nkat/l, using the molecular adsorption coefficient ε= 65,000 M cm^{-1} for monitoring the oxidation of syringaldazine. Qualitatively, laccase activity was visualized on plates containing 0,02% (w/v) of either guaiacol, ABTS (2,2'-azino-bis(3-ethylbenzothiazoline-6-sulfonate)) or catechol. A direct detection of laccase activity in polyacrylamide gels was achieved by staining with 2 mM guaiacol or 5 mM syringaldazine in citrate buffer (pH 4.5).

Kinetic Studies. Michaelis constant (K_m) and maximum reaction velocity (V_{max}) were calculated for the reaction of laccase with syringaldazine from Lineweaver-Burk plots using substrate concentrations between 250 and 1000 μM at pH 4.5 and 25°C.

Results and Discussion

Isolation and Cultivation of the Laccase Producing Fungus. From all isolates, the most promising laccase producer was chosen and identified as *Polyporus sp.*. Early studies showed that the presence of glucose in medium repressed laccase production by this organism (data not shown), in a similar manner as in *Trichoderma sp.(32)*.

The basidiomycete *Polyporus sp.*, cultivated on bran-based media, gave a peak of laccase activity after 8 days of cultivation. The appearance of laccase coincided with decrease of total protein content in medium. The kinetic profile for a *Polyporus sp.* incubation is shown in Figure 1, and it agrees with that of *P. anceps* (*3*). Traces of xylanase and mannanase activities were detected from the fourth day onwards, remaining constant during the entire period of fermentation. Cellulase activity reached a maximum of one filter paper unit after one week, only when the fungus was grown on sulfite pulp (Kappa 43) supplemented with 1% (w/v) yeast extract, and was not detected otherwise (data not shown).

When a potential inducer of laccase, 2,5-xylidine, was added to cultures of *Polyporus sp.*, no detectable increase of the laccase activity was observed, in contrast to the results reported previously for *Polyporus versicolor* (*17*), P*olyporus anceps* (*3)* and other basidiomycetes (*31*), but in agreement to results obtained with Basidiomycete PM1 (*41)* and *Rhizoctonia praticola (42)*. Other possible inducers gallic acid, 2,6-dimethoxyphenol, organosolv lignin and guaiacol also did not increase laccase levels.

The activity detected on *Polyporus sp.* culture filtrate after eight days of cultivation was at least twice as high as that detected after 10 days of non-induced cultivation of *Trametes versicolor* and ten times higher than for *Fomes annosus (42)* and *Polyporus anceps (3)* cultivated under the same conditions.

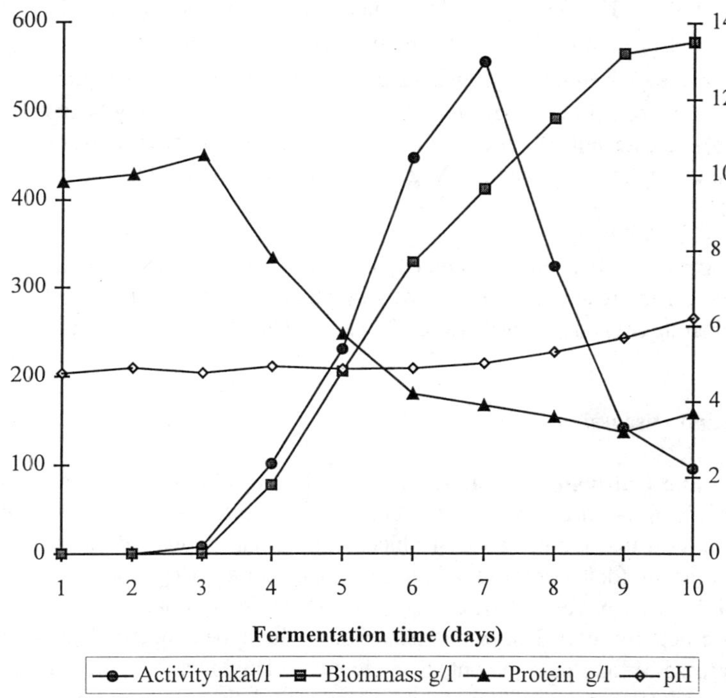

Figure 1. Kinetic profile of *Polyporus sp.* fermentation.

Table 1. Purification of the extracellular laccase from *Polyporus* sp.

Purification Step	Activity (nkatals / ml)	Protein (mg / ml)	Specific Acti. (nkatals / mg)	Yield (% protein)
Culture Filtrate (9 days)	13.35	3.02	4.42	100
Diafiltration 6x conc.	43.13	6.40	6.73	21
Preparative SDS-PAGE	26.64	0.09	296	14

The results obtained with the enzyme purification sequence are summarized in Table 1. When the most abundant laccase form produced by the organism was purified to apparent electrophoretic homogeneity, a protein of 45 kDa molcular mass was obtained. After several months storage at 4°C, further electrophoretic analysis revealed a second weaker band at aproximately 90 kDa. This result indicates the laccase protein from *Polyporus sp.* is a dimeric protein (two subunits of 45 kDa each).

Enzyme Size and Isoelectric Point Determination. SDS-PAGE estimated a molecular mass of 45 kDa for the main band with laccase activity, which is about the same size as the one from *P. chrysosporium* (46.5 kDa). At least two minor protein bands with larger size also showed laccase activity. The samples remained active after electrophoresis in the presence of SDS, and allowed a direct detection of laccase activity by staining. Activity staining was also possible after protein staining with Coomasssie brilliant blue dye. This fact agrees with what has been reported for *B. cinerea*, whose laccase tolerates SDS concentrations up to 4% (*1*).

Analytical isoelectric focusing PAGE of the main laccase band gave an isoelectric point value of 3.5 using Biorad standards.

Estimates of Laccase pH Optimum. The pH activity profile is in accordance to that expected with the substrate employed, see Figure 2. Activity is detectable from pH 3 until about pH 7, reaching a maximum at the pH value of 4. The differences in activity values caused from changing the type of buffer were within the range of the experimental error (inferior to 1%), and in contradiction to what was reported for *P. anceps* and *P. sanguineus*. For these organisms, pH optimum and activity differ when the buffer is changed. Using catechol as substrate, the *Polyporus sp.* laccase was also active in the same pH range from 3 to 7 (data not shown).

Effect of Temperature on Laccase Activity. Retention of activity after exposure to different temperatures gave the results presented in Figure 3. The laccase shows a higher activity at room temperature, 25°C:

Inactivation of Laccase by Solvents and Inhibitors. Retention of activity when inhibitors or organic solvents are present in standard assay medium is given in Table 2. The results corroborate those reported for the inducible form of laccase from *Trametes versicolor(40)* except for sodium azide. For this compound, the laccase from *Polyporus sp.* still exhibits a 20% activity at 30 mM, while the one from *Trametes* is completely inhibited. The enzyme remained active in concentrations of up to 80% ethanol and 40% methanol for several days at room temperature (data not shown).

Enzyme Kinetics. Enzyme activity versus substrate concentration for syringaldazine, using 10μl of purified laccase (2.7 nkat) was measured. From Lineweaver-Burk plots the K_m (0.5μM) and the V_{max} (21μM min^{-1}) values were estimated by extrapolation.

Conclusions. The main laccase produced by *Polyporus sp.* possesses general characteristics of the laccases of basisdiomycetes already studied, within respect to

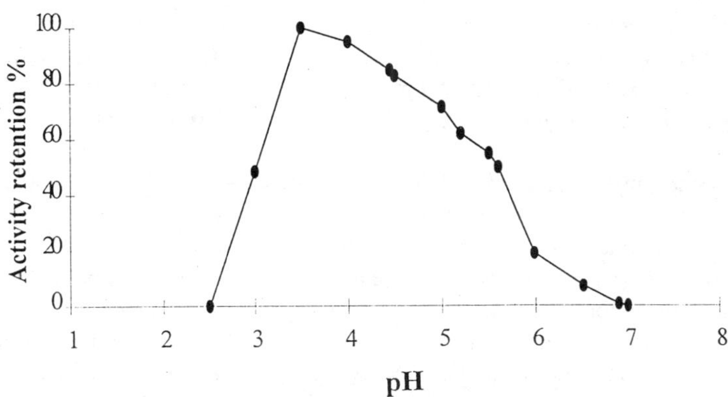

Figure 2. Influence of pH on the *Polyporus sp.* laccase activity.

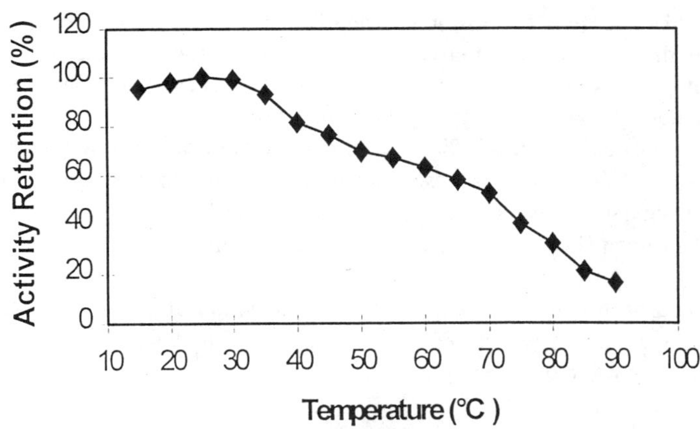

Figure 3. Influence of temperature on the *Polyporus sp.* laccase activity.

Table 2 Influence of Solvents and Inhbitors on the *Polyporus sp.* laccase activity

Substance	Concentration in reaction mixture	Remaining activity (%)
Ethanol	50%	80
Methanol	50%	40
Acetone	50%	67
Sodium Azide	0.03 M	20

molecular mass, isoelectric point and pH behavior. The optimum temperature for this laccase of 25°C, lower than the value for laccases previously studied is probably related to the characteristics of the organism. Our enzyme tolerates relatively high concentrations of SDS and retains its activity in ethanol and methanol solutions. Further studies will follow with the objective of optimizing enzyme production for obtaining a more competitive product for commercial use in bleaching and effluent treatments.

References

1. Slomczynski, D.; Nakas, J. P.; Tanenbaum, S. W., *Appl. and Environm. Microbiol.* **1995,** *61(3),* 907.
2. Fukushima, Y.; Kirk, T. K., *Appl. and Environm. Microbiol.* **1995,** *61(3),* 872.
3. Petroski, R. J.; Peczynska-Czoch, W.; Rosazza, J. P., *Appl. and Environm. Microbiol.* **1980,** *40(6),* 1003.
4. Milstein, O.; Nicklas, B.; Hüttermann, A., *Appl. Microbiol. and Biotechnol.* **1989,** *31,* 70.
5. Huang, H. P.; Cai, R. X.; Du, Y. M.; Zeng, Y. N., Analytical Letters 1995, 28(1), 1.
6. Guo, M.; Hu, S.; Wang, H.; Zhou, X., *Fenxi Kexue Xuebao* **1994,** *10(4),* 16.
7. Dzhafarova, A. N.; Skorobogat'ko, V.; Stepanova, E. V.; Yaropolov, A. I:, Jr. *Prikl. Biokhim. Mikrobiol.* **1995,** *31(1),* 128.
8. Yoshida, H., *I. J. Chem. Soc* **1883,** *43,* 472.
9. Leatham, G.; Stahmann, M. A., *J. General Microbiology* **1981,** *25,* 147.
10. Clutterbuck, A. J., *J. General Micrbiology* **1990,** *136,* 1731.
11. Worral, J. J.; Chet, I.; Hüttermann, A., *J. General Micobiology* **1986,** *132,* 2527.
12. Bourbonnais, R.; Paice, M. G., *FEBS Letters* **1990,** *267,* 99.
13. Bourbonnais, R.; Paice, M. G.,. *Appl. Microbiol. Biotechnol.* **1992,** *36,* 823.
14. Ishiara, T., *Lignin degradation: microbiology, chemistry and potencial applications;* CRC Press: Boca Raton, Fla., 1980, Vol. 2., pp 17-31.
15. Williamson, P. R., *Journal of Bacteriology* **1994,** *176(3),* 656.
16. Tamaru, H.; Inoue, H., *Journal of Bacteriology* **1989,** *171(11),* 6288.
17. Fåhraeus, G.; Reinhammar, B., *acta Chem. Scand.* **1967**, *21(9),* 5.
18. Nishizawa, N.; Nakabayashi, K.; Shinagawa, E., *Journal of Fermentation and Bioengineering* **1995**, *80(1)*, 91.
19. Sannia, J.; Giardina, P.; Luna, M.; Rossi, M.; Buonocore, V., *Biotechnoogy Letters* **1986**, *8(11)*, 797.
20. Matcham, S. E.; Wood, D. A., *Biotechnoogy Letters* **1992**, *14(4)*, 297.
21. Murao, S.; Oyama, H.; Nomura, Y., Tono, T.; Shin, T., *Biosci. Biotech. Biochem.* **1993,** *57(2)*, 177.
22. Bekker, E. G.; Petrova, S. D.; Ermolova, O. V.; Eliasashvili, V. I.; Synitsyn, A. P., *Biochemistry-URSS* **1990,** *55(11),* 1506.
23. Banerjee, U. C.; Vohra, R. M., *Foila Microbiologica* **1991,** *36(4),* 343.
24. Rigling, D.; Van Alfen, N. K., *Appl. and Environm. Microbiol.* **1993,** *59(11),* 3634.
25. Haars, A.; Hüttermann, A., *Arch. Microbiol.* **1983,** *134*, 309.
26. Kofujita, H.; Ohta, T.; Asada, Y.; Kuwahara, M., *Mozukai Gakkaishi* **1991,** *37(6),* 562.
27. Goshadze, M. K.; Eliashvily, V. I., *Biochemistry-Moskow* **1993,** *58(12),* 1448.
28. Faure, D.; Bouillant, M. L.; Bally, R., *Appl. and Environm. Microbiol.* **1994,** *60(9)*, 3413.

29. Oda,Y.; Adachi, K.; Aita, I.; Ito, M.; Aso, Y.; Igarashi, H., *Agric. Biol. Chem.* **1991,** *55(5)*, 1393.
30. Sirinivasan, C.; D'Sousa, T. M.; Boominathan, K.; Reddy, C. A., *Appl. and Environm. Microbiol.* **1995,** *61(12)*, 4274.
31. Bollag, J.-M.; Leonowicz, A., *Appl. and Environm. Microbiol.* **1984,** *48(4)*, 849.
32. Flegel, T. M.; Assavanig, A.; Amornkitticharoen, B.; Ekpaisal, N.; Meevootisom, V., *Appl. Microbiol. Biotechnol.* **1992,** *38,* 198.
33. Aramayo, R.; Timberlake, W.E., *Nucleic Acids Research* **1990,** *18(11),* 3415.
34. Devries, O. M. H.; Kooistra, W. H. C. F.; Wessels, J. G. H., *Journal of General Microbiol.* **1986,** *132,* 2527.
35. Bollag J.-M.; Shuttleworth, K. L.; Andreson, D. H., *Appl. and Environm. Microbiol.* **1988,** *52(12)*, 3086.
36. Novo Nordisk A/S, Den.; Novo Nordisk Biotech, Inc., *Patent, WO 9507988 A1 950323,* **1995**.
37. Valtion Teknillinen Tutkimuskeskus, Finland, *Patent, Wo 9201046 A1 920123,* **1992**.
38. D'Sousa, T. M; Reddy, C. A.; Raghukumar, C.and S.; Chinnaraj, A.; Chandramohan, D., *Botanica Marina* **1994,** *37(6),* 515.
39. Laemmli, U. K., *Nature (London)* **1970,** *227*, 680.
40. Leonowicz, A.; Grywnowicz, K., *Enzyme Microb. Technol.* **1981,** *3*, 55.
41. Coll, P. M.; F-Abalos, J. M.; Villanueva, J. R.; Santamaria, R.; Pérez, P., *Appl. and Environm. Microbiol.* **1993,** *59(8)*, 2607.

Chapter 21

Toner Removal by Alkaline-Active Cellulases from Desert Basidiomycetes

Hassan K. Sreenath, Vina W. Yang, Harold H. Burdsall, Jr., and Thomas W. Jeffries

Institute for Microbial and Biochemical Technology, Forest Products Laboratory, Forest Service, U.S. Department of Agriculture, One Gifford Pinchot Drive, Madison, WI 53705

Two isolates of wood-decay basidiomycetous fungi from the Sonoran Desert were selected from a preliminary screening of an extensive collection of wood-decay fungi. The isolates produced extracellular, alkaline-active carboxymethyl cellulase (CMCase) and filter paper cellulase activity, especially on solid substrates. Cultivation on wheat bran gave a higher cellulase yield than cultivation on corn steep liquor or paper pulp. The fungi produced alkaline-active cellulases when cultivated at pH 7.0 and 8.0, but not at pH 9.0 or above. Surfactants did not affect enzyme production level during cultivation on wheat bran. The fungi grew at 37 °C, but enzymes were produced at 27 °C. Isolate 2 produced more CMCase activity than did isolate 1 at both pH 7.0 and 8.5. In both isolates, enzyme activity was higher at pH 7.0 than at 8.5. Isolate 1 CMCase was stable at 50 °C and isolate 2 CMCase, at 60 °C. The crude alkaline enzymes could deink mixed office wastepaper that was 70% coated with toners. More ink was removed after addition of enzymes to paper pulp in the presence of surfactant.

Recycling of office and other wastepaper is an effective way to extend our nation's timber resource. For this reason, recycling has been given a high priority in Forest Service pulp and paper research. One of the greatest challenges in paper recycling is removal of contaminants; some of the most problematic contaminants are polymeric inks and coating. Toners such as those used in laser and xerographic copy machines are thermally fused to the surface of the printed page, and because they are nylon-based polymers that do not disperse, they are difficult to separate from fiber stocks. This is unfortunate because office copy paper is made of high-value bleached chemical pulp.

In 1991, Kim *et al.* (*1*) showed that crude cellulases applied to pulps could facilitate the deinking process. In subsequent studies, other researchers (*2–5*) showed that enzymatic deinking is most effective when cellulases are used during high-consistency fiberization in the presence of non-ionic surfactants. This step removes residual fibers from the toner surfaces. Subsequent floatation and washing steps remove toner or ink particles.

This chapter not subject to U.S. copyright
Published 1996 American Chemical Society

The most effective cellulases are those that exhibit activity on filter paper at neutral and alkaline pH. Activity at pH 8.5 is particularly important because calcium carbonate is often used as a filler and brightener in recycled fibers, and pulping processes leave the fiber with an alkaline pH. Most fungal cellulases are active in the acid region, and as the use of recycled fibers increases, the cost to neutralize the carbonate becomes excessive. Thermal stability is also important for effective processing.

Previous studies showed that the complete cellulase complex from the ascomycetous fungus *Humicola insolens* is highly effective in removing toners from office waste papers at neutral pH (*6*), but at the alkaline pH (8.5) found with most recycled fibers, deinking activity of this enzyme preparation is negligible. We therefore sought novel cellulases that could remove toners at alkaline pH. Most known alkaline-active cellulases are from *Bacillus,* but they do not produce a complete cellulase complex. *Bacillus* and other cellulases are presently used in the detergent industry (*7–9*).

The objective of the present work was to determine whether basidiomycetous fungi from alkaline thermal environments would produce useful cellulase complexes. In preliminary studies, we screened 43 strains of wood-decay Sonoran Desert basidiomycetes for production of alkaline-active cellulases. The screening procedure consisted of growing the strains on defined liquid alkaline media containing carboxymethylcellulose (CMC) as the sole carbon source and visualizing its enzymatic activity by zymogram. In this work, we report the solid-substrate cultivation of two selected fungal isolates for the production of alkaline cellulases. This paper also describes some characteristic properties and application of these crude enzyme preparations in enzymatic deinking.

Materials and Methods

Media. The fungal strains were maintained on yeast malt agar and potato dextrose agar (PDA) (Difco, Detroit, MI). (Note: Trade or firm names are for the convenience of the reader and do not represent the endorsement of the U.S. Department of Agriculture of any product or service.) In the preliminary screening of fungal isolates, both Norkran (*10*) and Ghosh and Kundu (*11*) liquid cultivation media were employed. The Norkran medium consisted of (per liter) yeast extract, 0.5 g; $NH_4H_2PO_4$, 2 g; KH_2PO_4, 0.6 g; K_2HPO_4, 0.4 g; $MgSO_4{\bullet}7H_2O$, 0.5 g; and CMC (low viscosity; Sigma, St. Louis, MO), 1.32 g. After autoclaving, 10% sterile Na_2CO_3 solution was added to adjust the pH to 7.3. The Ghosh and Kundu medium consisted of (per liter) KH_2PO_4, 2 g; $(NH_4)_2SO_4$, 2 g; $MgSO_4{\bullet}7H_2O$, 0.3 g; $CaCl_2$, 0.3 g; $FeSO_4{\bullet}7H_2O$, 5 mg; $MnSO_4{\bullet}H_2O$, 1.6 mg; $ZnSO_4{\bullet}7H_2O$, 1.4 mg; $CoCl_2$, 2 mg; and CMC, 1.6 g. The pH of the Ghosh and Kundu medium was adjusted the same as that of the Norkran medium.

Screening and Cultivation of Strains. Forty-three strains of Sonoran Desert wood-decay fungi from the culture collection of the Forest Products Laboratory were selected for initial screening. The strains were grown on fresh yeast maltose agar plates. After the fungi had attained confluent growth, spores and mycelia were washed from the surface with 3 to 4 mL of sterile water and used as an inoculum. The cell suspension (0.2 mL) was added to 0.8 mL of both Norkran (*10*) and Ghosh and Kundu (*11*) media in 2-mL microfuge tubes. The tubes were placed in a microfuge fermentation tilt-rack at a 45° angle (*12*) and shaken at 225 revolutions/min, 30 °C for 6 days. To identify cellulase producers, the 3,5-dinitrosalicylic (DNS) acid color reagent (*13*) was used to estimate reducing sugars released from CMC into the media on days 2, 3, 4, and 6. Further attempts to scale up cellulase production were conducted in shake flask and in solid-substrate culture (SSC).

Shake Flask Cultivation. Twenty-five milliliters of CMC liquid Ghosh and Kundu medium in a 125-mL Erlenmeyer flask was inoculated with 5 mL of fungal inoculum and shaken at 225 revolutions/min at 30 °C for 6 days. Samples from days 4 and 6 were centrifuged and the supernatant was saved for DNS assay. To enhance CMCase production, the Ghosh and Kundu medium was supplemented with skim milk, 0.2% (*14*), corn steep liquor, 2.6%, or wheat bran, 1%.

Production of Alkaline Cellulases. The fungal isolates were cultivated on wheat bran, corn steep liquor, and paper pulp in SSC (*15*). One-hundred grams of solid substrate was adjusted to 60% moisture (wet weight basis) with an aqueous solution of $MgSO_4 \cdot 7H_2O$ (60 mg), K_2HPO_4 (1.2 g), and Na_2CO_3 (1.2 g), at pH 7.0. Corn steep liquor (3%) was used as an experimental variable to provide nitrogen and other nutrients. Fifty grams of wet substrate in a 500-mL conical flask was autoclaved at 15 psig at 121 °C for 1 h. After cooling, the content of the flask was inoculated with 5.0 mL of fungal culture from the isolates grown on PDA plates. The flasks were incubated at 27 °C for 10 days. The moldy bran was then air-dried, packed into a cylindrical glass column, and extracted with 125 mL of tap water. The extracted fluid was centrifuged at 3000 x g for 20 min, and the supernatant solution was used as crude enzyme extract.

Enzyme Assays. For determination of CMCase activity, soluble CMC (degree of substitution (DS) = 0.65 to 0.90), 2% (w/v), was prepared by suspending 2 g of CMC in 100 mL of 100-mM K_2HPO_4/KOH buffer, pH 7, or glycine buffer, pH 8.5, and stirring the suspension until completely dissolved. Crude enzyme preparations were appropriately diluted to obtain maximal activity consistent with a linear response.

The reaction mixture (0.5 mL, containing 0.25 mL of substrate, 1% (w/v) final concentration, and 0.25 mL of diluted enzyme in buffer, pH 7.0 or 8.5) was incubated at 50 °C for 30 min. The reaction was stopped by adding 1.0 mL DNS color reagent to estimate reducing sugars. Enzymatic activity was expressed as international units (IU) (μmol glucose) per min^{-1} mL^{-1}. All units were corrected with suitable substrate and enzyme controls under identical assay conditions.

Filter paper culture enzyme (FPase) activity was measured (*16*) in buffer (see previous description) at pH 7.0 and 8.5. To determine the reduction in viscosity caused by added alkaline cellulases, 10 mL of reaction mixture containing 1% CMC solution was incubated with 20 mg of enzyme protein at 50 °C for 1 h. After inactivating the enzyme at 95 °C for 5 min, the hydrolyzed samples were filtered and cooled to 25 °C, and viscosity was measured with an Ostwald viscometer (*17*). The reaction mixture without enzyme prepared under the same conditions was used as control. The protein contents of the crude enzyme preparation were estimated with the bicinchoninic acid reagent (*18*).

Buffers used for studying the effect of pH on enzyme activity were 100 mM of phosphate buffer (pH 6 to 8.5), glycine buffer (pH 8 to 10), and 3-cyclohexylamino-1-propane sulfonic acid (CAPS) buffer (pH 9.5 to 11) (Sigma, St. Louis, MO). For thermostability studies, the crude enzyme was pretreated for 30 min at various temperatures prior to CMCase assay.

Electrophoretic Analysis. Isoelectric focusing (IEF) was performed on a Bromma 2117 Multiphor horizontal slab-gel system (LKB, Sweden) using Servalyt precotes (pH 3 to 10, Serva, Heidelberg, Germany). Protein bands were revealed by staining with Serva violet. Zymogram analysis of CMCase and xylanase activity in IEF was performed as reported previously (*19, 20*). Substrate gel (1 mm) was prepared with 0.1% (w/v) soluble CMC in phosphate buffer, pH 7.0, containing agar and filtered dye (Congo red, 0.1%). A zymogram was obtained by overlaying the dye-bound substrate gel on native IEF gel after incubation at 50 °C for 1 to 3 h.

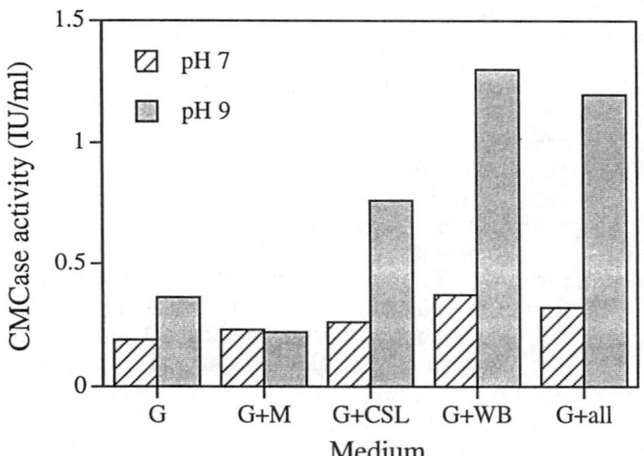

Figure 1. Enhancement of CMCase on isolate 2 grown on Ghosh (G) liquid medium, pH 9. Medium was supplemented with skim milk (M), corn steep liquor (CSL), wheat bran (WB), and all supplements (all). CMCase assays were performed on day 10 enzymes with pH 7 phosphate (dotted bar) and pH 9 glycine buffer (solid bar).

Table I. Production of Alkaline Cellulases and Isolate Activity at pH 7.0 and 8.5

Iso-late	*Sub-strate*	*Growth*	*CMCase (IU/mL)*		*Fpase (IU/mL)*		*Reduction in Viscosity (%)*	
			7.0	*8.5*	*7.0*	*8.5*	*7.0*	*8.5*
1	wheat	+++	0.78	0.60	0.70	0.95	29.8	19.3
	corn	+++	0.77	0.66	0.78	0.78	12.0	19.9
2	wheat	+++	1.11	0.97	0.39	0.38	46.4	29.0
	corn	-/+	0	0	0	0	0	0

Deinking Trial. Recycled paper pulps were collected from a mixture of photocopier waste papers that were coated with toners. Deinking trials were conducted as reported previously (*6*). However, surfactant solution (WITCO 5175-26A), 0.02%, was added to the pulp during the fiberization stage. Enzyme treatment was carried out at 14% final consistency at pH 8 to 8.5, 50 °C. The enzymes from these isolates and SP 342 Cellulase (Novo Industries, Danbury, CT) were added at three levels: 18, 36, and 72 IU CMCase/200 g oven-dry pulp. Shredded paper was added to water and surfactant in a custom-made, stainless-steel, water-jacketed Hobart mixing bowl. The initial solids content (consistency) was 17% (w/w). The suspension was mixed (fiberized) until paper particles were no longer visible (5 min). At that point, diluted enzyme was added. Enzyme treatment was conducted for 20 min to examine its effect on deinking. All pulping runs were followed by flotation in a 2 L-capacity laboratory flotation unit. The residual ink specks were counted on a Optomax ink scanner (Hollis, NH) and reported as parts per million. The size of the ink specks ranged from 0.02 to 2 mm. The specks were counted from five handsheets (made from enzyme-treated pulp) using 10 scans on each sheet. Suitable water and heat-killed enzyme controls were floated in each batch. The residual ink on handsheets made from enzyme-treated pulp was compared to that on sheets made from heat-killed enzyme.

Results

Identification of Isolates Producing Alkaline-Active Cellulases. Of the 43 fungal strains tested in the preliminary screening, only two strains released high amounts of reducing sugar within 4 days of cultivation. These strains are designated simply as isolates 1 and 2. Isolate 1 grew much slower than isolate 2 on both yeast malt agar and PDA. Isolate 1 does not form spores whereas isolate 2 is a spore-forming strain.

Effect of Shake Flask Culture on Enzyme Production. Low growth accompanied by low levels of CMCase production was observed with both isolates in liquid CMC medium. Alkaline CMCase production improved for isolate 2 in the liquid medium when corn steep liquor or wheat bran, but not skim milk, were incorporated (Figure 1).

Effect of SSC on Enzyme Production. In contrast to results with liquid culture, both growth and production of alkaline cellulases were good in SSC. Enzyme production was improved in SSC with regard to various solid substrates, pH of cultivation, cultivation and incubation time, and effect of surfactants. The best conditions were pH of harvested culture fluid in the range of 8.5 to 9.0 and use of wheat bran supplement. Column extraction was more efficient than batch-wise extraction in a beaker or flask. On wheat bran, both isolates produced alkaline CMCase, FPase, (Table I). Corn steep liquor supported growth and enzyme production by isolate 1 but not isolate 2. Paper pulp did not support growth or enzyme production (data not

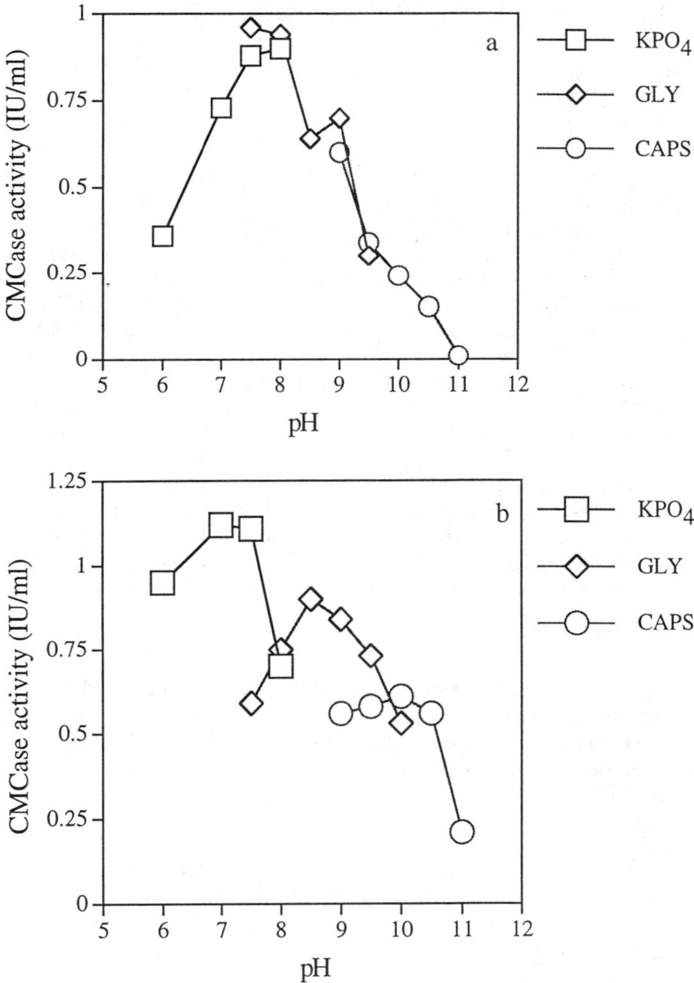

Figure 2. Effect of pH on CMCase activity of isolates 1 (a) and 2 (b). GLY is glycine buffer.

Table II. Effect of SSC pH Production of Alkaline Cellulase on Wheat Bran

Fungus	*SSC pH*	*Growth*	*CMCase (IU/mL)* *pH 7.0*	*pH 8.5*
Isolate 1	7.0	+++	0.97	0.58
	8.0	+++	0.46	0.51
	9.0	–	0	0
	10.0	–	0	0
Isolate 2	7.0	+++	1.20	0.79
	8.0	+++	0.71	0.46
	9.0	+++	0	0
	10.0	+++	0	0

Table III. Effect of Incubation Time on Alkaline Cellulase Production

Incubation	*CMCase (IU/mL)*			
	Isolate 1		*Isolate 2*	
(days)	*pH 7.0*	*pH 8.5*	*pH 7.0*	*pH 8.5*
0	0	0	0	0
8	0.77	0.70	1.20	0.78
10	0.98	0.89	2.25	1.12
12	0.79	0.60	1.11	0.79
28	0.54	0.17	0.68	0.37

shown). In each isolate, CMCase activity was higher at pH 7.0 than at pH 8.5. At pH 8.5, isolate 2 had 50% more CMCase activity than isolate 1. Isolate 2 produced low amounts of FPase and high amounts of CMCase. CMCase of isolates 1 and 2 caused 19% and 29% reduction in CMC viscosity at pH 8.5, respectively.

The production of alkaline CMCase was influenced by the pH of the SSC medium. Each isolate produced higher CMCase when cultivated at pH 7.0 (Table II). Neither isolate grew on wheat bran when cultivated at pH 9 or 10. The optimum period of cultivation on wheat bran was 10 days (Table III). Increasing cultivation beyond 10 days dramatically reduced alkaline CMCase activity in both isolates. The addition of surfactants to wheat bran cultivation medium did not alter growth or production of alkaline CMCase by isolate 2 (Table IV). However, corn steep liquor was used in most of our SSC studies because it supported favorable growth and higher enzyme production.

Effect of pH, Temperature, and Reaction Time on Crude CMCase. CMCase activity was detected over a wide pH range, from pH 6 to 11. Isolate 1 had optimum activity at pH 8; optimum activity of isolate 2 ranged from pH 7 to 9 (Figure 2). Thermostability was examined by pretreating the crude enzyme at various temperatures for 30 min prior to CMCase assay. Isolate 2 was stable up to 60 °C whereas isolate 1 was only stable to 50 °C (Figure 3). The CMCase activity of isolate 2 was progressive for at least 30 min at pH 8.5, 50 °C (Figure 4).

Electrophoretic Analysis. Isolate 2 enzyme showed a stronger clearance zone on CMC gel than did the isolate 1 enzyme (Figure 5). Hence, it was of interest to

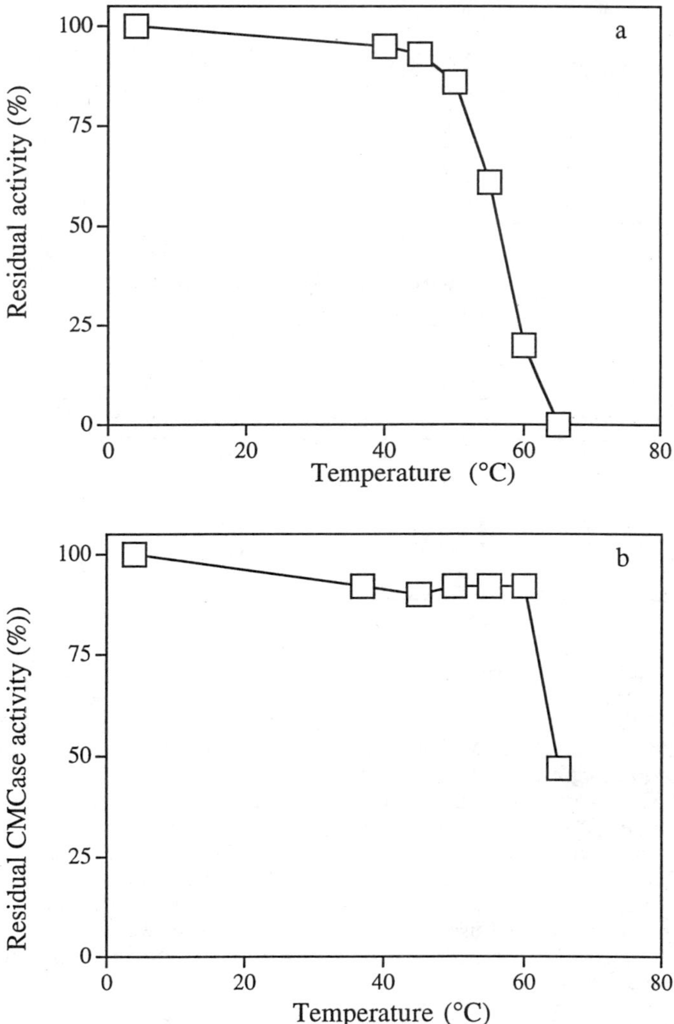

Figure 3. Thermostability of isolates 1 (a) and 2 (b). Crude enzymes were preincubated at various temperatures for 30 min prior to CMCase determination at pH 7. Residual activity is presented as a percentage of the original activity without heat treatment.

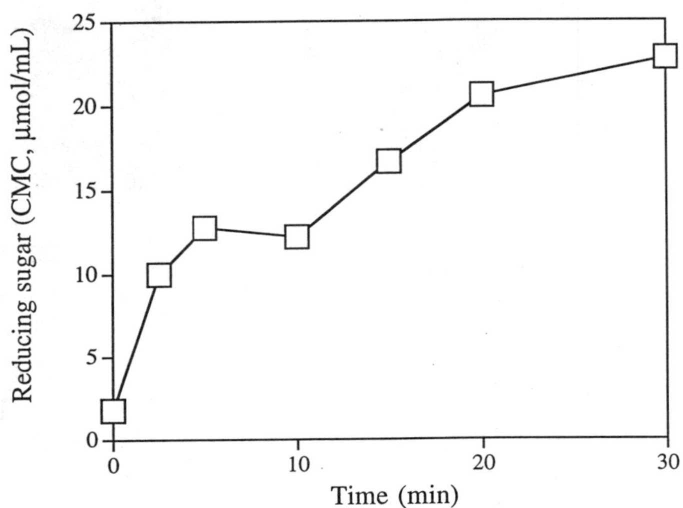

Figure 4. Effect of reaction time on CMCaseactivity of crude enzyme preparation from isolate 2 at pH 8.5, 50 °C.

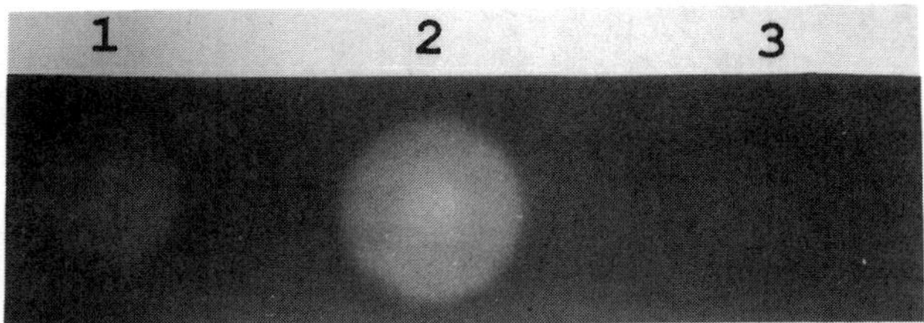

Figure 5. Effect of crude enzymes on CMC gel (pH 7.0) containing 0.1% Congo red. (1) Clearance zone shown by isolate 1 enzyme; (2) clearance zone shown by isolate 2 enzyme; (3) background effect by control sample cultivated at same condition but not inoculated with fungus.

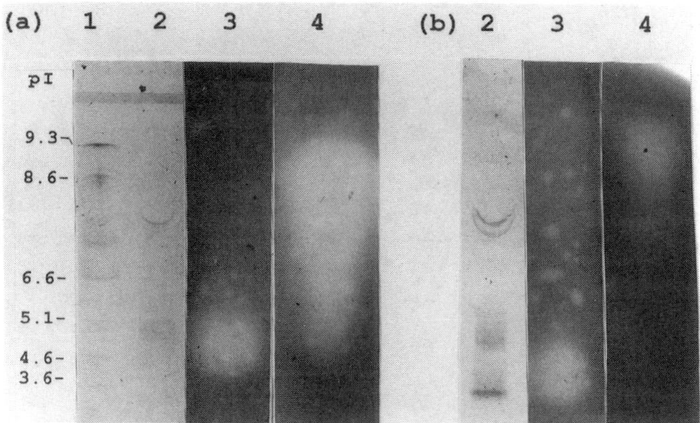

Figure 6. Isoelectric focusing (IEF) and zymogram of crude enzymes of isolate 1 (a) and isolate 2 (b). (1) Position of pI marker proteins; (2) IEF gel (pH 3–10) stained with Serva violet; (3) CMCase zymogram corresponding to lane 2; (4) xylanase zymogram corresponding to lane 2.

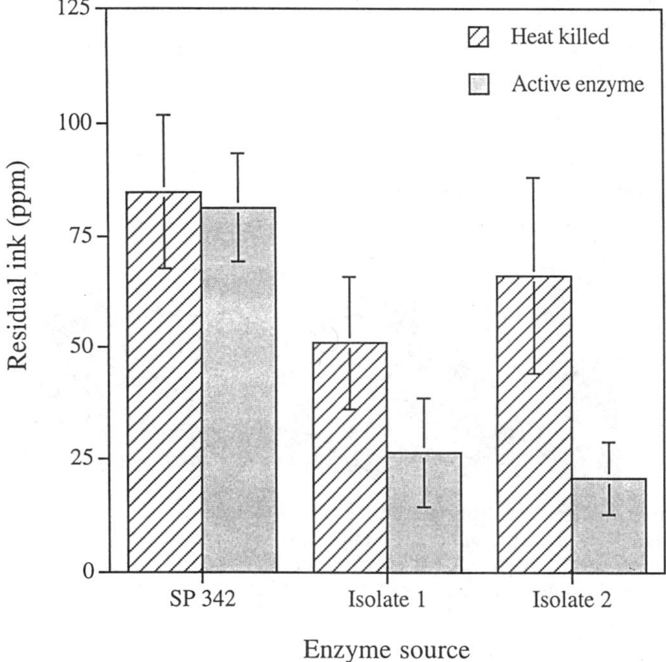

Figure 7. Reduction in residual ink from enzymatic deinking at pH 8.5 by SP 342 cellulase, isolate 1, and isolate 2 enzymes. Heat-killed enzyme served as control. Error bars represent standard deviation from replicate trials.

determine if multiple isozymes were formed. Isoelectric focusing and protein staining of each preparation revealed multiple protein bands in crude enzymes from both isolates (Figure 6). Zymograms showed that the pI values were 4.8 and 4.1, for the cellulases of isolates 1 and 2, respectively and 9.0 for the xylanases. Surprisingly, multiple cellulase isozymes did not seem to be present.

Table IV. Effect of Wheat Bran Surfactants on Production of Alkaline Cellulase by Isolate 2

Substrate	*Growth*	*CMCase (IU/mL)*	
		pH 7.0	*pH 8.5*
Wheat bran	+++	2.15	1.11
Wheat + corn liquor	+++	2.26	1.12
Wheat + Tween 80	+++	2.20	1.12
Wheat + triton X-100	+++	2.44	1.20

Effect of Deinking. Enzyme treatment of recycled wastepaper increased ink removal at pH 8.5 in the presence of surfactant (Figure 7). Enzyme from isolate 2 reduced residual ink on handsheets by about two-thirds as compared to that on heat-killed controls. At the same enzyme levels, enzyme from isolate 1 reduced residual ink count by about 50%. In comparison, cellulase from *Humicola insolens* showed no significant reduction in the ink particle count at pH 8.5 even though this fungus was proven to be very effective in pilot plant trials when the pH was adjusted to 7.9 or lower (*6, 21*).

Discussion

In preliminary screening work, only a few of the 43 fungal isolates cultivated on cellulose medium grew and secreted cellulolytic enzymes. In secondary screening studies, we repeatedly found that the fungal isolate 1 grew slowly on PDA plates as well as on CMC liquid medium; isolate 2 exhibited much better growth. Slow growth could be due to the presence of a growth-limiting substrate like cellulose as an organic carbon source in the fermentation medium (*22*) or could be intrinsic to the strain. On SSC, these isolates grew well and produced significant levels of alkaline cellulases (Table I). It should be noted that although the activity levels reported are low in comparison to the acid or neutral cellulase activities of *Trichoderma* (*23*) or *Humicola* (*20, 21*), they are reasonable when compared to levels observed with other alkaline active cellulases (*6–9, 24*). Typically, even the best producers of alkaline-active cellulase seldom produce more than a few international units of CMCase activity per milliliter. It is unclear whether this is due to the low turnover number of alkaline-active enzymes or to a low level of enzyme expression.

SSC is a simple technique that is easy to apply in small scale, and it has many advantages compared to submerged fermentation (*25,26*). SSC is particularly useful for enzyme production by fungi. Most filamentous fungi and especially basidiomycetes perform much better in solid substrate than in liquid cultivation probably because they are adapted to growth on solid surfaces. The carbon source in wheat bran, a typical solid substrate used in SSC, is primarily hemicellulose and cellulose. On the negative side, SSC is generally much more difficult to scale up than is submerged culture.

When cultivated at pH 7.0 by SSC, the fungal isolates produced higher alkaline activity on CMC compared with cultivation at pH 8.0 (Table II). Similarly, in

Bacillus maximum yield of the alkaline endo-1,4-β-glucanase was reached when the cultivation medium was adjusted to pH 7.5 and the enzyme activity decreased over pH 8.0 (*24*). The surfactants supplemented in SSC medium did not show any negative effect on enzyme production (*27*) nor did they enhance activity. Both alkaline CMCases from isolates 1 and 2 differed in their activity with regard to optimum pH and temperature (Figures 2 and 3).

The electrophoretic mobility of proteins and activity clearance on CMC gel of the isolate enzymes followed almost identical patterns. Alkaline enzyme activity on soluble CMC, viscosity reduction of CMC, and clearance zone activity on CMC gels on zymogram confirm the cellulolytic nature of the isolates. In deinking applications, the alkaline activities of the fungal cellulases are advantages because neutralization of the recycled fiber is not necessary.

Acknowledgments

The authors wish to acknowledge Freya Tan of the Forest Products Laboratory for the deinking analysis and the United States Agency for International Development (Project 10.299) for funding of this work.

Literature Cited

1. Kim, T. -J.; Ow, S. S. -K.; Com, T.-J. *Proc. Tappi Pulping Conf.* TAPPI Press, Atlanta, GA, **1991,** 1,023 pp.
2. Prasad, D. Y.; Heitmann, J. A.; Joyce, T. W. *Nordic Pulp and Paper Res. J.* **1993,** *2,* 284–286.
3. Jeffries, T. W., Klungness, J. H.; Sykes, M.; Cropsey, K. *TAPPI Recycling Symposium,* TAPPI Press, Atlanta, **1993,** 183–188.
4. Zeyer, C.; Joyce, T. W.; Heitmann, J. A.; Rucker, J. W. *Tappi J.* **1994,** *77,* 169–177.
5. Sykes, M. *TAPPI Recycling Symp.,* TAPPI Press, Atlanta, GA, **1995,** 61–64.
6. Jeffries, T.W.; Klungness, J.H.; Sykes, M.S.; Rutledge-Cropsey, K.R. *Tappi J.* **1994,** *77* (4), 173-179.
7. Kawai, S.; Okoshi, H.; Ozaki, K.; Shikata, S.; Ara, K.; Ito, S. *Agri. Biol. Chem.* **1988,** *52,* 1425–1431.
8. Ito, S.; Shikata, S.; Ozaki, K.; Kawai, S.; Okamoto, K.; Inoue, S.; Takei, A.; Ohta, Y.; Satoh, T. *Agric. Biol. Chem.* **1989,** *53,* 1275–1281.
9. Shikata, S.; Saeki, K.; Okoshi, H.; Yoshimatsu, T.; Ozaki, K.; Kawai, S.; Ito, S. *Agric. Biol. Chem.* **1988,** *54,* 91–96.
10. Eriksson, K. E.; Hamp, S. G. *Eur. J. Biochem.* **1978,** *90,* 183–190.
11. Ghosh, B. S.; Kundu, A. B. J. *Ferment. Technol.* **1980,** *58,* 135–141.
12. Sreenath, H.K.; Jeffries, T.W. **1996**. *Biotechnol. Tech. 10* , 239-242.
13. Miller, G. L. *Anal. Biochem.* **1959,** *31,* 426–428.
14. Patil, S. G.; Patil, B. G.; Bastawde, K. B.; Gokhale, D. V. *Biotechnol. Lett.* **1995,** *17,* 631–634.
15. Sreenath, H. K.; Joseph, R. *Chem. Mikrobiol. Technol. Lebensm.* **1983,** *8,* 46–51.
16. Mandels, M.; Andreotti, R.; Roche, C. *Biotechnol. Bioeng. Symp.* **1976,** *6,* 21–33.
17. Sreenath, H. K. *Lebensm. Wiss. Technol.* **1993,** *26,* 224–228.
18. Smith, P. K.; Krohn, R. I.; Hermanson, G. T.; Mallia, A. K.; Gartner, F. H. *Anal. Biochem.* **1985,** *150,* 76–85.
19. Robson, L. M.; Chambliss, G. H. *J. Bacteriol.* **1986,** *165,* 612–619.
20. Elegir, G.; Szakacs, G.; Jeffries, T. W. *Appl. Environ. Microbiol.* **1994,** *60,* 2609–2615.

21. Heise, O.U.; Unwin, J.P.; Klungness, J.H.; fineran, W.G., Jr.; Sykes, M.; Abubakr, S. *Tappi J.* **1996,** *79* (3), 207-212
22. Mandels, M.; Sernberg, D.; Andreotti, R. *Symposium on Enzymatic Hydrolysis of Cellulose;* Bailey, M.; Enari, T. M.; Linko, M., Eds.; Finnish National Fund for Research and Development (SITRA): Helsinki, 1975, pp. 81–109.
23. Gadgil, N. J.; Daginawala, H. F.; Chakrabarthi, T.; Khanna, P. *Enzyme Microb. Technol.* **1995,** *17,* 942–946.
24. Lusterio, D. D.; Suizo, F. G.; Labunos, N. M.; Valledor, M. N.; Ueda, S.; Kawai, S.; Koike, K.; Shikata, S.; Yoshimatsu, T.; Ito, S. *Biosci. Biotech. Biochem.* **1992,** *56,* 1671–1672.
25. Babu, K. R.; Satyanarayana, T. *Process. Biochem.* **1995,** *30,* 305–309.
26. Nandakumar, M. P.; Thakur, M. S.; Raghavarao, K. S. M. S.; Ghildyal, N. P. *Enzyme Microbiol. Technol.* **1996,** *18,* 121–125.
27. Al-Asheh, S.; Duvnjak, Z. *Biotechnol. Lett.* **1994,** *16,* 183–188.

Chapter 22

Mill-Scale Application of Enzymatic Pitch Control During Paper Production

K. Hata, M. Matsukura, H. Taneda, and Y. Fujita

Central Research Laboratory, Nippon Paper Industries Company, Limited, 5–21–1 Oji, Kita-ku, Tokyo 114, Japan

Pitch problems are the most common in papermaking process in the world. The mechanism in the degradation of resinous materials in wood during seasoning of logs was studied because the seasoning might be the best way to prevent the problems. From the results of the study, an enzymatic pitch control method was developed. Lipase hydrolyzes triglycerides to fatty acids and glycerol, therefore other resinous materials may not be coagulated without triglycerides cores. When lipase was added into groundwood pulping process, the effect of the enzyme addition on the product quality was detected in the long term. The enzyme did not spoil the quality and rather gave a good effect on the product. After the establishment of this technology, several mills in Japan have started to use it.

In the pulp and paper industry, pitch is an unsolved problem caused by resinous components in wood. During the pulping process these components are released and stick to the tile and metal parts of the production line, pitch also accumulates and sticks to the roll and wire of the papermachine. It stains the felts and eventually reaches the dryer section. Pitch accumulation causes paper stain and web break on the papermachine which is a severe problem in paper production. Enzymes can help alleviate these problems.

Fundamental Research

Wood Extractives. Extractives from wood used in papermaking, especially organic extractives from soft wood, consist of various components. They exert a number of effects on pulping and papermaking process. These extractives are hydrocarbons, alcohol, phenols, fatty acids, stilbenes, flavonoids, tannins, lignans and various terpenes

0097–6156/96/0655–0280$15.00/0
© 1996 American Chemical Society

such as mono-, di- and tri-terpene (*1*). Among these compounds, fatty acids, stilbenes and di-terpenes are of specific importance in pulping and papermaking process. Therefore, extensive studies have been carried out by many researchers on fatty acids in wood such as *Pinus, Picea, Larix, Populus, Betula,* etc. (*2-4*). Fatty acids in wood are saturated and unsaturated linear compounds. The numbers of carbon in these fatty acids are even numbers of 6, 8, 10 and all odd and even numbers from 12 to 32. The most important are unsaturated fatty acids with 18 carbons, such as oleic acid and linoleic acid. Fatty acids are stored in wood in various forms. A few exist as free acids, and most acids take the ester form as triglycerides. In addition, some of them make esters with high molecular weight alcohol such as β-sitosterol (*5*). During the seasoning of wood, free acids, especially unsaturated fatty acids, are partially oxidized and then polymerized. Triglyceride itself is relatively stable to oxidation, but it can be hydrolyzed during seasoning to free acids, which are further oxidized and subsequently polymerized.

Terpenes also have detrimental effects on products and process in wood industries (*6*). For instance, they result in pitch problems in the pulp and paper industry and adhesion problems in the lumber industry. Di-terpene, which is obtained as non-volatile fraction from pine oleoresin, is a glassy material commonly referred to as rosin. A typical resin acid is abietic acid, which is di-terpene. It is a three-ring cyclic compound. Pine oleoresin is a mixture of many wood resin acids. Isomerization takes place following heat or acid treatments and turns oleoresin into abietic acid. Abietic acid and its precursors are also known to cause pitch problems in the pulping of pine trees. Hardwood, especially tropical hardwood, contains four-ring and five-ring tri-terpenes, which also give rise to pitch problems during pulping (*7*).

In Japan, red pine (*Pinus densiflora*) is the most important wood for manufacturing of groundwood pulp (GP). Red pine GP has high opacity and printability. Therefore, it is indispensable for manufacturing newsprint and light weight paper. However, red pine GP contains large amounts of resinous substances which also cause pitch problems. It is well known that the content of the resinous materials in pine is more than that in spruce or fir.

Previously, two methods have been applied to control pitch problems. The first is seasoning of material wood before pulping (*8*). Seasoning requires raw wood logs to be piled up and allowed to stand outdoors at least for several months. It is the most commonly used method around the world because the resinous substances are decomposed during the seasoning period. However, it decreases the brightness of the resulting pulp greatly and requires a vast area of storage yard. Thus, this process is time consuming and quite expensive as well especially in Japan.

The second method used to reduce the accumulation of pitch is the adsorption and dispersion of the pitch particle with chemicals in papermaking process (*9-10*). This is accomplished by adding fine talc, dispersant and other kinds of chemicals. However, it is difficult to decrease pitch deposit to a lower level. Pitch deposits cause problems not only in a product quality but also in operation stability of the papermaking process.

Identification of Substances Causing Pitch Trouble. In 1986, an investigation to find the solution to pitch problems was started. In order to elucidate the mechanism of pitch deposit, a rapid analytical method for wood resin was established (Figure1).

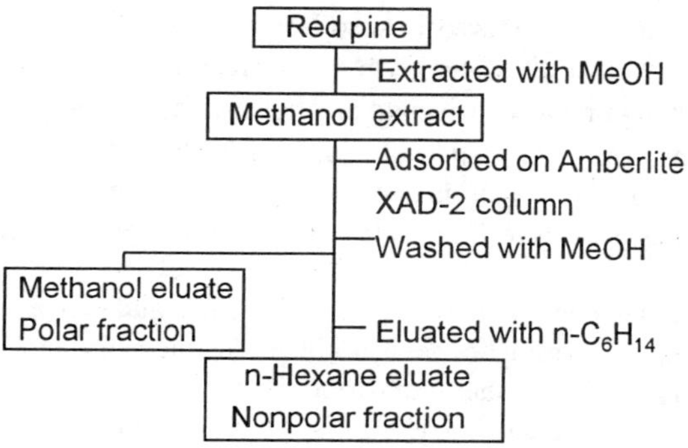

Figure 1. Fractionation of extractives by Amberlite XAD-2 column.

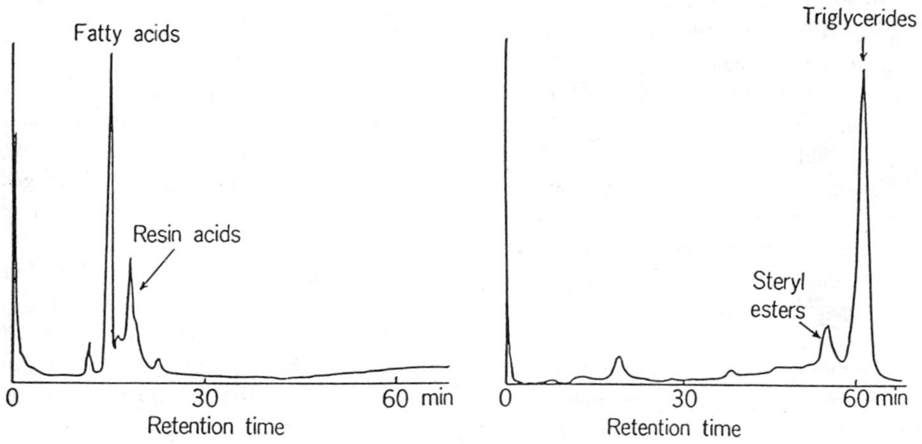

Figure 2. Gas chromatogram of polar fraction and nonpolar fraction from red pine.

Resinous components in red pine as well as the deposited pitch were fractionated by the new method and analyzed by gas chromatography. The changes in resinous components during the seasoning period and the contents of resin in fresh wood were also investigated in detail to understand the mechanism of seasoning process. These investigations presented following results (*11*):

1) Resinous component could be fractionated into polar and non-polar fractions. (Figure 2).
2) Fresh wood contained more non-polar compounds, especially in winter (Figure 3). The main component of non-polar fraction was triglycerides (TG).
3) 96% of the fatty acids that composed TG were oleic acid and linoleic acid (Figure 4).
4) TG was rapidly decomposed and reduced during seasoning (Figures 5-6).
5) Deposited pitch in the papermaking process always contained a lot of TG (Figure 7).

Based on these results, TG was estimated to be the key compound for pitch problems. In general, non-polar compounds such as TG may easily adhere onto the hydrophobic surface, such as papermachine rolls by Van der Waal's forces and build up to become the pitch deposits. TG could become the core and many other resinous substances could tend to accumulate to the structure of the core, thus they may cause pitch trouble. If TG in pulp slurry could be converted to less adhesive components, pitch deposits would decrease. The conversion of TG to free acid and alcohol would enable us to use fresh wood with less probability of pitch troubles.

Application of Lipase. The possibility of applying the lipase as a hydrolyzing enzyme of ester linkages of TG was investigated. The enzymatic reaction is selective under the mild condition and could not affect the environment nor a paper quality. In 1856, lipase was found in pancreatic juice by Bernard (*12*). It is well known as one of three prime digestion enzymes with amylase and protease. Lipase widely exists in animals, plants and micro-organisms. Recently, it has been produced in an industrial scale, and has come to be used not only for medical use but also for various purposes, such as an additive in detergents. In 1958, Desnuelle has proved that the lipase could only react on undissolved solid state substrate (*13*).

To prevent pitch deposits, the lipase from micro-organisms was used. Because it is commercially produced, relatively stable, and does not need co-factors(*14*). Lipases isolated from animals or plants, on the other hand, require co-factors to activate its catalytic action. For instance, pancrease needs cholic acid, and castor lipase needs recinolic acid polymer with some kind of protein (*15*).

Enzymatic Properties of Lipase from Micro-organisms. In general, lipase shows a diverse substrate specificity depending upon its resources. Chemical kinetic analysis is necessarily to determine its characteristics. The lipase from micro-organism easily hydrolyzes water insoluble TG, and then the fatty acids are liberated by the reaction. This enzymatic reaction is remarkably accelerated by calcium ion (*16*). The protein in the enzyme is not directly activated. The reaction mechanism is as a follow. When the concentration of fatty acid reaches a certain level, the calcium soap of fatty acid is formed. Emulsification of the substrate surface is accelerated by newly formed fatty

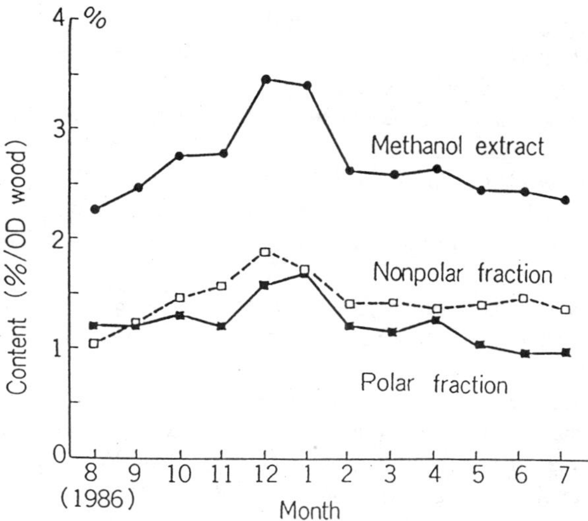

Figure 3. Changes of methanol extract yields from red pine.

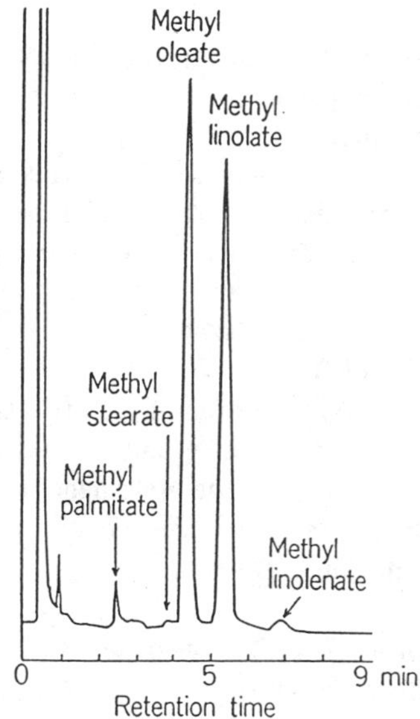

Figure 4. Gas chromatogram of fatty acid methyl esters obtained from red pine trigylcerides by saponification.

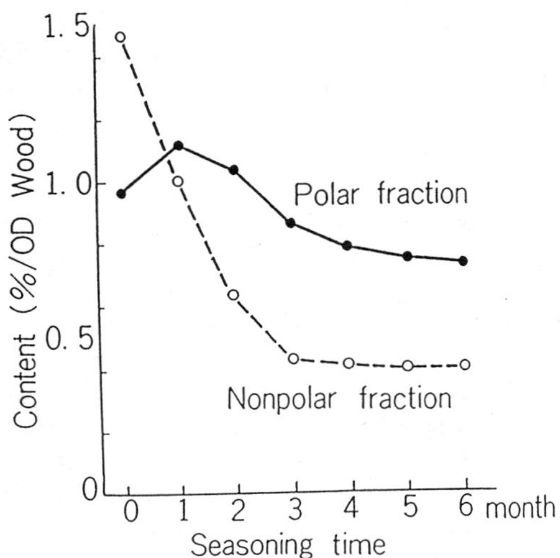

Figure 5. Effect of seasoning time on polar and nonpolar fraction yields from red pine methanol extract.

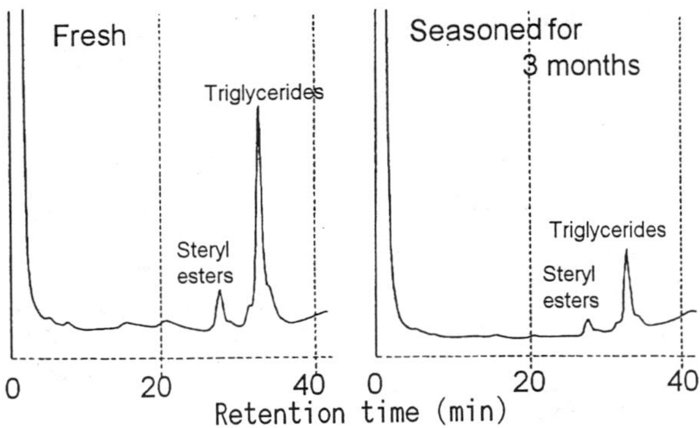

Figure 6. Gas chromatograms of nonpolar fraction from fresh and seasoned wood.

acid soap and this reaction accelerates the adsorption of lipase on the substrate surface.

As with other enzymes, inactivation of lipase can be classified into two categories. The first results from a change of molecular structure, and second caused by detrimental factors which hinder the reaction. Some of the adverse factors can be avoided but some of the effects are irreversible. In the case of lipase, metal ions and surfactants disturb the adsorption of lipase on substrate surface. For example, ferric ion is highly harmful to enzyme adsorption, and only 8 ppm of ferric ion delays the adsorption reaction by 90% (*17*). It is interesting to note that this adverse phenomenon decreases with time, and finally coincides with a system which does not contain detrimental factors.

Addition of ferric ion in the middle of the degradation reaction of oils and fats has adverse effects. The mechanism is thought to occur as follows. As the slow decomposition of oils proceeds, accumulated free fatty acids preferentially react with ferric ions and produce ferric soaps.

Alum is added in a latter step of the papermaking process in Japan and the formed aluminum soap can enhance the activation of the lipase enzyme. On the other hand, in the USA, alum is usually added at the former refining step in the groundwood fiber line, and the activation of lipase might be temporarily hindered. Beside metal ions, various materials such as alcohol, methyl oleate, surfactant, protein and so forth, have detrimental effects on lipase activity. Their effects are difficult to evaluate. It has also been thoroughly studied on the substrate specificity for structural isomer and position isomer, but it will not be touched on this issue.

Effect of Lipase Treatment on Prevention of Pitch Deposition. In 1987, laboratory simulation of pitch deposits was conducted. Resinous materials extracted from red pine wood and groundwood pulp (GP) were treated by lipase and their adhesiveness to the hydrophobic surface were determined using a specially designed instrument (*18*). Three kinds of lipase were selected from many lipase preparations after examining their activities and cost (Table Ⅰ).

Table Ⅰ. Principal properties of selected lipase

Lipase	A	B	C
Origin	*Candida cylindracea*	*Aspergillus sp.*	*Candida rugosa*
Molecular weight	53,000	-	64,600
Isoelectric point	4.3		4.3
Optimum pH	7	7	7
Optimum temperature	30℃	60℃	50℃
pH stability	3-8	4.5-8	3-8
Temperature stability	37℃	20-60℃	45℃

The pitch deposits increased when the ratio of nonpolar compounds to polar compounds increased. This result showed that the nonpolar compounds deposited more easily than polar compounds. It also became clear that the pitch deposits were reduced significantly after lipase treatment (Table II). The above result suggested that lipase hydrolyzed adhesive TG in the pulp suspension and therefore reduced the chance of pitch deposition. Thus, evidently the nonpolar compounds in the resinous component had higher adhesiveness to hydrophobic material and seemed to play an important role in pitch deposition. In other words, TG was a key compound in pitch deposition because the enzymatic hydrolysis of TG reduced pitch deposition significantly. TG was hydrolyzed to glycerol and fatty acids with lipase and the resulting glycerol dissolved into water. Fatty acids existed in the form of aluminum salt in the presence of alum. They were dispersed into pulp slurry or fixed on the surface of fibers (Figure 8).

Table II. Effect of lipase treatment on amount of deposits

Enzyme	Resinous materials	Pitch deposits (mg) Pitch-Water suspension		Pitch deposits (mg) Pitch-Pulp suspension	
	Polar/Nonpolar fraction	Control	Lipase treatment	Control	Lipase treatment
Lipase A	7/5	137	53	46	trace
	5/5	204	79	50	trace
Lipase B	5/5	196	10	38	5

Figure 9 shows the gas chromatogram of the pitch extracted from red pine GP after the treatment of lipase. In comparison with the untreated sample, evidently TG was hydrolyzed and fatty acids were increased. Again the resulting fatty acids were fixed on the surface of fibers as an aluminum salts.

Application to Papermaking Process. Since the effect of lipase on decreasing pitch deposition was confirmed, the enzyme was applied in the actual papermaking process. To select optimum conditions for lipase treatment in the mill, the following factors were investigated in a short period mill trial: the effect of enzyme concentration, reaction temperature, reaction time, and the agitation mode on the hydrolysis of TG. The following results were obtained from the investigation:

1) It was necessary to get a strong mixing system to get an effective reaction.
2) With sufficient mixing, lipase 10,000U/kg of GP. could hydrolyze more than 80% of TG in the surface pitch, which could be extracted by hexane, within two hours.
3) No effect of lipase treatment on the brightness and strength of pulp was observed.

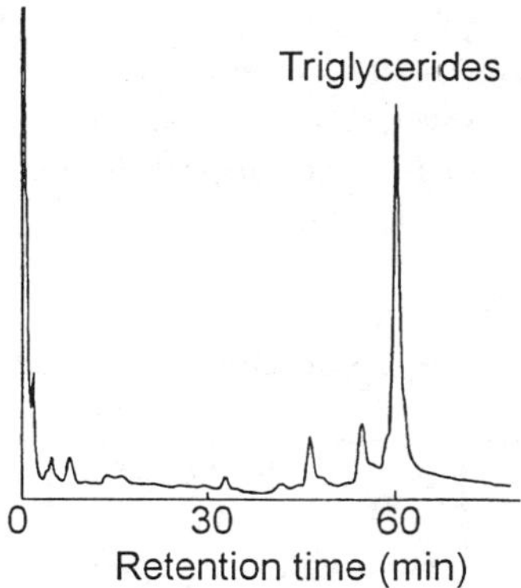

Figure 7. Gas chromatogram of nonpolar fraction from deposits on the center roll.

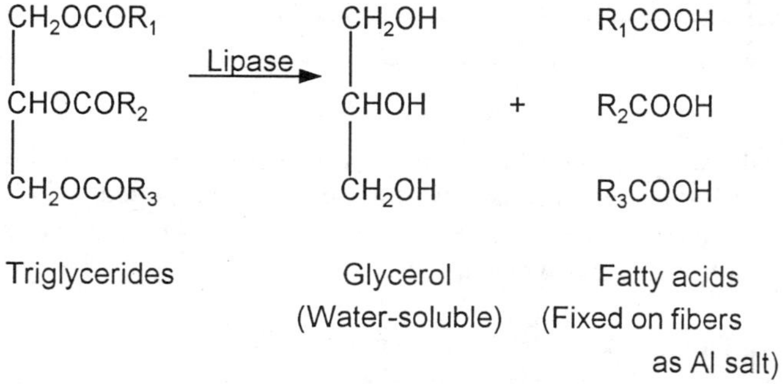

$$\begin{array}{l} CH_2OCOR_1 \\ | \\ CHOCOR_2 \\ | \\ CH_2OCOR_3 \end{array} \xrightarrow{\text{Lipase}} \begin{array}{l} CH_2OH \\ | \\ CHOH \\ | \\ CH_2OH \end{array} + \begin{array}{l} R_1COOH \\ \\ R_2COOH \\ \\ R_3COOH \end{array}$$

Triglycerides — Glycerol (Water-soluble) — Fatty acids (Fixed on fibers as Al salt)

Figure 8. Degradation of Triglycerides by Lipase.

Long Term Mill Trial. Based on these results, the first long term mill trial was conducted using large papermachine in the middle of 1988. In this mill, the papermachine using GP always had serious problems because large amounts of red pine wood was used as a raw material for GP. Normally 50% of unseasoned wood and 50% of seasoned wood, --which was seasoned for six months,-- were consumed. Therefore, GP had a high content of resinous materials and TG weighs amounted to 30% of the total extractives. In order to lessen the pitch problems, the seasoning period had been extended and the consumption of fine talc and dispersant had been also increased. As an attempt to prevent cost increase, lipase was added to the groundwood pulping line just before the post refiner(Figure 10). The operation conditions of papermachine were following;

Machine condition
Papermachine: Bel-Baie II, wire width 5080mm
Paper product: Yellow Telephone Directory paper(YTD) 34g/m^2
Pulp: Red pine groundwood pulp 15-40%, De-inked pulp,
Softwood bleached kraft pulp
Machine speed: 830 m/minute
Production rate: 200-270 ton/day
Enzyme: Lipase A 75-125 ppm/GP
Lipase B 500-750 ppm/GP

The reaction time of the enzyme means the time that it takes GP to travel from the addition point to the mixing chest. The reaction times for yellow telephone directory paper (YTD) and newsprint paper were 180 to 210 minutes, and 60 to 80 minutes, respectively.

The primary test was done in order to understand the proper dosage for long term mill trial and to know the hydrolysis rate of TG by lipase in the actual papermaking process. 125 ppm addition of Lipase A on the first day of the trial reduced the content of TG significantly. As a result, 74% of the TG was hydrolyzed in the first trial. After getting a good result, the lipase treatment had been carried out for one month. The following subjects were compared between the usual operation and the lipase treatment operation in major products, such as newsprint and YTD.

Content of the Surface Pitch and TG. The results are shown in Figure 11. The content of the surface pitch, which was extracted by hexane, was 1.5 to 2.0 % of oven-dried GP, and the content of TG was 16 to 26 % of surface pitch. Apparently the lipase hydrolyzed 70% of TG until reaching the mixing chest inlet. Furthermore, the accumulation of the pitch in the recycled whitewater (stock inlet, save all) decreased to a lower level after the lipase treatment (Figure 11).

First Pass Retention of Pulp and Pitch. As shown in Figure 12, the first pass retention (FPR) of pulp did not change with lipase addition. However, the FPR of the pitch increased from 5 to 9 %, to 12 to 19 % in YTD and from 9 to 14 %, to 13 to 24 % in newsprint. As the lipase hydrolyzed TG, the pitch was dispersed into the pulp slurry and then fixed into the fibers smoothly. Lipase also prevented the accumulation of pitch in the recycled whitewater system (Figure 12).

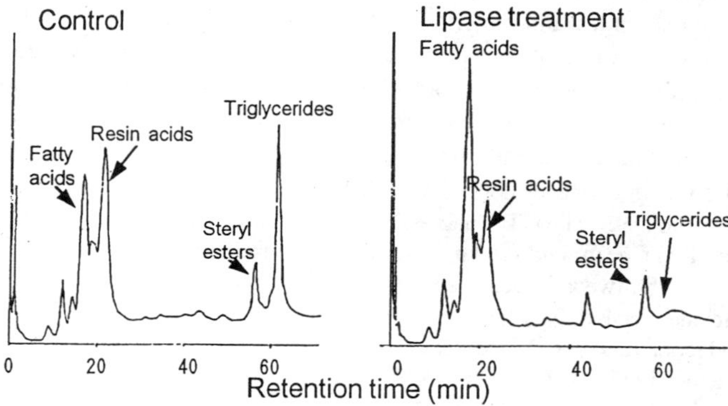

Figure 9. Gas chromatograms of n-hexane extract from lipase treatment red pine groundwood pulp.

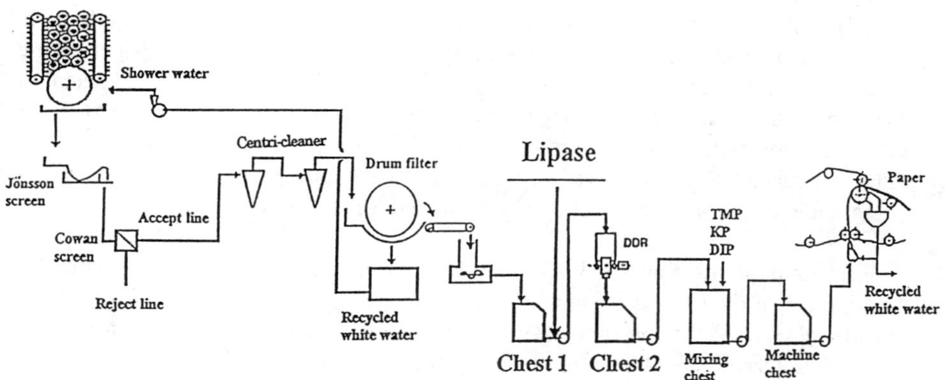

Figure 10. Outline of groundwood pulp fiber line to papermachine and addition of lipase.

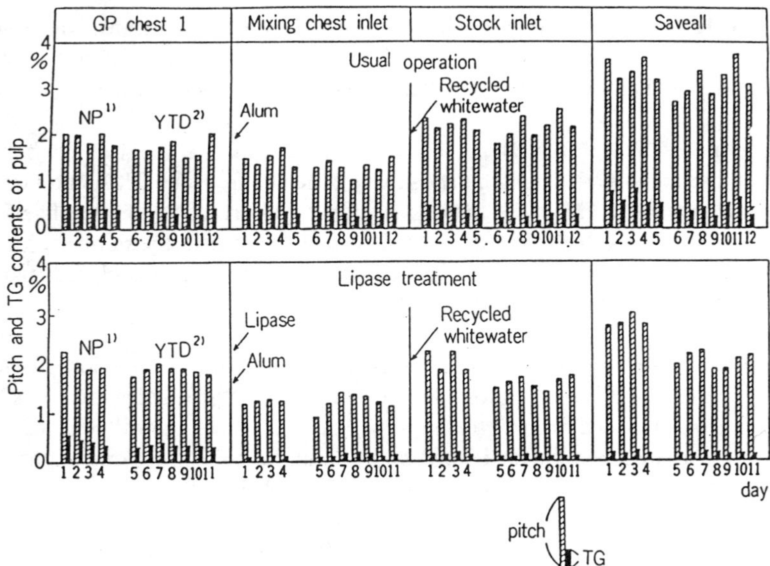

Figure 11. Amount of pitch and TG
1) Newsprint
2) Yellow telephone directory

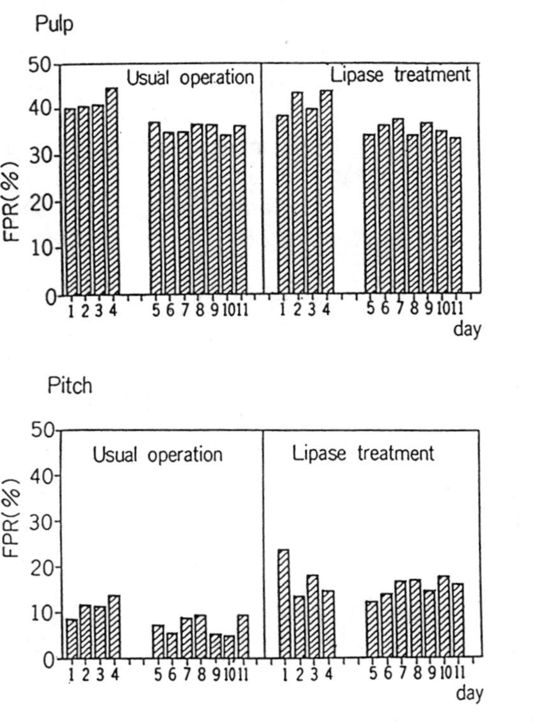

Figure 12. First pass retention of pulp and pitch.

Pitch Deposits on the Wall of the Machine Chest. Pitch deposit was observed as a black piling on the wall of the machine chest during the usual operation. However, pitch deposit could rarely be observed after a one-month trial with the addition of lipase.

Amount of Wet Pitch Deposit. In order to evaluate the wet pitch, which deposits in the press sections, pitch deposit was collected from the felt suction box and measured every day. Figure 13 shows the dramatic decrease in the weight of pitch deposit with the lipase treatment, compared to that of pitch in the normal operation. The above results strongly proved that TG in the pitch was hydrolyzed and then converted to less sticky compounds.

Number of Defects in Paper Web. As shown in Figure 14, the long term data collected by the spot detector showed that the number of defects, holes and spots larger than 1.5 mm was reduced from 61 to less than 19 as a daily average with the addition of lipase. Comparing long term data between the normal and the lipase operations, it was clear that the quality of products was improved (Figure 14).

Dynamic Friction Coefficient of Paper (19-20). When a paper roll is printed on a web offset printing press, it is very important to prevent runability problems such as snaking and wrinkling. This runability performance is especially a concern in newsprint rolls. The dynamic friction of paper is thought to be related to these problems and dynamic friction coefficient (DFC) is regarded as a quality control parameter at some Japanese paper companies. When DFC is low, the web tends to snake on the printing press. Therefore, when DFC of a newsprint drops to a lower level, white carbon (amorphous silica gel $Si0_2$ x H_20) is usually added to the paper furnish to increase DFC.

As lipase treatment was incorporated into the production process, there was an increase in the newsprint DFC and a decrease in the amount of white carbon dosage. In order to increase a certain DFC level during newsprint production, about 2% of white carbon was added in the production process. However, by incorporating the lipase treatment of GP, the dosage of white carbon could be decreased to 1%. With the lipase treatment of GP, the increase of DFC was also confirmed in laboratory experiments (Figure 15).

The other long term mill trial yielded the following results (16):

1) The addition of lipase decreased the pitch deposits on the center roll, and the frequency of cleaning decreased to the level of summer (Table III).

2) Increase of pitch deposit was not apparent by using 50% unseasoned wood. It was clear that a stable operation was performed(Figure 16). Therefore, with the use of lipase there was a decrease of TG content in fresh wood and profits with the reduction in cost of seasoning and bleaching chemicals. Based on the two mill trials, the addition points of lipase should be before the addition of alum (Figure 16, Table III).

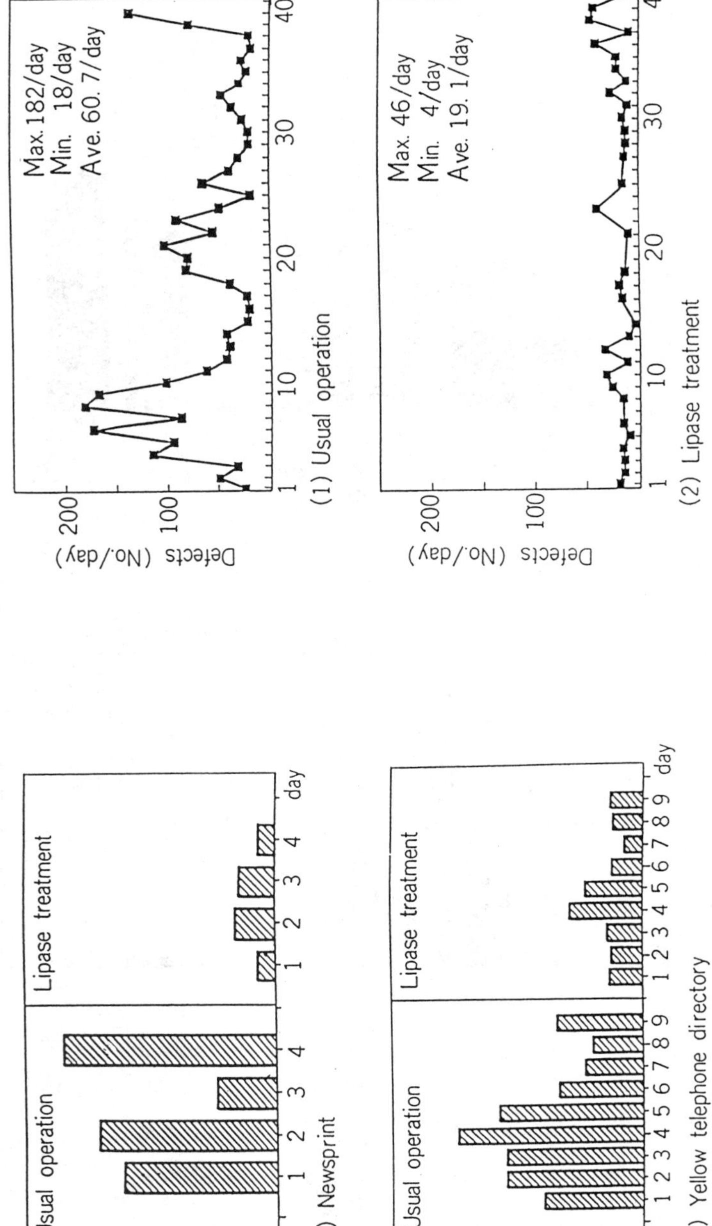

Figure 13. Amount of deposits on the No. 3 press suction box.

Figure 14. Number of defects by spot detector.

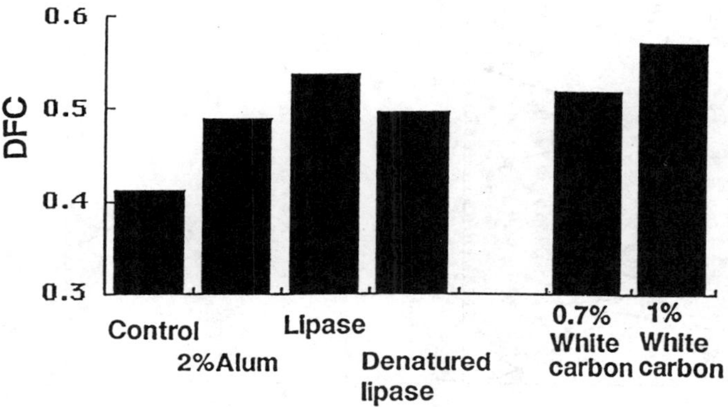

Figure 15. Effect of lipase treatment on DFC.

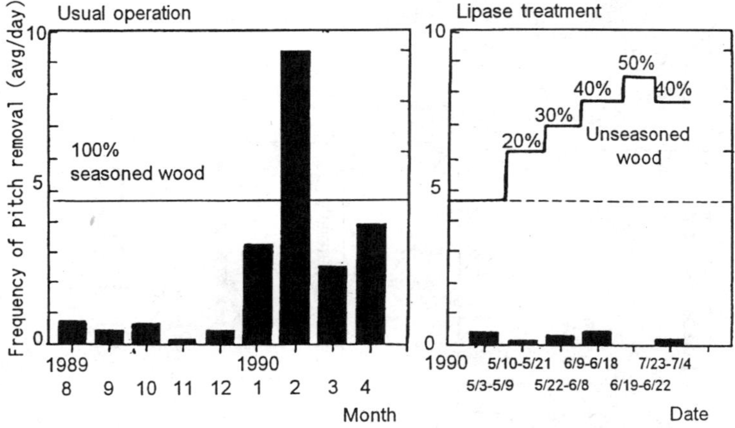

Figure 16. Effect of lipase treatment on frequency of deposit removal on the center roll.

Table Ⅲ. Frequency of deposits removal on the center roll

Day	Seasoned wood /Unseasoned wood	Frequency of deposits removal (av g/day)
3/1～12	100/0	2.5
3/13～31*	100～70/0～30	0.21
4/1～15	100/0	3.37

* Lipase treatment

Conclusions

This newly developed enzymatic pitch control method is the first successful case in using enzyme technology as a solution to pitch troubles. This technology has been tested and proved in actual papermaking process and it has been applied in the mills for industrialization. The following are the research results obtained and characteristics and advantages of this technology:

* The principal cause of pitch accumulation is triglyceride (TG).
* The method using TG hydrolyzing enzyme (lipase) is developed in order to control and prevent pitch deposits.
* The effectiveness of this technology has been proven in our laboratory and further technological know-how has been gained through long term mill trials.
* It solves pitch problems without altering the papermaking process and it requires no equipment investment.
* It is economical and easy to apply in mills. No more than a small-scale pump is required for the process application.
* It uses only a small amount of enzyme and it is regarded as a "clean technology".
* This technology improves paper quality, increases operation stability, allows shorter periods of seasoning, increases pulp brightness, reduces land space, and reduces cost allowing the use of more unseasoned wood.
* With the improvement in DFC, snaking and wrinkling in the web offset press decreases and the runability becomes stable.
* By using enzyme that requires only a short time for its effect, this technology is easily applied to the papermaking process. For this reason, this can be accepted as the first successful case in applying enzyme in the actual papermaking process.
* The enzymatic pitch control has been applied in the four big paper mills of three paper companies in Japan. This technology has contributed significantly to operation stability and to the decrease in costs. In those paper mills, 120,000-140,000T/Year of GP have been treated with lipase. The total amount of enzyme treated GP is corresponded to approximately 20% of whole GP production in Japan.

Literature Cited.

1. *Mokuzai Kagaku;* Migita,N.;Yonezawa,Y.;Kondo,T., Eds.; Kyoritsu Press: Tokyo, Japan, 1968; Vol.1, pp 436.
2. Hillis,W.E., *Wood extractives and their Significance to the Pulp and Paper Industries;* Academic Press: New York, NY, 1962; pp 367,405.
3. Buchanan,M.A.;Sinnett,R.U. Tappe,J.A. *Tappi* **1959**, *42*, 578.
4. Clermont,L.P. *Pulp & Paper Mag. Can.* **1961**, *62*, T-511.
5. King,F.E.;Turd,L. *J. Chem. Soc.* **1953**, 1192.
6. *Mokuzai Kagaku;* Migita,N.;Yonezawa,Y.;Kondo,T., Eds.; Kyoritsu Press: Tokyo, Japan, 1968; Vol.1, pp 466.
7. Kuroki,A. *Enshurin Hokoku.* **1958**, *30*, (Kyushu University).
8. Rydholm,S.A. *Pulping Process;* Interscience Publishers: New York, NY, 1965; pp 1029.
9. Clark,J.d'A. *Pulping Technology;* Miller Freeman Publications: San Francisco, CA, 1978; pp 607.
10. Rydholm,S.A. *Pulping Process;* Interscience Publishers: New York, NY,1965; pp 1025.
11. Fujita,Y.; Awaji,H.; Matsukura,M.; Hata,K. *Japan Tappi Journal.* **1991**, *45*, 905.
12. Bernard, C. *r,held. Sceancea de l'Acai.Sci.* **1956,** *I.*
13. Desnuell,D. *Biochim. Biophys. Acta.* **1958**, *30*, 513.
14. Fukumoto, J.; Iwai, M.;Tsujisaka,Y. *J. Gen. Appl. Microbiol.* **1963**, *9*, 353.
15. Borgström, B.; Erlanson-Albertsson, C. *Lipase;* Elsevier: Amsterdam, NY, 1984; pp 151.
16. Iwai,M.;Tsuujisaka,Y. *J. Gen. Appl. Microbiol.* **1964**, *10*, 87.
17. Iwai,M.;Tsujisaka,Y.;Fukumoto,J. *J. Gen. Appl. Microbiol.* **1970**, *16*, 81.
18. Irie,Y.;Matsukura,M.;Hata,K.;Usui,M. *Tappi Papermakers Conference Proceedings;* TAPPI PRESS: Atlanta, GA, 1990, pp1.
19. Fujita,Y.;Awaji,H.;Taneda,H.;Matsukura,M.;Hata,K.;Shimoto,H.;Sharyo,M.;Sakaguchi,H.;Gibson,K. *Tappi.* **1992**, *74*, 117.
20. Fujita,Y.; Awaji,H.; Taneda,H.; Matsukura,M.; Hata,K.; Shimoto,H.; Sharyo,M.; Abo,M.; Sakaguchi,H. In *Biotechnology in the Pulp and Paper Industry;* Kuwahara,M.; Shimada,M., Eds.; Uni publishers: Tokyo, Japan, 1992.

Chapter 23

Possible Applications of Cellobiose Oxidizing and Other Flavine Adenine Dinucleotide Enzymes in the Pulp and Paper Industry

Paul Ander[1,3], Geoffrey Daniel[1], Bert Pettersson[2], and Ulla Westermark[2]

[1]Swedish University of Agricultural Sciences, Department of Forest Products, S–750 07 Uppsala, Sweden
[2]Swedish Pulp and Paper Research Institute, STF1, Box 5604, S–11486 Stockholm, Sweden

Properties of the cellobiose oxidizing enzymes cellobiose:quinone oxidoreductase (CBQ) and cellobiose dehydrogenase (CDH) are outlined and comparison made with veratryl alcohol oxidase (VAO). Their importance for lignin and cellulose degradation by white-rot fungi is discussed as well as their possible use for applications in the pulp and paper industry. These applications include iron(III) and quinone reduction, studies on yellowing of paper, and (ligno)cellulose fibre characterization. Shortening of fibres and mechanical weakening of fibres by enzymes takes place most rapidly at specific sites of high acessibility, i.e. "nodes", along the fibre axis. The strongest attack on nodes is obtained with CDH + exoglucanase or with exo- and endoglucanases together. Experiments with exo-, endoglucanases and CDH have shown that neither of the cellulases can perform cleavage on their own. It is proposed that exoglucanases first release cellobiose, which is then used for oxidoreductive attack by CDH on pulp fibres.

Wood and lignocellulose are degraded by a mixture of enzymes attacking the different components lignin, cellulose and hemicellulose (*1*). At present the most important lignin-degrading enzymes are lignin peroxidase, Mn-peroxidase and laccase which are produced mainly by white-rot basidiomycetes. They degrade lignin by catalyzing one-electron oxidation of lignin and its degradation products. The resulting radical products undergo a variety of non-enzymatic coupling,

[3]Current address until Nov. 20th, 1997: Department of Mycology, Institute of Biochemical Technology & Microbiology, University of Technology Vienna, Getreidemarkt 9/172, A–1060 Vienna, Austria

0097–6156/96/0655–0297$15.00/0
© 1996 American Chemical Society

substitution and fission reactions. The application of the above mentioned phenoloxidases for degradation and bleaching of pulp is, however, complicated. This is because they form phenoxy radicals in vitro which can couple and form polymers (2). Under certain conditions it is possible to degrade synthetic lignins with both lignin peroxidase and Mn-peroxidase, although polymerization is also common (*3-6*). It has been suggested that cellobiose oxidizing enzymes produced during cellulose degradation by white-rot fungi are involved in lignin degradation (*7-10*). The reason is that these enzymes can reduce phenoxy radicals, which might regulate lignin polymerization/depolymerization. The radical, quinone and Fe(III) reducing effect of CBQ/CDH combined with the effect of cellulose oxidation, make these enzymes of interest for studies on pulp fibres and the resulting effect on paper properties. Some preliminary results along these lines are reported here together with a short background.

The Cellobiose Oxidizing Enzymes

CBQ and CDH are produced on cellulose media by white-rot and other fungi (*1,7,8,11-14*) and they both oxidize cellobiose to cellobiono-1,5-lactone. Via cellulose binding and catalytic domains, there is also a slow oxidation of cellulose fibres at accessible regions along the fibre at so called "nodes". In the presence of cellulose, the heme/FAD-containing CDH (EC 1.1.99.18) is cleaved by proteases to give CBQ (EC 1.15.1) plus a heme domain (*15*). Since CBQ/CDH is produced mainly in high nitrogen, cellulose media it is difficult to correlate these enzymes with lignin degradation and the ligninolytic enzymes produced in low nitrogen, glucose cultures. Simultaneous production of CBQ/CDH and lignin peroxidase/Mn-peroxidase has, however, been found in high nitrogen, cellulose cultures during degradation of synthetic lignin (*9*), and it was proposed that a delicate balance exists between ligninolytic and cellobiose oxidizing enzymes. Evidence for reduction not only of a large number of lignin-related quinones but also of phenoxy radicals and cation radicals by CBQ/CDH have been presented (*7,8,11,16,17*, Ander, P.; Marzullo, L. *J. Biotechnol.*, in press).

Veratryl Alcohol Oxidase (VAO)

In connection with radical reducing enzymes it is also appropriate to mention the FAD enzyme veratryl alcohol oxidase which has been reported produced on glucose media by *Pleurotus, Bjerkandera* and *Coriolus* spp. (*18-20*). The enzyme oxidizes veratryl alcohol to veratraldehyde during reduction of oxygen to hydrogen peroxide. The pH optimum is higher than for lignin peroxidase, and is ca 6.0-6.5. VAO has so far been shown to inhibit laccase-catalyzed polymerization of guaiacol and ferulic acid, as well as to reduce dichlorophenol-indophenol (DCIP) and the phenoxy radical from acetosyringone. A mixture of laccase and VAO was found to be more efficient in degrading lignosulfonate than laccase alone (*19*). Indications for CBQ production by the above fungi on cellulose was already obtained about 20 years ago (*12,13*), and it may be that

white-rot fungi produce either cellobiose or veratryl alcohol oxidizing enzymes depending on whether mostly cellulose or glucose is present in the local vicinity of the fungal hyphae. Comparisons during cultivation on wood has not been done. Since other FAD sugar oxidases like pyranose 2-oxidase also have at least quinone reducing activity (Ander, P.; Marzullo, L. *J. Biotechnol.*, in press), it seems very important for the fungus to retain lignin-related phenols in a reduced state (*21*).

Iron(III) Reduction and CDH

Cellobiose dehydrogenase reduces Fe(III) to Fe(II) (*22*), and in the presence of hydrogen peroxide produced by sugar oxidases there will be formation of reactive hydroxyl radicals and/or iron(per)oxo complexes:

$$H_2O_2 + Fe(II) \dashrightarrow OH^{\cdot} + OH^{-} + Fe(III), \text{ alt. } [Fe(IV)\text{-}OH)]^{3+} \text{ or } [Fe(IV){=}O]^{2+}$$

Degradation of labelled synthetic lignin (DHP), carboxymethylcellulose (CMC) and xylan was studied in a CDH system containing cellobiose, Fe(III) and hydrogen peroxide (*10*). The results in Table I show that passage of radioactive lignin through a membrane with cut-off 500, inc reased from 3.6% in buffer, to 6.6% in the complete CDH system. For a 1000 cut-off membrane, passage of the

Table I. Degradation of ^{14}C-labelled synthetic lignin by cellobiose dehydrogenase (CDH) and filtration through membranes with cut off 500 or 1000.

Conditions: CDH 92 nM, cellobiose 0.5 mM, acetate buffer pH 4.0. Incubation 16 h at 30°C. Filtration by pressure through Amicon ultrafiltration membranes. Adapted from ref *10*.

	% Radioactive lignin through membrane, cut off 500
Only acetate buffer	3.60
Hydrogen peroxide (HP) 0.5 mM	3.77
Hydrogen peroxide+$FeCl_3$ 0.2 mM	4.35
CDH + cellobiose + $FeCl_3$	5.40
CDH + cellobiose + $FeCl_3$ +HP	6.60
	% Lignin through membrane, cut off 1000
Acetate+cellobiose+Fe+HP	8.85
CDH+cb+Fe+HP (complete system)	14.0

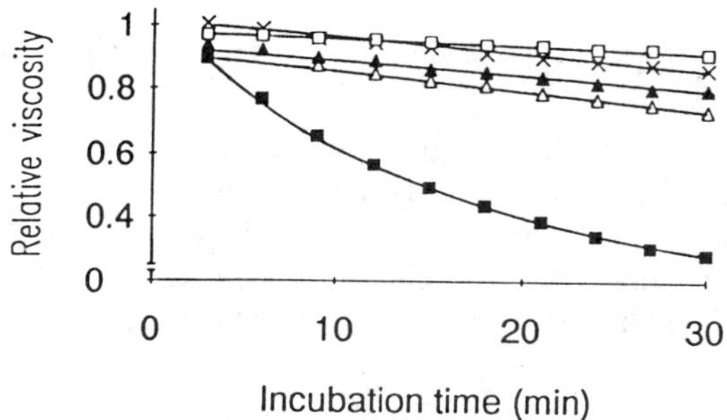

Figure 1. Viscosity decrease of carboxymethylcellulose (CMC) by the CDH system at pH 4.0. Conditions: CMC 0.4%, CDH 29 nM, cellobiose 2 mM, hydrogen peroxide 1.8 mM, ferricyanide 0.2 mM. (■) Complete system (+cb, HP, Fe); (Δ) without ferricyanide; (□) without CDH and cellobiose; (▲) without hydrogen peroxide and ferricyanide; (×) only CDH (-cb). (Reprinted with permission from reference 10. Copyright 1995 Springer-Verlag GmbH & Co. KG.)

lignin increased from 8.85% to 14.0 % in the complete system (*10,23*). It may be noted that the lignin used had a large portion with a M_n over 20.000 (*1, page 232*) probably making it more resistant to degradation than other DHPs. CMC degradation, as measured by relative viscosity, is shown in Figure 1 (*10*). Thus this system may be used by different fungi for initial attack on wood components and may also be used for in vitro degradation of xenobiotic compounds.

Recently, Fe(III) reduction by CDH in the presence of different lignins was investigated (*23-25*). The principle for this reduction is shown in Figure 2. Fe(III) reduction by CDH+cellobiose was strongly decreased by spruce guaiacyl lignin and coniferyl alcohol lignin in the presence of added hydrogen peroxide. With birch (i.e. guaiacyl/syringyl lignin), no decrease but rather a small increase in Fe(III) reduction was obtained (*25*). The decrease and delay of Fe(III) was related to the amount of guaiacyl lignin, but was obtained only in the presence of hydrogen peroxide. The effect is mainly thought due to reoxidation of Fe(II) by remaining hydrogen peroxide. In this connection, reduction of 2-methoxy-*p*-quinone (MQ) and 2,6-dimethoxy-*p*-quinone (DMQ) by CDH+cellobiose was also investigated. MQ was most rapidly reduced and it is thought that release of this quinone from birch lignin will increase the apparent Fe(III) reduction in the presence of birch lignin. Since the decrease in Fe(III) reduction was directly related to spruce lignin concentration it may be that this CDH, cellobiose, Fe(III), hydrogen peroxide-system can be used for surface analyses of residual lignin in bleached and unbleached kraft pulps and for lignin in (chemo)thermomechanical pulps (*24*). Unbleached kraft pulps are known to contain 2.5-4.5 times more lignin on the fibre surface compared with the total amount of lignin in the fibre cell wall (*26*).

At about 10 pM concentration, dimethoxyquinone DMQ strongly inhibited laccases from both *Phlebia radiata* and *Trametes (Coriolus) versicolor* (*27*), although CDH reduced as much as 0.41 mM DMQ (*23*). This suggests that the cellobiose oxidizing enzymes are important for decreasing the toxicity of quinones. Since the above lignin-related quinones are known to be involved in photo-yellowing of lignin-containing paper (*28*), CDH could possibly be used together with laccase and peroxidases for studies on mechanisms of quinone formation and yellowing of pulp and paper (*23,24*). Other FAD oxidases like veratryl alcohol oxidase and glucose oxidases my also be used for such studies.

Cellobiose Oxidizing Enzymes and Cellulose Degradation

Since both CBQ and CDH have a binding domain for cellulose (*15,29-31*), it is logical to believe that they are important for cellulose degradation and probably function in concert with cellulases as first indicated by Eriksson et al. (*32*) and later by Bao and Renganathan (*33*) and by us (see below). Recent results by Katagiri et al. (*34*) suggest that CBQ/CDH is more important in cellulose degradation than lignin degradation. However, since they measured cellobiose oxidizing enzyme activity with DCIP they could not distinguish between CBQ and CDH. They cultivated *P. chrysosporium* and *T. versicolor* on hardwood unbleached kraft pulp on low and high nitrogen and measured Mn-peroxidase

(MnP) and CBQ/CDH. As shown in Figure 3, it was found that brightening of the pulp correlated well with MnP production but not well with CBQ/CDH production. Cellulose degradation, however, correlated well with CBQ/CDH activity and this is shown in Figure 4 for the two fungi at both low and high nitrogen concentrations (*34*). The cultivations were done using solid state fermentation and such conditions may not be the same as during wood degradation in nature. CBQ/CDH production during primary metabolism on high nitrogen media does, however, not support their importance for lignin degradation. Furthermore, quinone and phenoxy radical reduction by CBQ/CDH and its possible relation to cellulose degradation has not been thoroughly investigated. Henriksson et al. (*10*) suggested that reduction of phenoxy radicals is a way of preventing precipitation of phenol or lignin polymers on the cellulose surface, which could otherwise decrease cellulose degradation.

Cellulase, CDH and Fibre Analyses

Recently the potential of enzymes including CDH were evaluated for the modification of pulp fibre cell walls. The approach involved the incubation of fibres and pulps with a variety of commercially available enzymes (e.g. Celluclast, Novo A/S) and highly purified specific hydrolytic enzymes (e.g. exo-, endoglucanase, xylanase, mannanase, pectinase and β-glucosidase) and CDH. Their morphological action on a variety of softwood fibres, e.g. O_2 bleached-, whole bleached- and unbleached sulfate pulp fibres, were observed using light- and electron microscopy.

Results have shown the specific attack and characteristic cleavage of fibres into short lengths bearing fibrillar ends by the simultaneous action of purified exo- and endo-glucanases from *Trichoderma reesei*. Cleavage and mechanical weakening of the fibres occurred most rapidly at specific sites visible in polarized light at so called "nodes" along the fibre axis (arrows; cf. Figure 5 with control Figure 8), with rate of attack related to enzyme activity. These nodes, usually appearing at a distance of 100-200 μm, are most easily seen after weak swelling of different pulps in LiCl/DMAC (*35*). The characteristic attack at nodes, of a hitherto unknown chemical composition, suggests a distinct variation in fibre structure which is more susceptible to cellulolytic attack, indicating that the nodes have cellulose as the main constituent. Interestingly, neither of the cellulases could perform cleavage on their own. Treatment with CDH+cellobiose (Figure 6) or CDH+endoglucanase caused swelling (perhaps caused by carboxyl group introduction) but not cleavage of the nodes, while CDH+exoglucanase resulted in cleavage (Pettersson, B. and Daniel, G., unpubl. results). The reason for the last result is probably that exoglucanase releases cellobiose, which is used by CDH for increased oxidative attack. Treatment with endo-, exoglucanase and CDH together also caused node attack. No visual attack of the nodes was, however, observed with any of the other purified enzymes apart from xylanase, where some delamination was noted for both the nodes and fibre walls (Figure 7). Currently the structural and chemical basis of the nodes is under study using a variety of immunological and microscopical methods.

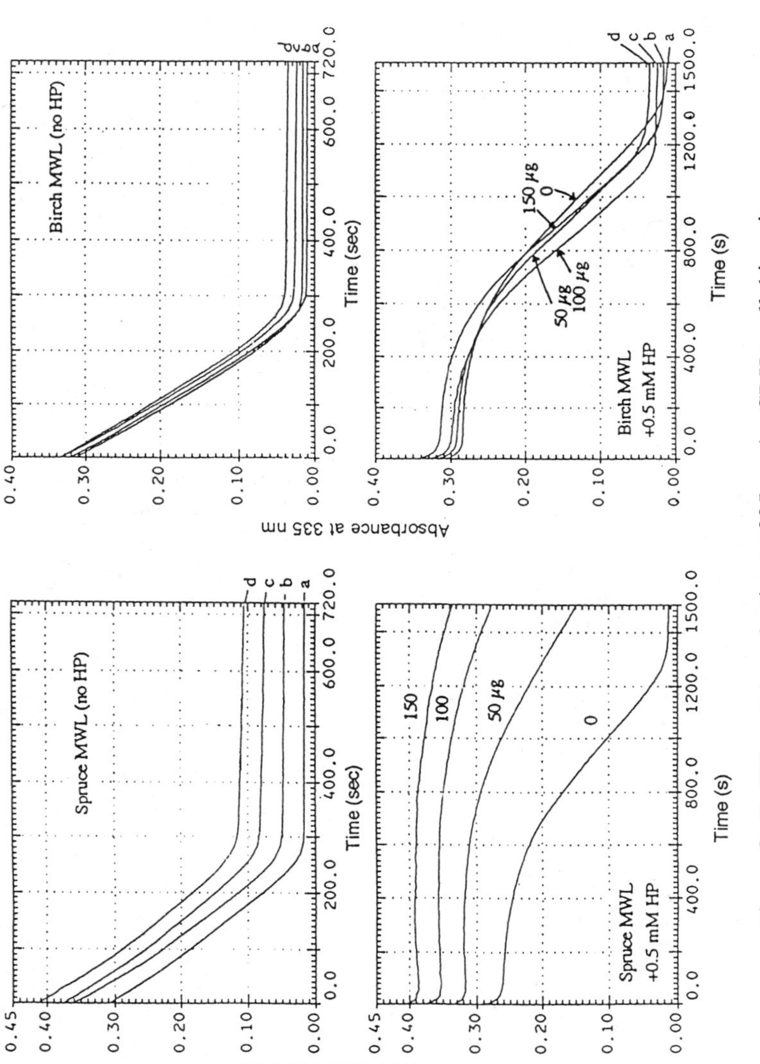

Figure 2. Fe(III) acetate reduction at 335 nm by CDH+cellobiose in the presence of spruce and birch milled wood lignin (MWL) ± 0.5 mM hydrogen peroxide (HP) in acetate buffer at pH 4.0. (Reprinted with permission from reference 25. Copyright 1996 Springer-Verlag GmbH & Co. KG.)

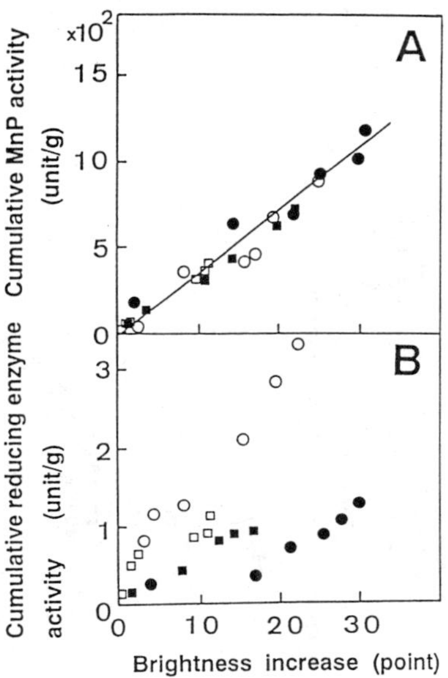

Figure 3. Relationship between the cumulative enzyme activity and brightness increase in the solid-state fermentation system. (•) *P. chrysosporium*, LN-HC; (o) *P. chrysosporium*, HN-HC; (■) *T. versicolor*, LN-HC; (□) *T.versicolor*, HN-HC. (**A**) Cumulative MnP activity. (**B**) Cumulative reducing enzyme activity. (Reprinted with permission from reference 34. Copyright 1995 Springer- Verlag GmbH & Co. KG.)

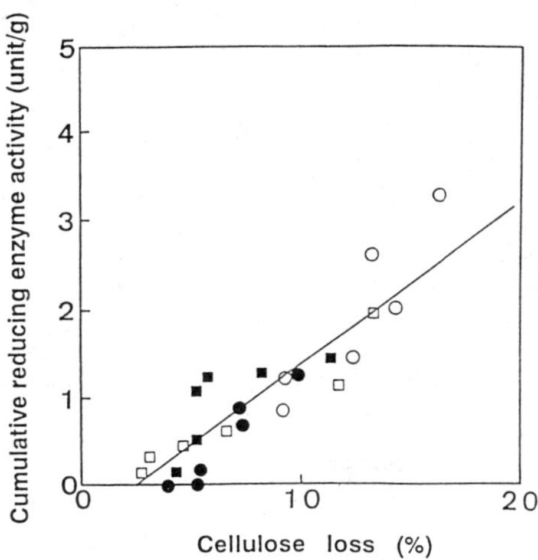

Figure 4. Relationship between the cumulative reducing enzyme activity and cellulose loss in the solid-state fermentation system. (•) *P. chrysosporium*, LN-HC; (o) *P. chrysosporium*, HN-HC; (■) *T. versicolor*, LN-HC; (□) *T. versicolor*, HN-HC. (Reprinted with permission from reference 34. Copyright 1995 Springer-Verlag GmbH & Co. KG.)

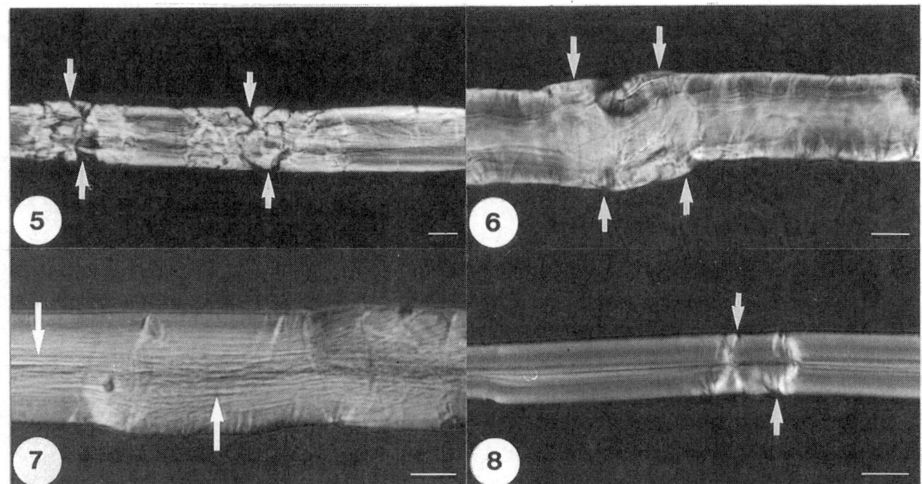

Figures 5-8. Polarized light micrographs showing pine holocellulose fibres after enzymatic treatments.
Figure 5. Characteristic cleavage of nodes (arrows) after cellulase (endo+exo) treatment in 50 mM NaAc pH 5.0 for 8 hrs at 40°C.
Figure 6. Swelling of nodes (arrows) after CDH+cellobiose treatment with 222 nM CDH in 50 mM NaAc, pH 5.0 for 12 hrs at 37°C.
Figure 7. Typical delamination (arrows) of fibres after xylanase treatment in 50 mM NaAc pH 7.0 for 18 hrs at 40°C.
Figure 8. Control, untreated fibre with nodes (arrows).
Note: After treatment the enzyme reactions were terminated by elevation of pH, and samples were washed 2x with distilled water before observation in a Leitz bright field light microscope equipped with polarized filters. Bars in Figures 5-8 = 20 μm.

Acknowledgments

Prof. Kurt Messner, TU, Wien for kind support to P.A. during the writing of this paper in his laboratory. Financial support to P.A. and B.P. from MoDo R&D, Sweden. Thanks are due to Dr. N. Katagiri, Shizuoka University, Japan for permission to use Figures 3 and 4.

Literature Cited

1. Eriksson, K.-E. L.; Blanchette, R.A.; Ander, P. *Microbial and Enzymatic Degradation of Wood and Wood Components*; Springer Verlag: Heidelberg, New York, 1990, p. 225-333.
2. Kirk, T.K.; Farrell, R.L. *Ann. Rev. Microbiol.* **1987**, *41*, 465-505.
3. Hammel, K.E.; Moen, M.A. *Enzyme Microb. Technol.* **1991**, *13*, 15-18.
4. Wariishi, H.; Valli, K.; Gold, M.H. *Biochem. Biophys. Res. Comm.* **1991**, *176*, 269-275.
5. Hammel, K.E.; Jensen, Jr., K.A.; Mozuch, M.D.; Landucci, L.L.; Tien, M.; Pease, E.A. *J. Biol. Chem.* **1993**, *268*, 12274-12281.
6. Bao, W.; Fukushima, Y.; Jensen, Jr., K.A.; Moen, M.A.; Hammel, K.E. *FEBS Lett.* **1994**, *435*, 297-300.
7. Eriksson, K.-E. L.; Habu, N.; Samejima, M. *Enzyme Microb. Technol.* **1993**, *15*, 1002-1008.
8. Ander, P. *FEMS Microbiol. Rev.* **1994**, *13*, 297-311.
9. Costa-Ferreira, M.; Ander, P.; Duarte, J.C. *Enzyme Microb. Technol.* **1994**, *16*, 771-776.
10. Henriksson, G.; Ander, P.; Pettersson, B.; Pettersson, G. *Appl. Microbiol. Biotechnol.* **1995**, *42*, 790-796.
11. Westermark, U.; Eriksson, K.-E. *Acta Chem. Scand.* **1974**, *B28*, 204-208.
12. Westermark, U.; Eriksson, K.-E. *Acta Chem. Scand.* **1974**, *B28*, 209-214.
13. Ander, P.; Eriksson, K.-E. *Physiol. Plant.* **1977**, *41*, 239-248.
14. Ayers, AR.; Ayers, SB.; Eriksson, K.-E. *Eur. J. Biochem.* **1978**, *90*, 171-181.
15. Habu, N.; Samejima, M.; Dean, J.F.D.; Eriksson, K.-E. L. *FEBS Lett.* **1993**, *327*, 161-164.
16. Ander, P.; Mishra, C.; Farrell, R.L.; Eriksson, K.-E. L. *J. Biotechnol.* **1990**, *13*, 189-198.
17. Samejima, M.; Eriksson, K.-E. L. *Eur. J. Biochem.* **1992**, *207*, 103-107.
18. Bourbonnais, R.; Paice, M.G. *Biochem. J.* **1988**, *255*, 445-450
19. Sannia, G.; Limongi, P.; Cocca, E.; Buonocore, F.; Nitti, G.; Giardina, P. *Biochim. Biophys. Acta* **1991**, *1073*, 114-119.
20. Marzullo, L.; Cannio, R.; Giardina, P.; Santini, M.T.; Sannia, G. *J. Biol. Chem.* **1995**, *270*, 3823-3827.
21. Schoemaker, H.E.; Meijer, E.M.; Leisola, M.S.A.; Haemmerli, S.D.; Waldner, R.; Sanglard, D.; Schmidt, H.W.H. In *Plant Cell Wall Polymers Biogenesis and Biodegradation*; Lewis, N.G.; Paice, M.G., Eds.; ACS Symp. Ser. *399*, Washington DC, 1989, Chapter 30.

22. Wood, P.M. *FEMS Microbiol. Rev.* **1994**, *13*, 313-320.
23. Ander, P. In *Biotechnology in the Pulp and Paper Industry. Recent Advances in Applied and Fundamental Research.* Proc. 6th Int. Conf. Biotechnol. Pulp Paper Ind. 1995, Vienna, Austria. Srebotnik, E.; Messner, K., Eds.; Facultas-Universitätsverlag, Wien, 1996, pp. 357-361.
24. Ander, P. *The 8th Int. Symp. Wood and Pulping Chem.*, June 6-9, 1995, Helsinki, Finland. Proc. Vol II, Poster presentations, pp.107-112.
25. Ander, P. *Holzforschung* **1996**, *50*, 413-419.
26. Heijnesson, A.; Simonsson, R.; Westermark, U. *Holzforschung* **1995**, *49*, 313-318.
27. Rogalski, J.; Józwik, E.; Hatakka, A.; Leonowicz, A. *J. Mol. Cat. A.* **1995**, *95*, 99-108.
28. Zhang, L.; Gellerstedt, G. *Third European Workshop on Lignocellulosics and Pulp. Extended Abstracts.* Hasseludden, Stockholm, Sweden, August 28-31, 1994, pp. 293-295. Organised by: Roy. Inst. Technol., Div. Wood Chemistry and STFI, Swedish Pulp and Paper Res. Inst., Stockholm, Sweden.
29. Renganathan, V.; Usha, S.N.; Lindenburg, F. *Appl. Microbiol. Biotechnol.* **1990**, *32*, 609-613.
30. Henriksson, G.; Pettersson, G.; Johansson, G.; Ruiz, A.; Uzcategui, E. *Eur. J. Biochem.* **1991**, *196*, 101-106.
31. Bao, W.; Lymar, E.; Renganathan, V. *Appl. Microbiol. Biotechnol.* **1994**, *42*, 642-646.
32. Eriksson, K.-E.; Pettersson, B.; Westermark, U. *FEBS Lett.* **1974**, *49*, 282-285.
33. Bao, W.; Renganathan, V. *FEBS Lett.* **1992**, *302*, 77-80.
34. Katagiri, N.; Tsutsumi, Y.; Nishida, T. *Mokuzai Gakkaishi* **1995**, *41*, 780-784.
35. Wallin, M. Examination of rigid elements in the fiber wall. Examensarbete på Kemitekniklinjen, CTH. Printed by STFI, Stockholm 1992. In Swedish, summary in English.

Chapter 24

Improvement of Wheat-Straw Pulp Properties with an Alkali-Tolerant Xylanase from *Pseudomonas* sp. G6-2

Jiachuan Chen[1], Jianyun Yang[1], Yinbo Qu[1], Peiji Gao[1], Baomin Wang[1], Qi Zhang[2], and Quihua Yang[2]

[1]State Key Laboratory of Microbial Technology, Institute of Microbiology, Shandong University, Jinan 250100, China
[2]Department of Chemical Engineering, Shandong Institute of Light Industry, Jinan 250100, China

An alkali-tolerant bacterium, which produced an extracellular alkali-stable xylanase, was isolated from soil and identified as a strain of *Pseudomonas*. The conditions of treating hypochlorite-bleached wheat straw pulps with this xylanase were investigated. Results showed an obvious improvement of brightness, water filterability and fragility of the pulp after enzymatic treatment .

Hemicellulases from microorganisms have been investigated because of the possibility of degrading hemicellulose from agricultural residues into xylose and other monosaccharides. However, there is an increasing interest in applying hemicellulases, mainly xylanase, to the pulp bleaching process to reduce the amount of chlorine used in bleaching processes as well as improve the quality of pulp produced (*1-4*).

Wheat straw is a major source of fibers for paper industry in China and many other developing countries. The main problems of straw pulps are poor water filterability, pasting the nets and rolls of paper machine and resultant paper products with low opacity, high fragility and low strength. As a result, straw pulps can only be used to produce paper of low grades. A part of wood pulps must be added in order to produce paper of high grades. The problems are probably related to high content of hemicellulose in the straw pulps. Therefore, straw pulp properties may be improved when the hemicellulose on the surface of fibers is partially removed by xylanase. Alkaline xylanases are preferable for this process because of the alkaline nature of most pulps. In this paper, the isolation of an alkalophilic bacterial strain, *Pseudomonas* sp. G6-2, the optimization of its alkali-tolerant xylanase producing conditions and the effects of treating hypochlorite-bleached wheat straw pulps with the xylanase on pulp properties were reported.

The results showed that the papermaking properties of nonwood-based pulp can be improved distinctly by xylanase treatment. This gives the hope that the high grade paper may be produced by this technique from agricultural residues instead of wood.

0097–6156/96/0655–0308$15.00/0
© 1996 American Chemical Society

This is very important for ecosystem protection in a world view, and is important for economic balance in the forest-inadequate countries such as China.

Isolation and Identification of Alkalophilic Xylanase Producing Bacteria

Isolation of the Bacteria. A small amount of soil was suspended in sterilized water and inoculated into an flask containing enriching media (peptone, 0.2%; xylan, 2%; NaCl, 0.5%; pH 9.0). After incubation at 37℃, 120 rpm for 2 days, the culture broth was spread on alkaline xylan agar plates (xylan, 1%; KNO_3, 0.1%; $MgSO_4$, 0.05%; NaCl, 0.05%; K_2HPO_4, 0.05%; $Fe_2(SO_4)_3$, 0.001; pH 9.0) which were then incubated at 37 ℃ for 2 days. From about 2000 colonies isolated from soil samples, an alkalophilic strain, designated as strain G6-2, was selected for xylanase production accordding to its high extracellular xylanase activity.

Characterization and Identification of the Bacteria. Morphological properties and other taxonomic characteristics of the bacteria were studied according to the methods described in Bergey's Manual of Determinative Bacteriology (*5*).

The strain G6-2 was an aerobic, gram-negative, motile and rod-shaped bacterium with polar flagellation. It grew well under pH 5.0-pH 10.0 and the growth rate was the highest at pH 8.5. The maximum temperature for growth was 42 ℃, and optimal temperature was about 37℃ at pH 8.5. The strain reduced nitrates, hydrolyzed starch and gelatin, utilized D-xylose, DL-arginine and phenylalanine, synthesized fructan from sucrose and demonstrate activities of arginine dihydrolase and oxidase. It is clear that the bacterium belongs to the genus *Pseudomonas,* and is quite similar to *Pseudomonas pseudomallei.*

Optimization of Xylanase Production

Media and Culture Conditions. The medium used for xylanase production was as following unless stated: 4% wheat bran, 0.25% urea, 0.25% peptone, 0.5% NaCl, pH 8.5. Cultivation was carried out at 37℃, with a shaking speed of 130 rpm.

Assays of Enzymes. The culture supernatant was used as crude xylanase solution. Activity was assayed by measuring the amount of reducing sugars liberated from xylan by the 3,5-dinitrosalicilic acid method (*6*). The reaction mixture contained enzyme and 1.0% xylan (from oat spelt, Sigma Chemical Company) in 40 mM sodium barbital-0.4M HCl, pH 7.2, which was treated according to Bailey et al. (*7*). The reaction was carried out at 50℃ for 30 minute unless stated.

Cellulase activity was assayed by incubating 0.5 ml crude enzyme solution with 0.5% CMC-Na, or with 50 ± 5 mg filter paper (Whatman No. 1) in 2 ml 40 mM sodium barbital-0.4M HCl, pH 7.2 for 30 min. After incubation, the reducing sugar released from the reaction was measure by DNS method as described above.

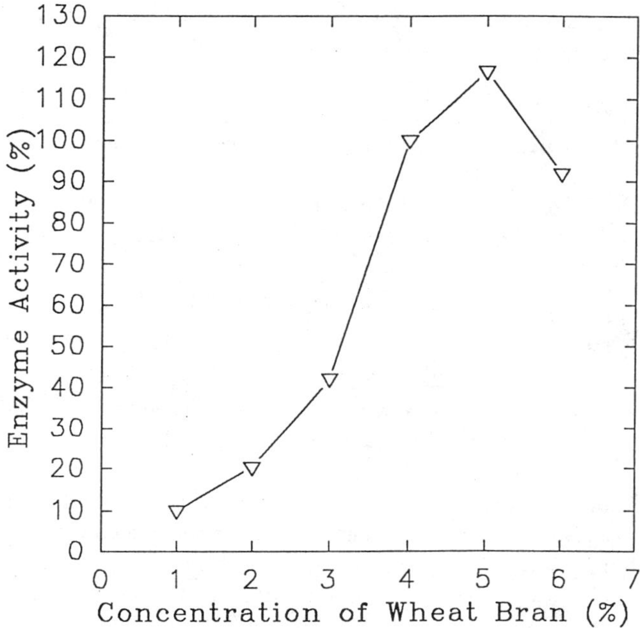

Fig. 1 Effects of concentration of wheat bran on the xylanase production

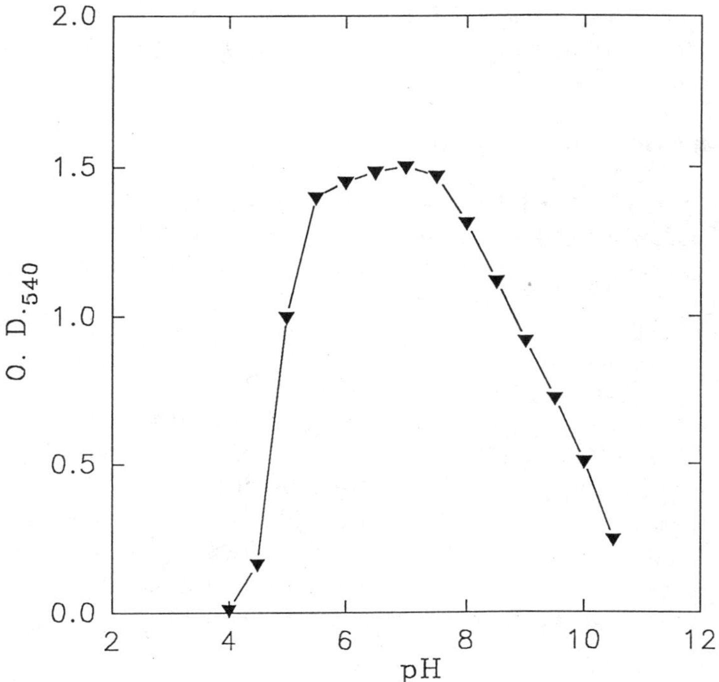

Fig. 2 Effects of pH on enzyme activity

One unit of activity corresponds to the equivalent of 1 μ mol xylose or glucose liberated per minute.

Effects of Culture Conditions. Effects of carbon sources, concentrations of wheat bran, nitrogen sources and aeration conditions on the xylanase production of *Pseudomonas* G6-2 were investigated (Table I, Fig. 1 and Table II).

Since the isolate is strictly aerobic, aeration may have a strong effect on xylanase production. In this study, the aeration condition was changed by using a three-baffled flask instead of a normal flask. The results showed that the enzyme production was greatly improved. Under the optimal conditions, *Pseudomonas* sp. G6-2 can produce more than 300 IU/ml xylanase within 96 hr.

Table I Effects of Carbon Sources on the Xylanase Production

	1% wheat bran 0.5% glucose	*1% wheat bran 0.5% xylose*	*1% wheat bran 0.5% starch*	*1% wheat bran*
Relative activity (%)	0	0	23.2	100
Final pH	<6	<6	8.8	8.5

Table II Effects of nitrogen sources on the xylanase production

	0.25 % peptone 0.25% urea	*0.5% $(NH_4)_2SO_4$*	*0.5% $NaNO_3$*	*0.5% urea*	*0.5% sodium glutamate*
Relative activity (%)	100	0	98.6	94.6	110.2
Final pH	8.5	<4	9.7	8.5	8.5

Characterization of the xylanase

Effects of pH on the Enzyme Activity and Stability. The reaction pH was adjusted with 0.1M citrate-sodium citrate buffer (pH 4.0-6.5); 40 mM sodium barbital-HCl (pH 7.0-9.5) and 0.2 M glucine-NaOH (pH 10.0-10.5). Other conditions were the same as those of the standard assay method. The xylanase was most active at pH 6.5-7.5 while remained about 80% of the optimal activity when the reaction pH value reached 8.5 (Fig. 2). Stability of the enzyme was investigated in buffer solutions of various pH values. It was stable at 4℃ for 24 h in the range of pH 4.0 to 10.5.

Zymogram analysis. Zymogram analysis was carried out on a 7.5%-polyacrylamide gel containing 0.1% sodium dodecylsulphate (SDS) and 0.1% xylan. After electrophoresis, renaturation treatment, incubation for 10 minute at 50 ℃, staining the gel with 0.1% Congo red for 30 minute, washing with 1.0M NaCl and placed in 5% acetic acid (*8*), zymograms of the culture supernatant of alkaliphilic *Pseudomonas* sp. G6-2 were obtained. It was shown that there were two major bands of xylanases. Both were still active at pH 10.5.

Effects of Temperature on the Enzyme Activity and Stability. The xylanase was most active at 60℃, while lost most of its activity when the incubation temperature

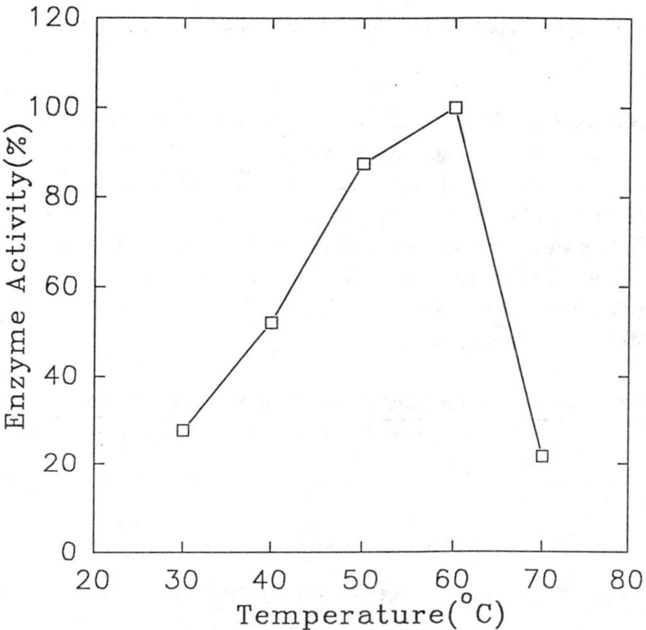

Fig. 3 Effects of temperature on enzyme activity

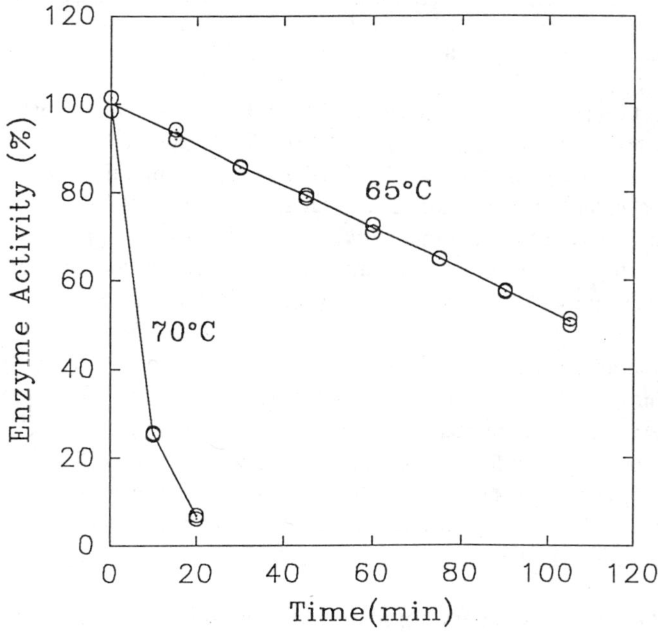

Fig. 4 Effects of temperature on enzyme stability

was raised to 70 ℃ (Fig. 3). Half lives of the enzyme were determined by incubating the enzyme solution at different temperatures. Samples were withdrawn at defined times and stored at 4 ℃ before measuring the activities. The half lives at 65 ℃ and 70 ℃ were about 105 min. and 5 min., respectively (Fig. 4). The enzyme lost its activity slightly while incubating at 60 ℃ for 2 h.

Cellulase Activity of the Enzyme. No cellulase activity was detected when using the filter paper method or the CMC-Na method.

Improvement of Straw pulp properties by xylanase treatment

Enzymatic Treatment of the Straw Pulp. The pulp used for this experiment was a hypochlorite-bleached wheat straw pulp from a mill in Shandong Province, China. The pulp has a brightness of 71.5% ISO and a DP of 760. Enzymatic treatments were done in polyethylene bags and kept at the desired temperatures by water baths. The treatment conditions were enzyme dosage 5 IU/g pulp, pH 7.5, pulp consistency 10 %, 30 ℃ for 90 min. unless stated. After treatment the pulp were washed with tap water and brightness, CED viscosity (DP) and physical properties were measured (*9*).

Effect of Enzyme Treatment Conditions on the Pulp Properties. The effects of enzyme dosage on the pulp properties are shown in Table III. After enzymatic treatments, beating degree and fragility were decreased, brightness and DP increased, but physical strength properties varied slightly with the exception of tear index. When enzyme dosage rose, the DP increased, fragility dropped, but the others changed little. Enzyme dosage of 1 IU/g is enough for the improvement of beating degree, brightness and strength properties, but in this case the fragility is still high. The dosage of 3-5 IU/g is suitable for comprehensive consideration.

Table III Effects of Enzyme Dosage on the Pulp Properties

EnzymeDosage, IU/g	*Control*	*1*	*3*	*5*	*10*	*15*	*20*
beating degree, ° SR	34	28.5	28	28	28	28	27.5
brightness, % ISO	71.5	74.0	74.8	74.8	75.0	75.0	75.0
DP	760	768	773	776	784	792	801
breaking length, km	3.66	3.65	3.66	3.68	3.69	3.69	3.72
burst index, KPa · m^2/g	2.24	2.26	2.26	2.27	2.29	2.28	2.30
tear index, mN · m^2/g	3.8	4.2	4.2	4.4	4.5	4.5	4.4
fragility*, %	20.5	15.8	12.0	7.8	7.2	7.0	6.4

* Fragility (%)= (tensile for unfolded sheet - tensile for the sheet folded twice on each side) × 100/tensile for unfolded sheet

Xylanase from strain G6-2 has higher enzyme activity under neutral or weak alkaline conditions. The effects of pH on the pulp properties were investigated. As can be seen in Table IV, it is effective when pH varies within the range of 6.5-8.5 for improvement of water filterability, brightness and fragility. The optimal pH is 7.5.

Under the above conditions, beating degree and fragility dropped, brightness and DP risen as reaction was prolonged, but these items had no obvious changes when time was over 90 min.(Table V)

Effects of reaction temperature (from 30 to 70 ℃) on the pulp properties were also investigated. As temperature rises, the reaction speeds up, but the enzyme activity may decreased gradually. So, no obvious effect was found between 30 to 50℃. When the temperature was over 50 ℃, beating degree and fragility became higher, but brightness and DP were lower. With enough time, high temperature is unnecessary.

Table IV Effects of Reaction pH on the Pulp Properties

pH	*Control*	*6.5*	*7.0*	*7.5*	*8.0*	*8.5*
beating degree, ° SR	34	29	28	28	28.5	29
brightness, % ISO	71.5	73.6	74.2	74.9	74.5	74.1
DP	760	767	773	781	772	767
breaking length, km	3.66	3.68	3.66	3.68	3.67	3.66
burst index, KPa · m^2/g	2.24	2.24	2.23	2.28	2.25	2.24
tear index, mN · m^2/g	3.8	4.0	4.1	4.3	4.1	4.2
fragility, %	20.5	12.2	9.8	7.5	8.2	9.0

Table V Effects of Reaction Time on the Pulp Properties

Time, min.	*Control*	*30*	*60*	*90*	*120*	*150*	*180*
beating degree, ° SR	34	30	29	28	28	28	27.5
brightness, % ISO	71.5	73.8	74.3	74.6	74.6	74.8	74.6
DP	760	765	772	784	785	791	796
breaking length, km	3.66	3.65	3.70	3.69	3.75	3.78	3.80
tear index, mN · m^2/g	3.8	4.1	4.3	4.6	4.6	4.5	4.6
fragility, %	20.5	14.4	10.8	7.2	7.1	6.7	6.8

To obtain effective dispersion of enzyme, it is necessary to select suitable pulp consistency, The effects of consistency on the pulp properties are given in Table VI.

Table VI. Effects of Consistency on the Pulp Properties

Pulp consistency, %	*Control*	*5*	*10*	*15*
beating degree, ° SR	34	27.5	28	29
brightness, % ISO	71.5	75.0	74.8	74.2
DP	760	790	782	770
breaking length, km	3.66	3.72	3.70	3.69
tear index, mN · m^2/g	3.8	4.5	4.5	4.2
fragility, %	20.5	7.4	7.3	8.4

The results indicate that under the high consistency (15%), water filterability, brightness and fragility of the pulps are not as good as under the 5-10% consistency.

Comparison of the Treatment Results between Xylanase G6-2 and An-76. Xylanase from *Aspergillus niger* An-76 is an acidic fugal xylanase(*10*). Comparison of the treatment results between the two kinds of xylanases is shown in Table VII. The results indicate that xylanase G6-2 is superior to xylanase An-76 in improving fragility, but the latter is better in increasing strength property (breaking length).

Table VII. Comparison of the Treatment Results between Xylanase G6-2 and Acidic Xylanase An-76

	Control	*G6-2*	*An-76*
enzyme dosage, IU/g	0	5	5
pH	-	7.5	5.0
beating degree, ° SR	34	27.5	28
brightness, % ISO	71.5	74.2	74.0
DP	750	789	778
breaking length, km	3.40	3.65	4.0
tear index, mN · m2/g	4.3	4.9	4.7
fragility, %	18.0	7.9	10.2
brightness after ageing*, % ISO	67.5	71.8	71.0
PC value	1.65	0.97	1.14

*Ageing conditions: 105 ± 2 ℃, 180 min.

Conclusion

1. The alkali-tolerant xylanase preparation from *Pseudomonas* sp. G6-2 is suitable for using in pulp industry.
2. The treatment of hypochlorite-bleached wheat straw pulps with xylanases improves the water filterability, fragility, brightness and brightness stability of the pulp.
3. The optimum conditions of treating the pulps with xylanase G6-2 are: enzyme dosage 3-5 IU/g. pH 7-8, consistency 5-10% and 30-40℃ for 90 minutes.
4. In comparison of xylanase G6-2 with An-76, it was found that the former is superior to the latter in decreasing fragility, but it is in reverse order in increasing strength property (breaking length).

Literature Cited

1. Viikari, L.; Ranua, M., Kantelinen, A.; Sundquist, J. and Linko M. In *Proceeding: 3rd Int. Conf. on Biotechnol. Pulp Pap. Ind.*, Stockholm, 1986, pp.67-69.
2. Viikari, L.; Kantelinen, A. and Ranua, M. In *Biotechnology in pulp and Paper Manufacture: Applications and Fundamental Investigations*; Kirk, T. K. and Chang, H. M., Ed; Butterworth-Heinemann, Boston, 1990, pp 145-151.
3. Viikari, L.; Ranua, M.; Kantelinen, A.; Linko, M. and Sundqvist, J. In *4th Int. Symp. on Wood and Pulping Chem., Paris, 1988. Proceedings*, Vol. 1, pp 151-154.
4. Kantelinen, A.; Sundqvist, J.; Linko, M. and Viikari, L. In *6th Int. Symp. on Wood and Pulping Chem., Melbourne, Australia, 1991. Proceedings,* Vol. 1, pp 493-500.

5. *Bergey's Manual of Systematic Bacteriology*, Vol. 1; Krieg, N. R., Ed.; Williams & Wilkins: Baltimore, MD, 1984.
6. Miller, G. L.; Blum, R.; Glennon, W. E. and Burton, A. L. *Anal. Biochem.* **1960**, *2*, 127-132.
7. Bailey M. J.; Biely, P. and Poutanen, K. *J. of Biotech.* **1992**, *23*, 257-270.
8. Mackenzie, C. R.; Bilous, D. and Johnson, K. G. *Biotechnol Bioeng.* **1984,** *26*, 590-594.
9. *Experiments in Pulp and Paper Making* (in Chinese).Chen Peirong, Qu Weijun and He Fuwang Ed. Chinese Light Industry Press. Beijing. 1990.
10. Chen H.; Gao P. and Wang Z. *Acta Microbiologica Sinica.* **1990**, *30*,351-357.

Author Index

Affiliation Index

Subject Index

W

X

Y

MW01641406